# HIMALAYA TO THE SEA

Plate tectonic collision, climate oscillation, glacial fluctuation, severe wind and water erosion – all have wrought dramatic change on the landscape of the Western Himalaya, one of the most dynamic and spectacular landscapes on Earth. Study of the region – from the Western Himalaya foothills and lowlands to the Arabian Sea – is of particular value to geology and geomorphology because of the size and frequency of events. That much of South Asia is relatively inaccessible has enhanced the significance of research in Pakistan and adjacent areas.

*Himalaya to the Sea* focuses on the general evolution of landforms in Pakistan but also represents an essential guide for predictive, protective and remedial measures to mitigate the natural hazards which plague the region and constrain development. The book describes regional erosion and sedimentation within the context of topographical evolution; more specifically, chapters deal with neotectonics, past and present glaciation, general mountain geomorphology and process mechanics, past and present fluvial processes and landforms, wind blown loess deposits, age dates, soils, marine terraces and archaeology.

This is the first integrated assessment of the geomorphology and Quaternary evolution of this region, from highlands to ocean. Presenting new research, methodologies and theory, this highly illustrated volume also provides the first comprehensive bibliography to the region.

**John F. Shroder jr.** is Professor and Chair of Department at the Department of Geography and Geology, University of Nebraska at Omaha.

# HIMALAYA TO THE SEA

## Geology, geomorphology and the Quaternary

*Edited by*
*John F. Shroder, jr.*

Routledge
Taylor & Francis Group

LONDON AND NEW YORK

First published 1993
by Routledge
2 Park Square, Milton Park, Abingdon, Oxfordshire OX14 4RN

Simultaneously published in the USA and Canada
by Routledge
711 Third Avenue, New York, NY 10017

First issued in paperback 2014

*Routledge is an imprint of the Taylor and Francis Group, an informa business*

Typeset in 10 point September by Leaper & Gard Ltd, Bristol

*British Library Cataloguing in Publication Data*
A catalogue reference for this book is available from the British Library

*Library of Congress Cataloging in Publication Data has been applied for.*

ISBN 13: 978-0-415-06648-8 (hbk)
ISBN 13: 978-1-138-86704-8 (pbk)

He knows what far snows melt
Along what mountain wall

The waters have risen
The springs are unbound –
The floods break their prison,
And raven around.

And presently the floods break way
Whose strength is in the hills

And belts of blinding sand show cruelly
Where once the river ran

> (Kipling: *Song of the Fifth River,
> The Astrologer's Song, The
> Floods* and *The Masque of
> Plenty*)

# CONTENTS

# PLATES

# FIGURES

# TABLES

# CONTRIBUTORS

| | |
|---|---|
| Mirza Arshad Ali Beg | Pakistan Council of Scientific and Industrial Research, Karachi. |
| John R. Bowman | Department of Geology and Geophysics, University of Utah, Salt Lake City, Utah. |
| Michael E. Brookfield | Land Resource Science, Guelph University, Ontario. |
| Douglas W. Burbank | Center for Earth Sciences, University of Southern California, Los Angeles, California. |
| Thure E. Cerling | Department of Geology and Geophysics, University of Utah, Salt Lake City, Utah. |
| Vincent S. Cronin | Department of Geosciences, University of Wisconsin-Milwaukee. |
| Edward Derbyshire | Department of Geography, University of Leicester. |
| Louis Flam | 48 Scott Terrace, Clifton, New Jersey 07013. |
| James S. Gardner | Department of Geography, Faculty of Environmental Studies, University of Waterloo, Waterloo, Ontario. |
| Michael D. Harvey | Water Engineering and Technology Inc., Fort Collins, Colorado. |
| Kenneth Hewitt | University of Waterloo, Waterloo, Ontario. |
| Jonathan A. Holmes | School of Geography, Kingston University, Surrey. |
| M. Asif Jah | Department of Earth Sciences, Cambridge University, Cambridge. d. 1991. |
| Gary D. Johnson | Department of Earth Science, Dartmouth College, Hanover, New Hampshire. |
| Norman K. Jones | Department of Geography, St Mary's University, Halifax, Nova Scotia. |
| David W. Jorgensen | Department of Geology, Washington and Lee University, Lexington, Virginia. |
| M. Javed Khan | Presentaly at Department of Geology, University of Peshawar, Peshawar. Based at Lamont-Doherty Geological Observatory of Columbia University, Palisades, New York. |
| Neil D. Opdyke | Department of Geology, University of Florida, Gainesville, Florida. |

| | |
|---|---|
| Lewis A. Owen | Department of Geography and Geology, Royal Holloway and Bedford New College, University of London. |
| D. Papastamatiou | National Technical University, Athens. |
| Jay Quade | Department of Geosciences, University of Arizona, Tuscon, Arizona. |
| H.M. Rendell | Geography Laboratory, University of Sussex, Falmer, Brighton. |
| S.A. Schumm | As Flam *and* Harvey. |
| John F. Shroder, jr. | Professor and Chair, Department of Geography and Geology, University of Nebraska at Omaha. |
| Rodman J. Snead | Department of Geography, The University of New Mexico, Albuquerque, New Mexico. |
| C. Vita-Finzi | University College London. |

# FOREWORD

Driven by the ongoing collision of the Indian subcontinent with Asia, the Himalayan range is one of the most dynamic mountain chains in the world. Recent isotopic studies indicate that not since the Pan-African orogeny over 500 myr ago has a continental collision of this scale and intensity occurred. Rapid, late Cenozoic rates of bedrock uplift in the Himalaya, in response to compressional stresses, have created and sustained the highest summits and the most massive mountain range on the earth's surface today. Intense denudation has accompanied this uplift and has removed over half of the topographically elevated crustal mass in most regions. Much of the mass lost from the mountains is carried by some of the largest and most sediment-laden rivers in the world into the Indo–Gangetic foreland basin. Beyond this, in the northern Indian Ocean, the Bengal and Indus fans accommodate an enormous volume of Cenozoic sediment derived from the Himalaya.

This volume focuses on the processes that have shaped the Himalayan landscape and on the products of the dynamic balance between uplift, erosion, transport and deposition adjacent to the soaring summits of the Himalaya and along the tortuous route to the sea more than 1000 km away. As described here, the Miocene–Holocene history of mountain uplift and of sediment delivery to and storage within the Himalayan foreland and the Indian Ocean submarine fans provides an overview of long-term sediment fluxes related to the topographic development of this collisional range. Within this overall framework, one might ask the following.

(a) What geomorphic processes have acted to denude the mountains?
(b) How much of that eroded debris is stored within the range itself and in what form?
(c) How rapidly is erosion of highlands occurring today and what is the role of glaciers as erosive and transport agents in this heavily glaciated terrane?
(d) To what extent has changing late Cenozoic climate influenced the geomorphic record?
(e) What is the magnitude and character of climate change that can be reconstructed from the stratigraphic record?

The papers in the first half of this volume attempt to address these questions as applied to the Himalaya, Karakoram and Hindu Kush ranges in northern Pakistan. They provide new insights upon the complex mosaic of asymmetric, climatically controlled processes that operate on alpine valley walls and deliver sediment to the Himalayan rivers; the irregular sediment storage within intra-montane basins and valleys; and the role of glaciers both as sculptors of the mountain landscape and as recorders of the episodic nature and variable pattern of major climate changes.

As the sediment-laden rivers exit from the mountains, they debouch into the Indo–Gangetic foreland and wend their way toward the sea. The foreland itself is a dynamic region where encroaching deformation irregularly disrupts the proximal basin, where rapid rates of basinal deposition persist owing to ongoing crustal loading in the adjacent ranges, and where rivers can shift courses dramatically in response to both subtle, tectonically driven topographic changes and autocyclic processes. Both the long- and short-term history of the deposition in the Indus foreland of Pakistan is examined in this volume. A number of provocative problems persist within this realm of low topographic relief.

(a) To what extent can the study of soils, loess, and isotopes reveal the changing nature of Pleistocene climates and the interplay between erosion and deposition within the foreland basin?

(b) Why do the courses of some major rivers, such as the Indus, appear to remain focused within a narrow quadrant of the foreland for millions of years?

(c) How can this apparent stability be reconciled with previously documented drainage geometries and with the clear evidence for unstable river courses during the Holocene?

(d) What has controlled the major shifts in course and the stream-capture events that dominate the recent record?

(e) How has tectonism interacted with fluvial and coastal processes to control depositional patterns and the development of marine terraces?

The papers in the second half of this volume address many of these questions. The style, timing and magnitude of loess deposition in northern Pakistan contrast strongly with that recorded by the famous loess deposits of northern China, and yet indicate a strong climatic control. Pedologic and isotopic studies of soils suggest that similar climatic conditions for soil formation persisted throughout much of the Quaternary in northern Pakistan. It is argued that the depositional axis of the Indus river was localized across the present Trans-Indus ranges for 5 myr during late Miocene and Pliocene times. In the context of these apparently persistent conditions, it is intriguing to consider the evidence for dramatic shifts in drainage patterns that have occurred in Holocene times as the Indus river and its tributaries have swept broadly across the foreland and as major avulsion and river-capture events have continuously reorganized the network of foreland rivers. The dynamic nature of deposition today within the foreland is well illustrated by the partitioning of zones of erosion and deposition along the course of the Indus, as well as by the strong contrasts in channel patterns that are controlled in a predictable manner by changing local gradients and base levels.

The collection of papers in this volume presents a new perspective on the exceptional array of geomorphic processes that have operated upon the Himalayan landscape since Miocene times. From the mountains to the sea, many of these processes appear to have acted at larger scales and at higher rates than are typical of most other regions. As a result, the imprint that has been left on the Himalayan landscape is often written larger and perhaps more clearly than that to which we are accustomed. These papers offer a fresh sampling of that landscape and will enhance our appreciation and understanding of the processes that have shaped this continuously evolving mountain range and the products that have been left in the wake of the ongoing Indo–Asian collision.

Douglas W. Burbank
Los Angeles, January 1992

# PREFACE

*Himalaya to the Sea* is a book that describes the late Cenozoic evolution of the geology and landforms of the Pakistan region. Information from the neighbouring areas of Afghanistan, China, India and Iran is included as needed to explain certain cross-border phenomena. During the 1960s and 1970s scientific analysis of the terrain of these neighbouring countries had proceeded well because of the reduction in hostilities in some of the often contentious border areas. Except for China and Tibet, western scientists were usually welcome in all of these countries at that time. Over these two decades, therefore, a great wealth of new information about geologic changes throughout the late Cenozoic became available for the mountains and plains of Pakistan. In the 1980s in the bordering countries, much access was restricted owing to revolution, invasion, and cultural and political sensitivities. Pakistan escaped most of these troubles and to this day has remained reasonably open to research. This book is a direct product of that openness.

Most of the landforms of the mountains and plains of Pakistan have been produced within the last 1–2 myr, which is the Quaternary period of the late Cenozoic era. Nevertheless, this book also contains considerable information about underlying geology and geomorphology of the region that is not solely Quaternary in age. Some geomorphology is older, having been formed in the mid- to late Tertiary period. This includes: the formation of the great depositional basins of Jalalabad, Peshawar, Campbellpur, Potwar and Kashmir; the course of the Indus river through the Himalaya and its deltaic growth into the 'Sindh estuary'; and the Dasht-i-Nawar volcanic caldera in Afghanistan and its Bain lahar deposit in Pakistan. Furthermore, the pace of geomorphic events in some parts of this region is remarkably more rapid than in many other areas of the world. Thus, some mountains have been thrust upward by as much as 8000 m in but a million years, as in the case of Nanga Parbat. In another case, in the Northwestern Frontier Province the original topography in Waziristan, over which the Bain lahar travelled after eruption from the Dasht-i-Nawar caldera ~ 2–2.2 myr ago, has been completely eroded away, although the caldera itself in Afghanistan still appears reasonably youthful. In addition, both the Bain lahar and the Jalipur tillite near Nanga Parbat are now bedrock that is folded up to 45°

in some places from the original flatlying attitude at time of emplacement. These enigmas, including the reasonably young sediments already lithified into rock, the really old landforms that have survived for so long and the very new ones that seem so much older, necessitate great care in dating and in interpretation lest the student of landform and Quaternary time blunder significantly. So little of the Quaternary or geomorphology of this part of the world has been studied in any detail at all that we may look to many coming years of fruitful effort.

Much of the geomorphology of the world is a product of high frequency–low magnitude events. The ebb and flow of tides and the movement of sediment in a meander loop are geomorphic processes that are common and not very spectacular. Low magnitude events, even though common enough, may require so many iterations to effect visible change that landform alteration is too slow to analyse successfully and may not allow collection of sufficient data for understanding of the overall landform shaping in any case. Certain low frequency–high magnitude events, such as the massive breakout floods across the channelled scablands of the USA or the Deccan flood basalts of India, may leave a large erosional or depositional record of a catastrophic but uncommon character. On the other hand, because of the high energy gradients available within Himalayan relief, some geomorphic processes there take on a moderate–high frequency, moderate–high magnitude character. Thus, major catastrophic floods on the Indus river from the breaching of ice and slope failure dams may have been relatively common in Quaternary times. Certainly large-scale glacier advances, retreats and surges have been common in the Himalaya in the past. Slope failure phenomena of almost all kinds are ever present as well. Major drainage changes through capture and avulsion seem to have occurred many times in the Punjab. The highly active Makran coast has uplifted abundant fault blocks, mud volcanoes and raised beaches as a result of moderate frequency–moderate magnitude events. The study of geomorphology in such places can be highly productive because the phenomena under investigation may be more readily active and available for analysis than elsewhere.

The rationale for the arrangement of papers in this book is, first, to present then in order from the mountains to the sea and, second, in so far as possible, to order them according to a sense of superposition, i.e. from older to younger. About half of the papers in this book deal with phenomena based in the mountains and half with phenomena in the lowlands. Two papers are concerned with geomorphic activities along the coast.

This book is about the general evolution of landforms in Pakistan but can also be viewed as a partial outline of potentials for the future. An understanding of the history of geomorphic process change and the long-term effects of a dynamic landscape is essential to help to guide regional design and planning in agriculture, development of underground water supplies, generation of power, transportation and construction. Improved assessment of natural hazards, such as floods, glacier advances and various forms of slope failure, will constrain development in Pakistan, even as the population continues to expand into already sorely stressed environments. Highways, settlements and waterworks are vital, yet the important Karakoram Highway and the massive Tarbela Dam are already threatened. Continued study of landform-generating processes in the western Himalaya will provide data for predictive, protective and remedial measures to mitigate hazards, a vital necessity according to the United States National Research Council's recent call for an International Decade of Hazard Reduction. It is hoped that this book, in some small measure at least, may bring together diverse material and focus attention on this dynamic and exciting landscape of Pakistan.

From the Himalaya to the sea is a journey of topographic and climatic extremes, matched only by the diversity of people and culture spread throughout vast regions. The course of water and sediment from the highest mountains on Planet Earth to the depths of ocean basin is a traverse of imagination and a delight of discovery for the scientist. The deep time of geologic imagery, the sweep of landform generation: to put these concepts into a frame of understanding is a task of great reward. Perhaps this book will entice others to have as much fun as I did in travelling this route.

Jack Shroder
1 May 1991

# ACKNOWLEDGEMENTS

This book was suggested originally by Ali H. Kazmi, former Director of the Geological Survey of Pakistan (GSP). He had recognized that, over the past two or three decades, a great wealth of new scientific material had become available concerning the late Cenozoic evolution of the mountains and plains of Pakistan. Also, from the 1960s to the 1980s, analysis of the terrain of neighbouring Afghanistan, China, India and Iran had also proceeded well. It was clear to Dr Kazmi and me that the time was ripe for the beginnings of synthesis of the wide variety of earth science for the late Cenozoic of the region that had been produced in recent years. His inspiration was essential to this project.

*Himalaya to the Sea* represents the work of many people, not the least of whom are the many hospitable Pakistanis who put in so much time, effort and scarce financial and logistical resources to the studies described. Rashid A.K. Tahirkheli, Qasim Jan, Javed Khan, Arif Ghauri and M. Saqib Khan all helped me immensely in the early stages in 1984 while I was a Fulbright Lecturer at Peshawar University. Bruce Lohoff, then Director of the US Educational Foundation in Pakistan (USEFP), and his staff in Islamabad were a great help, as was John Dixon, Director of USIA in Peshawar when I lived there. R. Lawrence, R. Yates and their students from Oregon State University were helpful in field support as well. Excellent editing was provided by Melissa Glick Wells and Susan Maher. M. Saqib Khan typed the references. Mary Trahman and Janet Dean, my secretaries, kept me from straying too far from the job and typed many of the letters and other materials. Carol Zoerb found many obscure references. Financial support for my part of this project was provided by the Smithsonian Institution, Fulbright, the National Geographic Society, the Kiewit Foundation, the University of Nebraska at Omaha and Peshawar University.

M.E. Brookfield was supported by the NSERC (Canada) and the National Geographic Society. He is grateful to the villagers and government officials in Pakistan, India, USSR, China and Tibet who provided hospitality, accommodation and assistance.

J. Holmes was supported by a research studentship (GT4/84/GS/85) from the Natural Environment Research Council (UK). He thanks L. Owen, W. Kick, W. Haeberli, E. Derbyshire and H. Osmaston for their assistance.

V. Cronin and G. Johnson were supported by NSF (USA) grants EAR 8018779, EAR 8407052, INT 8019373-A01 and INT 8308069, and by Dartmouth College, Texas A & M University, the University of Wisconsin–Milwaukee and Peshawar University. M. Hassan, G. Hussein, M. Khan, O. Khan, S. Khan, H. Rahman and R.A.K. Tahirkheli assisted in Skardu and L.E. McRae processed the palaoemagnetic samples.

J.S. Gardner and N.K. Jones were supported by NSERC (Canada) and the Snow and Ice Hydrology Project. F. de Scally, E. Mattson, M. Anwar and F. Rahmat helped in the field. K. Hewitt and H. Afzal provided other assistance.

J. Quade, T. Cerling, J.R. Bowman and M. Asif Jah are greatly indebted to J. Barry, K. Behrensmeyer, R. Dennel, M. Raza, I.M. Shah and the Geological Survey of Pakistan. Grants from the Smithsonian Institution and NSF (USA) BNS-8703304 provided funding.

H. Rendell carried out fieldwork under the auspices of the Potwar Palaeolithic Project of the British Archaeological Mission to Pakistan and was funded by the British Academy and Royal Society. Her laboratory work was funded by the Science and Engineering Council. She thanks particularly R.W. Dennell, P.D. Townsend, the Director of the British Mission, and the Director-General of Archaeology of the Government of Pakistan (GOP).

D. Papastamatiou and C. Vita-Finzi thank Dames and More Ltd, Ali H. Kazmi, former Director of the Geological Survey of Pakistan, and the Secretary of the Ministry of Petroleum and Natural Resources, GOP for permission to publish.

L. Flam is grateful to Muhammad Alim Mian (Director), Mohammad Akram, Ch. M. Rafiq, Abdul Hadi Ansari and the officers and staff of the Soil Survey of Pakistan. He also thanks Stanley A. Schumm, M.D. Harvey, M.H. Rizvi, Shabir Ahmed Khan Chandio, Abdullah Waryah, Babar Soomro, A.A.A. Razak Soomro, M. Rafique Mughal, G.L. Possehl, J.G. Shaffer, W.C. Jones, Cecile Jones and R. Woodhead. Flam was supported by the Fulbright Research Scholar Program and had much help from USEFP.

D.W. Jorgensen, M.D. Harvey, S.A. Schumm and L. Flam were helped extensively by Muzammil H. Qureshi, Secretary, Irrigation and Power Department, Government of Sindh, Karachi, and his staff; Mohammad Idries Rajput; Mahadev J. Mathrani; Iqbal Sheikh; and Zeeshan Ahmed Farooqi, Statistical Officer of the Research Division. Others who were of great assistance are Alisdar Paine in England, Larry Grahl, Richard and Isabella Von Glatz and Jim Moore of the US Consul General in Karachi, and Nawabzada Shabir Ahmed Khan Chandio; the Makhdoom of Hala and his representative Khalifa Mohammad Yusuf, Mahtab Akbar Rashdi, Rashid A. Shah, M. Mukhtiar Kazi, G.M. Sher, Mohammad Lund, Mohammad Ishtiaq Khan, and Ahmed Nabi Khan from the various departments and agencies of GOP. Funding was provided by the Smithsonian Institution grant SFC-50163700, and NSF grant BNS-8303707.

# 1

# HIMALAYA TO THE SEA: GEOMORPHOLOGY AND THE QUATERNARY OF PAKISTAN IN THE REGIONAL CONTEXT

*John F. Shroder, jr.*

## ABSTRACT

The geomorphic evolution of Pakistan and bordering areas is a result of ongoing tectonism associated with the movement of the Indo–Pakistani plate into the Eurasian plate. The raising of the Qinghai–Tibetan plateau and the Himalaya led to complex formation of the Indus and Punjab drainage systems that evolved through superposition, antecedence, fault offsetting, headward erosion, ponding and overflow, avulsion, capture and delta building to the present configuration. Across this vast tectonic and fluvial landscape, the erosional and depositional results of volcanism, glaciation, slope failure, aeolian activity and marine processes have left their strong imprint as well.

## INTRODUCTION

The geomorphology and Quaternary development of the Qinghai–Tibet Plateau, the western Himalaya, the Hindu Kush and down across the foothills and plains to the Indian Ocean are a vast aggregation of extremes; extremes of altitude and relief and of geomorphic process, driven by the fundamental alterations of a plate-tectonic world. This landscape of Southwest Asia is a product of plate-tectonic collision and uplift, of rapid erosion and thick sedimentation. Massive glaciers and torrents of water, ubiquitous slope failures and the wind, and quiet but relentless lake and ocean processes have all sculpted the dynamic landforms of the region, culminating throughout the late Cenozoic to produce the landscapes of today.

Pakistan embodies these extremes. More high mountain peaks occur there than in any other place in the world; there are five > 8000 m and more than 36 > 7000 m. At 8125 m, Nanga Parbat is the tenth highest mountain in the world and is still rising rapidly. The mighty Indus river, one of the world's great drainages, is the single most important natural resource in Pakistan, the source of

1

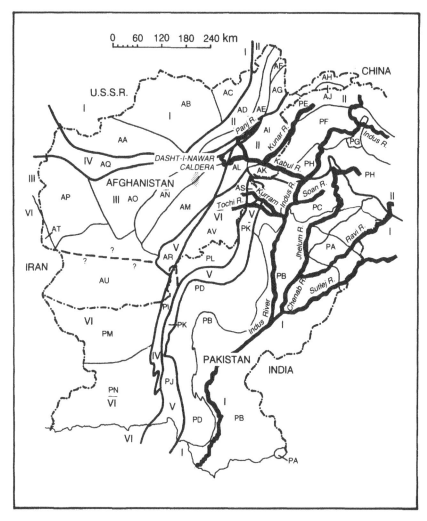

*Figure 1.1* Index map of Pakistan, Afghanistan and part of northwest India. Subdivisions are based upon plate-tectonic divisions (Shroder, 1984; 1989b) and include:

I Pre-collisional Eurasian and Indo–Pakistani plate margin terranes: plateaux, plains and basins;

II Interplate marginal collision zone mountain massifs:

III Iran–Afghanistan microcontinental plate basins and ranges;

IV Interplate marginal geosuture ridges and valleys;

V *Interplate marginal obduction zone basins and ranges;*

VI Interplate marginal convergence boundary basins and ranges.

Other divisions are keyed by letter to Table 1.1. Only the Indus river and its main tributaries or those discussed in the text are included here. A stippled pattern indicates the Dasht-i-Nawar caldera in Afghanistan.

*Table 1.1* Morphostructural zones of Pakistan, Afghanistan and part of northwestern India (after Shareq *et al.*, 1977; Kazmi and Rana, 1982; Shroder, 1984). Emphasis is on morphostructure of Pakistan.

| | | |
|---|---|---|
| I | Pre-collisional Eurasian and Indo–Pakistani plate margin plateaux and basin zones | |
| | AA | Murghab block plateau and ranges |
| | AB | Balkh block basin and plateau |
| | AC | Fore-Badakshan fold ranges |
| | PA | Sargodha–Shahpur and Nagar Parkar basement ridges and hills of circumalluviation |
| | PB | Foreland belt of Punjab and Indus alluvial plains |
| | PC | Himalayan fold-belt basins, plateaux and ranges |
| | PD | Sulaiman and Kirthar fold-belt ranges |
| II | Interplate marginal collision zone | |
| | AA–AL | Hindu Kush and Pamir ranges of Afghanistan |
| | PE | Karakoram (Tethyan) ranges |
| | PF | Kohistan volcanic and igneous ranges |
| | PG | Nanga Parbat–Haramosh massif |
| | PH | Himalayan crystalline and Schuppen (imbricate) ranges |
| III | Iran–Afghanistan microcontinental plate | |
| | AM–AP | Various fault-block basins and ranges |
| IV | Interplate marginal geosuture ridges and valleys | |
| | AQ | Hari Rud–Panjsher fault ridges and valleys |
| | PI | Ornach Nal-Chaman fault ridges and valleys |
| V | Interplate marginal obduction basins and ranges | |
| | AR–AS | Various fault and fold-basins and ranges |
| | PJ–PK | Various fault and fold-basins and ranges |
| VI | Interplate marginal convergence boundary basins and ranges | |
| | AT | Asparan range |
| | AU–PM | Chagai volcanic and igneous ranges and basins |
| | PN | Makran flysch ranges and basins |
| | AV–PL | Katawaz and Kara Khorasan flysch basins and fold ranges |

much of the country's irrigation water as well as almost all of its electric energy. Aside from the active tectonism of the region, the river and its tributaries are responsible for more of the landform generation there than any other process. For perhaps as much as 30 myr, the Indus river has been delta building south into a bay off the Indian Ocean which was formed by the collision between the Indo–Pakistani and the Eurasian plates (Figure 1.1).

The reasons for the great sweep of varied landscapes of this region are explained best by concepts of plate tectonics. Central Afghanistan and the main Indo–Pakistani plate appear to have been split from the Gondwanaland parent block (Antarctica, Africa, Australia), beginning for the Afghanistan microcontinent (Figure 1.1 and Table 1.1) as much as 250 myr ago, or the late Permian, and for the main plate ~130 myr ago, or the early Cretaceous. The

3

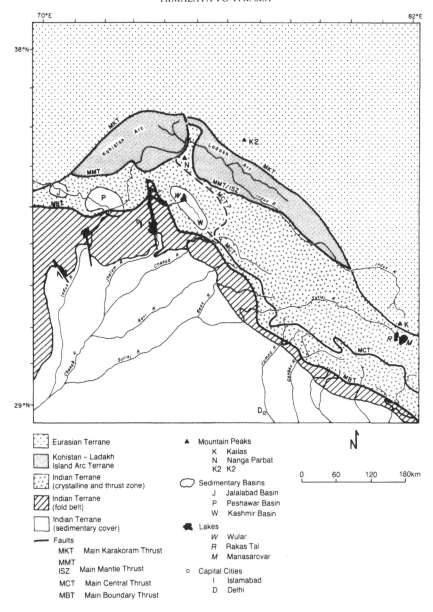

*Figure 1.2* The various mountain ranges of the western Himalaya are part of a single geologic complex formed after the collision of the Indo–Pakistani and the Eurasian crustal plates. The volcanic island arcs of Kohistan and Ladakh were caught between the Indo–Pakistani and Eurasian plates as they came together. Main faults are shown which have deflected drainages and along which great uplifts have taken place (after Gansser, 1980; Kazmi and Rana, 1982; Yeats and Lawrence, 1984; Zeitler, 1985).

fragments subsequently drifted northward to dock against volcanic island arcs and the Eurasian plate through the late Cretaceous and into the Cenozoic. The collision caused the metamorphism of the sedimentary rock on the edges of the Eurasian plate and the intrusion of the Karakoram batholith, beginning about 100 myr ago and continuing for more than 50 myr. Silica-rich volcanics associated with these intrusions were later eroded and then redeposited from north to south as continental molasse. The Eurasian terrane was upthrust southward along the melange zone of the Main Karakoram Thrust (MKT) as the Indo–Pakistani terrane was shoved northward beneath it (Figure 1.2). The MKT seems to connect southeastward with the Indus Suture Zone (ISZ) of the eastern Himalaya (Gansser, 1980; Zeitler, 1985). As with the Eurasian terrane, the leading edge of the Indo–Pakistani plate also has a variety of older metasediments and intrusives of the deformed continental margin and marine platform of Gondwanaland, overlain by the thick continental molassic sediments shed from the rising Himalaya. The first major deposition of these sediments began, perhaps 20 myr ago, south of the main Himalaya as the Rawalpindi and Siwalik Groups (Meissner et al., 1974), which were deposited on the leading edge of the Indo–Pakistani plate. Siwalik sedimentation continued up to less than 1 myr ago. The Indo–Pakistani terrane is delineated on its north by faults known as the Main Mantle Thrust (MMT) which pass from Pakistan eastward into India into the ISZ.

Ancient volcanic island arcs had also formed in the Tethys ocean straits between the Eurasian and Indo–Pakistani plates prior to 65 myr ago. The Kohistan and Ladakh island arc terranes are mainly highly deformed mafic to silicic intrusives, volcanics and sedimentary debris shed from the mafic-rich source areas. The beginning of collision of the Indo–Pakistani and Eurasian plates began perhaps 50–60 myr ago and the island arcs were thrust upwards, to be sandwiched between the two plates as they closed. The collision appears to have been most intense ~ 30 myr ago. Also, at about this time, the rocks of the Nanga Parbat–Haramosh massif seem to have moved north, perhaps as the Kohistan and Ladakh arcs were thrust south along the MMT fault over the metasedimentary and other rocks of the northern margin of the Indo–Pakistani terrane (Andrews-Speed and Brookfield, 1982). This fault has been locked since ~ 15 myr ago (Zeitler et al., 1982b; Farah et al., 1984a,b). The rocks of the Indo–Pakistani terrane were, in turn, thrust south over the late Tertiary molasse sediments of the Rawalpindi and Siwalik Groups along the Main Boundary Thrust (MBT).

The Himalaya mountain range forms a narrow rampart along the southern margin of the massive Qinghai–Tibet plateau. This barren and arid plateau now averages 4.5 km elevation but in the Cretaceous, ~ 70 myr ago, it was partly marine. Prior to 30 myr ago, tropical and subtropical forests existed there, but rapid uplift and denudational unroofing began ~ 20 myr ago (Harrison et al., 1992). The present altitude of the plateau was obtained by ~ 8 myr ago, so that the vegetation from 10–5 myr ago became deciduous and comparable to modern temperate forests. In this model, the changes in the vegetation, along with assorted tectonic and sedimentary evidence, implied to Ruddiman *et al.* (1989) that, as a result of the plate-tectonic collision, the Qinghai–Tibet plateau overall has risen more than 4 km, with perhaps 2 km of rise in the past 10 myr. Uplift of the great plateau was tied to worldwide climate cooling by forced diversion of air masses (Ruddiman and Kutzbach, 1989), enhanced chemical weathering, and decreased atmospheric $CO_2$ (Raymo *et al.*, 1988). Molnar and England (1990), however, inverted cause and effect in these ideas when they noted that externally forced, worldwide climate change in late Cenozoic instead could have led toward lower temperatures and increased alpine glaciation, with associated peturbations to precipitation and vegetative cover. These climate-forced changes could, in turn, have increased erosion rates, caused exhumation of rocks at the edge of the Qinghai–Tibet plateau through incision of deep valleys, and thereby unloaded the crust. Such crustal unloading would have produced isostatic compensation in the form of the rise of the modern Himalayan peaks without significant additional tectonism.

## EVOLUTION OF HIMALAYAN DRAINAGE SYSTEMS

Ever since the time of the first religious pilgrims and European explorers, the origin and evolution of the Himalayan rivers and their sediments have been the subject of debate (Oldham, 1893; Pilgrim, 1919; Pascoe, 1920a; Mithal, 1968). Twin parallel rivers were hypothesized to have originated in the eastern highlands of Tibet and to have emptied westward into the 'Sindh estuary', an arm of what remained of the Tethys Sea in the northwest, some time after ~ 60 myr ago when the first uplifts between northern India and Asia were forcing the seas out to east and west. An 'Indo–Brahm' river, thought to have deposited the Siwaliks on the south, and an ancestral Tsangpo–Upper Indus–Amu Darya on the north were believed to have been partly captured and reversed to produce the present drainage configuration of the region. Commonly cited, for example, were the strongly curved ' < -shaped' patterns of many tributaries that flow west or northwest out of the Himalaya, as though originally joining a west-flowing river at a slight angle and then later being forced to flow southeast through an apparent elbow of capture into the east-flowing Ganges. Rivers eroding headward from the Bay of Bengal were presumed to have captured the northwest-flowing Indobrahm to become the modern southeast-flowing Ganges, as well as capturing the Tsangpo–Brahmaputra that now joins the Ganges in the Bangladesh delta. These speculations were adequate for the times but the lack of radiometric dates, sedimentologic data or detailed information from the mountains allowed little additional progress. In the past two decades, however, extensive new explorations and discoveries have changed this. We now know, for example, that the curved pattern of the tributaries to the Ganges is an oversimplification because deflection also occurs in other directions as a result of deposition upstream in valleys formed behind downward rotating fault blocks

6

broken from the edges of the great boundary thrust wedges (Seeber and Gornitz, 1983).

Two other peculiar drainage features also caught the attention of observers from the outset. Most striking is the fact that the Indus and the Tsangpo–Brahmaputra rise near each other in Tibet, 100–150 km north of the main Himalaya watershed, and, after flowing parallel to the range in opposite directions along the rather straight watershed boundary for over 1000 km, both turn abruptly to break through it and escape southward. The other odd fact is that most of the remaining larger rivers of the Himalaya rise to the north of the highest peaks and cut deep and narrow transverse gorges south through the range. These rivers, however, do not deflect the linear watershed boundary, as they would do if they had captured their headwaters from across the range.

A number of possibilities have been suggested to explain the courses of the rivers that flow from the Himalaya. One is that the rivers are older than or antecedent to the mountains and have maintained their courses by active erosion across newly rising tectonic structures. A second is that the rivers are younger than the mountains and have cut down through an overlying cover of sediments to superimpose themselves across older structures buried beneath. A third is that the steep, active headwaters of early rivers cut down through drainage divides to capture parts of other rivers. A fourth is that periodic damming by ice, slope failure or faulting caused ponding and eventual overflow through low passes in different directions. Two other minor drainage controls occur where rivers are deflected as they cross faults which have lateral offset, and through avulsion where rivers switch out of their channels through aggradation differences. Despite the ambiguities of the structural and sedimentation history of the region, lengthy histories of drainage evolution through a mix of some of these controlling possibilities can be deducted for some of the rivers.

The Indus river rises on the north slopes of Mt Kailas (6714 m) in the Gangdise range of Tibet (Allen, 1984) (Figure 1.2). Mt Kailas, or Kangrinboche Feng to Tibetans, is sacred to both Hindus and Buddhists, probably in part because it is the source of the three great river systems of South Asia (Tsangpo–Brahmaputra, Ganges and Indus). The mountain is an isolated remnant of debris eroded from the volcanoes emplaced above the batholiths generated 40–60 myr ago by the plate collision. The Kailas deposits are likely to be some 20–40 myr old and are correlative with the Indus molasse that seems once to have covered the Ladakh are intrusives to the west (Honegger et al., 1982; Sharma, 1983; Sharma and Gupta, 1983). Some of the sedimentation was in alluvial fans and plains; other sediment types were produced by drainage impedance owing to mountain uplift across stream channels. As the Tethys seaway receded and an isthmus between the two plates grew in its place, at least part of the ancestral Indus appears to have come into existence near Mt Kailas to drain the new highland. Similarly, the Tsangpo may have formed and flowed east to erode further the volcanic region (Seeber and Gornitz, 1983).

Subsequently the ancestral Indus seems not only to have eroded through the

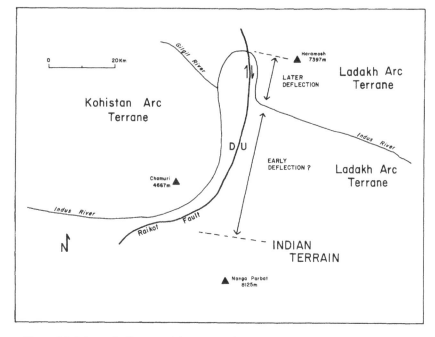

*Figure 1.3* Schematic diagram of the course of the Indus river between Haramosh and Nanga Parbat mountains and beside the Raikot fault trace along which the river shows an apparent two generations of deflection.

overlying sediments to superimpose itself upon the rocks of the Ladakh terrane beneath but also to have maintained erosion as an antecedent stream across other newly rising structures further downstream. Thus, by ~ 20 myr ago, the ancestral Indus had established itself across the Ladakh arc. The relationship of the river to the Kohistan arc and its greater course to the Indian Ocean are less clear at this time, although Kazmi (1984) cited evidence that at that same time the marine delta of an ancestral Indus (or other large river) was some 650 km north of its present location. In any case, near the Nanga Parbat–Haramosh massif, where the Indus first strikes the rocks of the Kohistan arc and crosses the active Raikot fault (Lawrence and Ghauri, 1983; Lawrence and Shroder, 1984; Madin, 1986), the drainage controls as a result of uplift and offset are pronounced.

On first glance at a small-scale map of the middle and upper Indus, the river seems deflected around the massif in an overall left-lateral sense; that is, the Kohistan block on the west appears to have moved south as the Ladakh block moved north and twisted the river across them (Figure 1.2). The collective apparent offset is ~ 30–40 km and might be the result of sinistral movement on the MMT prior to 15 myr ago, in the waning phases of suturing the island arc between the plates. This bend in the river, of course, may also be nothing more

8

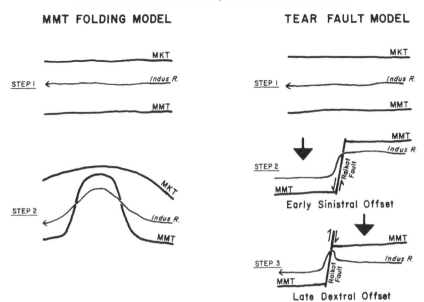

*Figure 1.4* Schematic diagram of two possible hypotheses for deflection of the Indus river across the Nanga Parbat syntaxial bending or offset of faults. In the MMT model, the MMT thrust sheet is supposedly folded into the syntaxial bend to produce apparent strike-slip fault offsets that would provide roughly equivalent offsets of the river. In the tear fault model, on the other hand, the two generations of movement on the Raikot fault produce the offset observed at present.

than a fortuitous relic of irregular superposition of the ancestral river across the structures. The likelihood of this seems low, however, given the 1000 km of linear flow along the structural grain upstream and then the abrupt deflection across the regional structures coincident with a major fault system between the two arc terranes.

In addition to the above apparent offset, close inspection of large-scale maps of the area reveals yet another picture of events. The Indus river flows west–northwest from Skardu toward the Nanga Parbat–Haramosh massif and then is deflected first *north* along the Raikot fault zone before turning abruptly west again out of the fault zone and then finally south into the larger and older deflection mentioned above (Figure 1.3). This younger deflection is right lateral and is opposite in sense of motion to the original deflection; that is, in the younger case the east block moved south relative to the west block moving north. Detailed mapping of the Raikot fault (Madin, 1986) also has revealed various lineations in the rocks along the fault that confirm the change in directions of motion.

The nature and locations of all of the fault sutures between the Eurasian, Indian and island arc terranes are controversial but the new ideas of a tear fault on the MCT, which pushed up the Nanga Parbat–Haramosh massif, better explain the younger deflections of the river than the older idea of the MMT fault plane that was later folded and exposed as the massif pushed up through it. In this latter case the Indus should be curved smoothly across the massif (Figure 1.4), but it is not.

From Nanga Parbat downriver to the mountain front, structural and other controls of the course of the Indus are known only poorly. At Thakot the lower gorge of the Indus above Tarbela Dam the river is strongly deflected to the east in what appears to be a deeply ingrown or entrenched meander loop. No structural or lithologic change in the bedrock is known at this site, so the basic reason for the deflection is not understood. In any case, in the sloping core of the meander loop occurs a thick series of pendant bars whose coarse cobble, pebble and sand load appears to have been deposited by periodic catastrophic floods.

From Thakot to Manshera, along the Himalayan foothill reaches of the Karakoram Highway, a geomorphic section is developed and exposed that reflects a period of relative tectonic stability during tropical soil development on a surface of rolling hills between 20 and 5 myr ago, followed by uplift and erosion of that surface (Lawrence and Shroder, 1985). The area of interest extends from the Indus river up the valley of Nandihar Khwar to a pass at ~ 1800 m, then across the Chattar plain to an escarpment where the surface declines to the Manshera basin. Some of this region is a relict landscape that is temporarily preserved between the deeply entrenched and rapidly downcutting Indus river to the west and the Kunar river to the east (Figure 1.5).

The valley of Nandihar Khwar above Thakot has a tectonically disturbed drainage system in which the lower segment of the stream has a steep gradient ($\sim$ 16 m km$^{-1}$) and a narrow V-shaped valley, whereas the upper part has a broad bottom, gentle sides and a low gradient ($\sim$ 8 m km$^{-1}$). Below Batgram a series of paired bedrock terraces record the continuation of the upper valley floor that has been incised owing to downcutting at the mouth of the stream by the Indus. The active uplift of this region proposed by Zeitler et al. (1982b) gives a most plausible cause for the active stream incision, with a projection of the upper valley gradient and its relict terraces suggesting an uplift of $\sim$ 300–400 m in the episode recorded here. The upper valley segment and the Chatter plain directly over the pass are relict fragments of an uplifted and tilted landscape covered by thick (4–10 m) residual soils which locally preserve china clay deposits of some economic potential. The upper part of the soils are thick clay-rich saprolites which commonly preserve granitic textural features. At depth, as yet unweathered core stones occur in the saprolite and these pass downward or laterally into disintegrating, jointed granite. Such relict deep saprolitic weathering, characteristic of low relief, tropical weathering profiles, is also present on several other rock types in Swat (Moosvi et al., 1974; Rosenberg and Khan, pers. comm., 1983) and elsewhere along the foothills into India.

Metamorphism of the granites and overlying sediments of the region reached a peak prior to 30 myr ago (Kazmi et al., 1984; Maluski and Matte, 1983) and

10

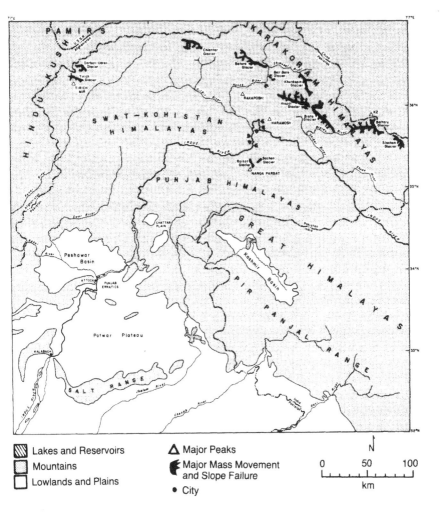

*Figure 1.5* Map of the western Himalaya showing a few of the larger glaciers and slope failures studied by the writer, as well as other features mentioned in the text.

uplift of the area is dated by fission-track ages at 18–20 myr (Zeitler *et al.*, 1982b). These dates appear to record an early episode of thrust faulting that emplaced the granites and metasediments over little metamorphosed units to the south. Development of the Chattar soil began as soon as a surface of low relief developed over these structures. Raymo *et al.* (1988) have proposed that uplift of plateaux and mountain ranges in late Cenozoic has increased the rate of chemical erosion on a globally averaged basis. Heavy monsoons, which develop at the margins of highlands, unleash intense rainfall, and uplift-related tectonism also exposes fresh rock to weathering. The steeper slopes created by uplift cause faster runoff which removes erosion products and intensifies weathering (Ruddiman and Kutzbach, 1989). Just north of the Chattar plain is the Indus Kohistan Seismic Zone (IKSZ of Seeber and Armbruster, 1979) and northeast of this is a northwest-trending topographic step along which the terrain to the northeast rises to peaks > 8000 m (Gornitz and Seeber, 1981). This zone is considered to overlie a blind ramp fault in the Himalayan sole thrust and is perhaps an extension of the faults in the Pir Panjal range. The uplift of this range and of the area north of the IKSZ is dated at 4–5 myr by a change in the source of palaeo-magnetically dated Siwalik sediments (Raynolds, 1981) and by the similarly dated initiation of Karewa sedimentation (Burbank and Johnson, 1982, 1983). Tilting, uplift and erosion of the extended Chattar weathering surface was probably initiated at this time and continues to the present. Rapid uplift of the Pir Panjal and IKSZ began ~ 350,000 yr ago (Burbank and Raynolds, 1984). The tentative chronology proposed by Lawrence and Shroder (1985) suggested that prolonged weathering between 20 and 5 myr ago developed the residual saprolites that were somewhat eroded during the tectonism of the last 5 myr and then rapidly dissected in the last 350,000 yr.

The Chattar plain is an older landscape remnant within the Hazara inter-montane basin which also includes the Manshera and Abbotabad depositional basins and plains. The Havellian Group of sediments that fills the Manshera and Abbotabad basins is normally polarized and correlated with the Brunhes chron ( < 0.73 myr to present) and is possibly in part a sandur or product of deposition of glaciofluvial sediment and wind-blown loess derived from small glaciers in the nearby uplands (Pivnik, 1988). Some of this section is fine grained and may also be derived from the erosion and redeposition of saprolites of an eastern extension of the Chattar plain. Lenses of gravel occur high in the section, and the top is an extensive gravel sheet. Some of these gravels may also be derived in part from the core stones and bedrock exposed by erosion of the residual soils.

In general, uplift in the Plio–Pleistocene initiated final erosion of the Chattar and other upland saprolites and this process may also be the source of some of the sediment in the Manshera, Peshawar, Campbellpur and Haro intermontane basins. Furthermore, Lawrence and Shroder (1985) have suggested that the extensive silts of the norther Potwar plateau may also be in part of such an origin, although most workers consider them loess derived from extensive reworking of glacially derived Indus river silts.

Beck and Burbank (1990) noted that the fluvial stratigraphy of the Himalayan foreland basin at the front of the range seems to show strong influence by apparently repeated continental-scale diversion of large river systems between the (modern) Indus and the Gangetic forelands. In the Kohat–Potwar area south of Tarbela and Attock, Abbasi and Friend (1988) discovered in the Indus Con-glomerate a blue-green hornblende characteristic of the Kohistan island arc lithologies. Thus, ~ 15 myr ago in that area, an ancestral Indus-like river was depositing a thick section ( ~ 1500 m) in the area of the modern Indus. Other palaeocurrent and lithologic evidence from this region also indicate that a large sandy river similar to the modern Indus debouched from the mountain front at the same longitude as the modern river and flowed southward toward the Arabian Sea from at least 13.5–11.5 myr ago (Beck and Burbank, 1990). Enhanced subsidence of the Gangetic foredeep, owing to accelerated loading by the Himalaya at 11 myr, seems to have diverted this large, south-flowing Indus-like river eastward into the Gangetic system and the Bay of Bengal. In the Jhelum area, just prior to about 5 myr ago, the molasse sediments of the upper Siwalik Group have a lower unit of clast assemblages and palaeocurrent structures that suggest a major river flowing east to northeast (Raynolds, 1981). Pivnik (pers. comm., 1992), however, has found no evidence of such drainage reversal in the area. Initial uplift of the Salt range of northern Pakistan by ~ 5 myr ago and continued development of related folds and faults convinced Beck and Burbank (1990) that these structures diverted the Indus-like river back toward the Arabian Sea by ~ 4.5 myr ago. In the Jhelum area the old east-flowing river was replaced by a southward flowing ancestral Jhelum or Soan river after ~ 4.5 myr ago. This river was responding to migration of the Himalayan foredeep south at ~ 2 cm yr$^{-1}$. From ~ 2.4 to 0.7 myr ago the molasse coarsens upward into conglomerates that advanced south ~ 3 cm yr$^{-1}$ as a time-transgressive phenomenon in response to outward displacement of the Pir Panjal orogenic front (Reynolds and Johnson, 1985).

This progradation of the ancestral Jhelum fan gravels likely would have included lateral migrations as well. Lateral movement of Himalayan rivers through avulsion across their fans is not unusual, as for example the Arun–Kosi river in eastern India that shifted laterally over 100 km and through as much as 75° in overall flow direction in only two centuries (Gole and Chitale, 1966). Such extreme lateral movement, over 15,000 times more rapid than the Jhelum progradation, could have been involved in this apparent shift from an east–northeast flow to a later southerly direction. Certainly having a palaeo-Indus in approximately its present position 15 myr ago (Abbassi and Friend, 1988), shifting it to drain into the Ganges ~ 5 myr ago (Burbank and Raynolds, 1984; Beck and Burbank, 1990), and then back again to the modern location seems unlikely. Parsimony in hypothesis construction would indicate a greater likelihood that the evidence for an east-flowing river in the Jhelum area might be better explained by reworking of Indus sediments and later avulsion on the Jhelum fan. In addition, deflection into the S-shaped bend of the modern Jhelum river could have been facilitated by the ongoing thrust-driven counterclockwise rotation of over 20° of part of the Potwar plateau and Salt range that occurred between 1.8 myr ago and the present (Opdyke et al., 1982; Burbank and Raynolds, 1984; Yeats et al., 1984).

Aside from the basins already mentioned, the other nearby major sedimentary basins produced in the last few million years in the front of the western Himalayas include the Kabul, Jalalabad and Kashmir basins (Figures 1.2 and 1.5). The Peshawar basin, through which the Kabul and Indus rivers flow, began by ~ 3 myr ago through differential uplift of the Attock range and ponding of

13

the pre-existing fluvial system (Burbank and Tahirkheli, 1985). Quaternary lake sediments in the Peshawar basin are intersected in many places by discordant clastic dikes which indicate fracturing and injection possibly associated with seismicity associated with episodic tectonism (Haneef *et al.*, 1986). Widespread intermontane deposition in the basin was terminated ~ 600,000 yr ago about the same time that the Attock range uplift accelerated, after which catastrophic floods periodically inundated the basin to produce cyclicly bedded sediments (Burbank, 1983b). Shroder *et al.* (1989) have suggested that many of these floods were produced by self-dumping, ice-dammed lakes caused by glacial blocking of the Indus river in Pleistocene, especially around Nanga Parbat. Some of the Punjab erratics (Figure 1.5) were emplaced near Attock as a result of these floods. More modern, but still catastrophic floods (Hewitt, 1982; Shroder, 1989c) from failure of ice and slope-failure dams have swept down the Indus for much of recorded history.

In its course south–southwest from Attock, the Indus passes through the south end of the Campbellpur basin, skirting to the west of the Potwar plateau. The river turns at nearly a right angle to pass west through the Salt range and across the Kalabagh tear fault along which the west edge of the Salt range is being thrust to the southeast (Yeats *et al.*, 1984; Yeats and Lawrence, 1984). From there the Indus turns south again and spreads into a broad braid plain for its remaining 1150 km course to the sea. The Kalabagh fault has 12–14 km of right lateral offset (McDougall, 1987 and pers. comm., 1987), which also accords with the above-mentioned palaeomagnetic studies of the Middle Siwaliks that indicate counterclockwise rotation of the region. Deflection of the south-flowing Indus west across the Kalabagh fault and then south again is supported also by the new discovery of extensive high-level gravels characteristic of a Himalayan source. The river apparently was carried south and west away from the gravels as fault movement progressed, so that in the last 0.5 myr it occupied the lower part of the Soan river system (McDougall, 1989).

The Salt range itself is being thrust south over the sediments of the Jhelum river plain (Yeats *et al.*, 1984). The boundary between the plain and the steep south flank of the Salt range is abrupt and consists of a series of scallops convex to the south. The steep, south-facing slopes constitute an eroded fault scarp, but pediments have not developed there. Instead only alluvial fans are as yet concentrated at sharp re-entrants between adjacent scallops. The longer erosion time required for pediment development has not yet occurred because of the recency of the faulting. Even so, Holocene alluvium appears to be undeformed, suggesting that the recurrence interval of faulting may be measured in thousands of years.

High rates of erosion ($1-15$ mm yr$^{-1}$) of fluvial strata have persisted over intervals of 0.2–1.5 myr in the late Cenozoic foreland basin of northern Pakistan (Burbank and Beck, 1991). The Potwar plateau, Soan river and Salt range areas have been eroded in association with thrust fault uplift and crustal shortening. The Cenozoic fluvial molasse strata were eroded rapidly during Pleistocene from above the underlying older carbonates of the Salt range, which now seems to be eroding at much slower rates in this semi-arid climatic regime.

Amidst the Plio–Pleistocene molasse sequence of the Siwalik Group, exposed in the Bhittani and Shinghar ranges of the Salt range to the west of the Indus, occurs the Bain diamictite. This was originally considered a glacial unit (Morris, 1938), but now is recognized as the product of at least two volcanic lahars ~ 2.16 ± 0.4 and 2.89 ± 0.6 (fission track) myr ago (Khan *et al.*, 1985; Pivnik, pers. comm., 1992). The distinctive, cliff-forming unit forms hogbacks and monoclinal ridges of a matrix-supported heterogeneous lithology with clasts ranging from boulders to clay. The largest boulders are volcanic and tend to be consolidated acidic pyroclastics. The most abundant clasts are limestones and greywackes. The most diagnostic of transport direction are clasts of mudstone, shale, slate and quartzite from outcrops in South Waziristan and Afghanistan (Stuart, 1922; Coulson, 1938; Tahirkheli, pers. comm., 1984).

Khan *et al.* (1985) pointed out numerous reasons why the Bain could not be the product of glaciation or debris flows, and why it should be considered the result of a volcanic lahar, emplaced down the existing drainages of the time. The large Dasht-i-Nawar caldera in nearby Afghanistan (Figure 1.1), source of much of the volcanic ash in the Siwaliks (G. Johnson, pers. comm., 1989), seemed the most likely source and is known to have been active at the time when the Bain diamictite was deposited (Bordet, 1970, 1972a, 1972b; Boutiere and Clocchiatti, 1971).

The distance from the caldera across the irregular Ghazni–Gardez plains of Afghanistan and down the large Kurram river valley in Pakistan to the distal end of the Bain deposit is about 400 km (Figure 1.1). A more direct route down the Tochi valley is ~ 300 km. In any case, the topography of 2 myr ago would have to have been less tectonically disturbed or eroded than now or the lahar would have been deflected or stopped in that distance. In comparison with the probable distance of transport, the Bain lahar is 2–3 times longer than any other known deposit of this type. The volume of the Bain diamictite is perhaps $30 \times 10^9$ m$^3$, which is compatible with the probable pre-explosion stratovolcano size at Dasht-i-Nawar. G. Johnson (pers. comm., 1992) has noted new evidence of two flows instead of one. Given the dramatic circumstances implicit in this geomorphologic reconstruction, further detailed field work is required. The civil war in Afghanistan and inaccessible Pushtoon tribal territories, however, make such effort unlikely in the foreseeable future.

In spite of its large size, the Bain deposition seems to have exerted no lasting effect on the course of the nearby Indus river, if indeed the two ever were in close temporal or spatial proximity.

## OROGENY AND DENUDATION IN THE HIMALAYA

Relationships between uplift and erosion in the western Himalaya are important to understanding the great relief and the highly dynamic geomorphic processes there. Variations in mountain landforms commonly reflect differences between slope processes and fluvial or glacial incision of valleys, rather than between overall or general rates of uplift and erosion. Nevertheless, consideration of general regional effects produces interesting questions about local relief characteristics.

Theoretical considerations (Ahnert, 1970) show that, in order to have a balance between rate of erosion and a 'typical' rate of mountain uplift (1 cm yr$^{-1}$), a relief of $> 50$ km would be necessary. Of course, the rates of erosion probably increase exponentially with altitude but the fact that the Himalaya are at least six times lower than the theoretical maximum is an indication that balance is likely only achieved by discontinuous pulses of rapid uplift, alternating with longer periods of quiescence or slow uplift. Numerous measurements made at various places have shown this disparity between present rates of denudation and orogeny (Schumm, 1963), but only a portion of this disparity can be attributed to the fact that process rates are partly a function of measured time interval (Gardner *et al.*, 1987). In general, the modern rates of uplift ( $\sim 7$–8 mm yr$^{-1}$) are about eight times greater than average maximum denudation. Differences in the forms of mountain slopes are the result of different rates of channel incision and slope erosion. Where rocks are very resistant, incision will be greater than slope erosion and a narrow canyon will be formed. The highest rates of any erosion processes occur in glacial and periglacial regions, with glacierized erosion rates of $\sim 0.6$ mm yr$^{-1}$ and mountain rivers of $\sim 0.4$ mm yr$^{-1}$ (Corbel, 1959).

The Arun–Kosi river system that drains the Everest area fills in reservoirs at a 1 mm yr$^{-1}$ denudation equivalent (Khosla, 1953), and the Hunza river tributary to the Indus carries about a 1.8 mm yr$^{-1}$ denudation equivalent (R.I. Ferguson, 1984). The total volume of sediments from the Himalayan source regions over the last 40 myr is $8.5 \times 10^6$ km$^3$ (Menard, 1961). This is equivalent to a regional long-term denudation rate of $\sim 0.2$ mm yr$^{-1}$. The short-term present-day erosion rate, however, seems to be at least 1 mm yr$^{-1}$; five times greater than the long-term rate. Searle (1991), for example, has noted a time averaged exhumation rate for the southern Karakoram of 0.95 mm yr$^{-1}$.

At various hot closure temperatures deep within the Earth, certain minerals anneal or close radioactively generated fission tracks. Below such closure temperatures, fission tracks accumulate in the cooler rock as a function of time, enabling determination of length of time since the mineral was at that temperature. Many researchers have attempted to determine uplift rates in the Himalaya by assuming past uniform geothermal gradients at depth and then using fission tracks accumulated in rocks now on the surface to calculate an original closure depth and the time for uplift from that depth to the surface. Zeitler (1985), for example, first used fission tracks to determine that in the last 10 myr the western Himalaya seemed to be uplifted some 3–6 km regionally and $> 10$ km locally. Nanga Parbat (8125 m), Haramosh (7397 m) and Rakaposhi (7790 m) apparently experienced uplift during the past 7 myr at rates increasing from

$< 0.5$ mm yr$^{-1}$ to greater than several mm yr$^{-1}$. According to Zeitler (1985), in the past 2 myr alone the uplift rate of the Nanga Parbat–Haramosh massif, across which the Indus maintains a great trench, seems to have increased from 2.5 mm yr$^{-1}$ to 5 mm yr$^{-1}$.

England and Molnar (1990), however, have pointed out recently that the uplift of a land surface is really equal to the total overall uplift of the rock minus the exhumation that has occurred. In fact what Zeitler (1985) had actually measured was a rate of exhumation. Yet Zeitler and Chamberlin (1991) have recently discovered exceptionally young leucogranites, as little as 2.3 myr old, which are now high in the Raikot valley on Nanga Parbat. The melting and emplacement at depth of these young leucogranites, during the rapid late Tertiary denudation of the Nanga Parbat region, suggests that the granites may have been produced by decompression melting from removal of overburden load through major fluvial or glacial erosion. Following their emplacement, the granites must have been raised quickly toward the rapidly eroding surface overhead until they emerged in the recent past. These new cooling age dates thus also support Zeitler's (1985) original accelerating uplift path at Nanga Parbat, which may have reached unroofing rates approaching 10 mm yr$^{-1}$ in the last few million years.

In the Indus valley directly below Nanga Parbat the sediments at the base of the Jalipur formation, now shown to be deposited from glacial ice (Shroder *et al.*, 1989), seem to be preserved beneath the Raikot fault. The upper Jalipur valley fill has Nanga Parbat clasts with fission-track dates showing it to be younger than 1–2 myr, yet the lower Jalipur tillite contains almost no rock types from the mountain itself. Although the idea is controversial (see Owen, 1988a, and Shroder *et al.*, this volume), apparently the mountain was not high enough to generate glaciers at this time. This immense mountain seems to have risen in only the past few million years, at a rate perhaps sufficient to affect the growth of glaciers on its flanks.

Vertical relief between the Indus and the top of Nanga Parbat is nearly 7000 m in 21 horizontal km. On this huge mountain face, the Raikot glacier (Figure 1.5) descends in a series of tumbled ice cliffs at an angle of about 51° to an altitude of 5000 m. From there it moves at a more gentle slope of about 7° to its terminus at 4250 m. The glacier was discovered recently to be eroding the area at a rate of about 4 mm yr$^{-1}$ (Gardner, 1986 and pers. comm., 1986), which nearly balances the fission-track uplift rate of 5 mm yr$^{-1}$ (Zeitler, 1985).

Near-equilibrium conditions between uplift and erosion may occur in the nearby Indus river gorge as well. Such conditions are not widespread, however, because the Nanga Parbat and Haramosh summits loom far above both glacier and gorge. Although the mountains are deeply incised by many other glacier and river systems, these processes do not act upon the entire ground surface, and the fact that this major relief still stands shows that great rock strength allows the maintenance of steep slopes even in such seismically active areas.

# GLACIERIZATION OF THE WESTERN HIMALAYA

Glaciers of northern Pakistan are some of the largest and longest mid-latitude types on Earth (Figure 1.5). Himalayan glaciers have been studied by a wide variety of scientists for over a century and a half (see Shroder *et al.*, this volume, for further references). The tremendous accumulations of mass movement and snow avalanche debris that falls to the glacier surfaces in the Himalaya (Hewitt, 1988), coupled with the intense summer heat that melts the ice rapidly in the deep arid valleys, produce great thicknesses of tills on glacier surfaces, as well as massive moraines down valley. For this study, field survey of a number of glaciers in Swat, Kohistan, Nanga Parbat and the main Karakoram provided basic data for downstream assessment of past glacial history (Shroder *et al.*, 1989), as well as modern behaviour (Shroder *et al.*, 1984), and provision of ground truth for simple photogrammetric analysis of satellite imagery at the behest of the US Geological Survey (Shroder, in press a). Emphasis was directed toward glaciers with unusual advance, retreat or surge histories (Table 1.2). Only a few examples of the larger data base are presented here.

Sachen glacier (Figure 1.5) on the east slopes of Nanga Parbat, like many in the Himalaya, is fed largely by ice and snow avalanches from the cliffs on the mountain above. This produces an unusual configuration of a glacier at about 3500–4000 m altitude with little direct connection to its source area 2000 m above on the mountain where snow and icefields accumulate above 6000 m. Nevertheless, this source of nourishment is steady enough to maintain the digitate front of the glacier with remarkably little change over the last 50 yr since its first survey in 1934 (Finsterwalder, 1937), although our recent rephotography and tree-ring dating do show some downwasting at the north terminus.

Sachen glacier is remarkable in also having several rock glacier-like forms that have broken out through the lateral moraines and into the deep ablation valleys

*Table 1.2* Criteria used on satellite imagery of the western Himalaya for recognition of resolvable types of unusual differential or rapid glacier movement followed by less rapid movement or retreat.

| | |
|---|---|
| 1. | Strongly convoluted medial or lateral moraines (5 glaciers; including Hispar Glacier and 2 in Shimshal valley). |
| 2. | Medial or lateral moraines convoluted or offset by extensive crevasses or ogives (13 glaciers). |
| 3. | Tributary ice overriding or displacing main glacier ice (7 glaciers). |
| 4. | Marked depression of ice surface below lateral moraines or tributary glaciers (14 glaciers). |
| 5. | Lateral and medial moraines slightly sinuous, or lobate, or offset by crevasses or ogives (76 glaciers). |
| 6. | Recent major melting or retreat of ice front leaving light-coloured scars behind (9 glaciers). |
| 7. | Extensive area of debris-covered downwasting ice or stagnant ice (106 glaciers). |
| 8. | Linear moraines and accordant tributaries (47 glaciers). |

alongside the glacier. Rock glaciers (Giardino *et al.*, 1987), a polygenetic landform involving mostly slow movement of ice and rock fragments, as well as water under pressure in some cases, abound in the Himalaya and Hindu Kush. The enigmatic examples at Sachen glacier appear most unusual in that their motion is close to perpendicular to the main direction of flow of the parent ice glacier.

The Batura glacier (Figure 1.6) in Hunza is ~ 59 km long, making it one of

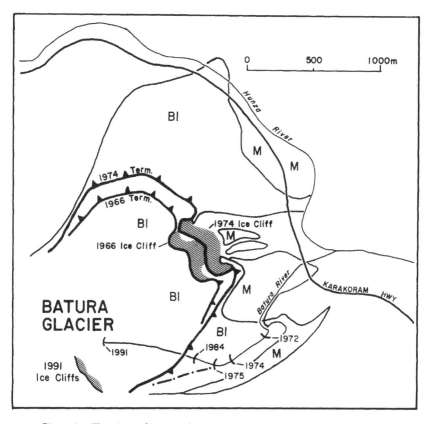

*Figure 1.6* Terminus of Batura glacier in 1991 showing changes since 1975.

BI Buried stagnant ice covered with gray till ( ~ 1885–1991);
M Yellowish moraine about two centuries old;
KKH Karakoram Highway.

The prominent ice cliff of 1966, which advanced to the position noted in 1974, was only a steep till-covered slope in 1984 and 1991. Dates on the Batura river indicate positions of main ice-cave portal; dot and dashed line indicates position of collapsing secondary cave (after Batura Glacier Investigation Group, 1979, 1980; Perrot and Goudie, 1984; and Zhang, 1984).

the eight largest glaciers in the middle and low latitudes. Although it is not a hazardous surging type of glacier, it has been much studied in the past two decades (Batura Glacier Investigation Group, 1980; Shroder *et al.*, 1984) because of some unusual characteristics and its close proximity to the Karakoram Highway between Pakistan and China.

The climate of the deep valleys of the Karakoram near Batura is dry, with only ~ 10 cm of precipitation on average. Annual snowfall on the upper reaches of the Batura is greater, but still only 100–130 cm. The annual 0 °C isotherm is near 4200 m and the equilibrium line occurs at ~ 5000 m. The glacier is therefore cold in its upper reaches, and temperate or warm based in its middle and lower parts where two-thirds of the glacier is covered with thick debris, except for a thin ( ~ 700 m wide) strip of white ice that extends to within ~ 4 km of the terminus.

The Batura terminus itself extended in 1984 to about half a kilometre from the vital Karakoram Highway (Figure 1.6). Desire by engineers to know more about the potential threat to the highway by the glacier led to a series of studies about fifteen years ago of climate, tree rings, ice flux, ablation and velocity (Batura Glacier Investigation Group, 1979, 1980). Results indicated that the glacier would advance close to the highway into the 1990s and then decline thereafter for 20–30 yr. Resurvey in 1984 showed that the predicted advance had not begun and instead the frontal ice cliff downwasted and the ice cliff above the main meltwater channel backwasted (Shroder *et al.*, 1984). By 1991, a large area of ~ 1 km² of newly stagnant ice had been isolated behind the old frontal ice cliff by the continued retreat of the ice portal and extension of the Batura river into the ice behind the old frontal cliff. No further hazard is anticipated but monitoring continues.

Apart from the well studied glaciers of Alaska and the Yukon, the western Himalaya seem to have more glacier surges than anywhere else. Over twelve documented cases of rapid glacier advance are known from there (Hewitt, 1969; Wang *et al.*, 1984), and plentiful twisted moraine loops show other examples (Shroder, in press a). Recent observation in Alaska of subglacial tunnel collapse leading to hydrostatic 'floating' and rapid surges (Kamb *et al.*, 1985) suggests that plentiful meltwater in the hot mountain valleys may be a cause of many surges in the western Himalaya.

With so many glaciers in the world in retreat and fresh water in increasingly short supply, and in part because of the need to gather data pertinent to suspected world climatic warming, the US Geological Survey is using satellite imagery to assess the world glacier ice mass. To this end, 168 glaciers of Pakistan were assessed (Shroder, in press a) to gain increased understanding of these remote and difficult areas (Table 1.3). The largest and most prominent of glaciers with strongly convoluted moraines indicative of surge behaviour are in the valleys of Shimshal, Hispar and Braldu, although several also occur in directly adjacent China north of K2 mountain. The Shimshal valley is notorious for having glaciers that dam the main river periodically and cause floods.

*Table 1.3* Late Quaternary chronology in southern Pakistan and northwestern India (after Allchin and Goudie, 1978; Allchin *et al.*, 1978).

| Phase | Date | Evidence and effects |
|---|---|---|
| Harrapan wet phase | 3000–1800 BC 5000–3800 yr BP | Pollen evidence of humidity. Archaeological sites in some currently dry areas. Indus civilization. |
| Pre-Harappan less moist phase | 7500–3000 BC 9500–5000 yr BP | Pollen analysis of Rajasthan lakes shows *slight* elimination of rainfall. |
| Early Holocene moist phase | 10,000–9500 yr BP | Freshwater lakes inundated dunes in places. Some dune weathering and drainage incision. Extensive human settlement. |
| Major dry phase | > 10,000 yr BP – Later Upper Palaeolithic | Dunes extend over lake basins, Middle Stone Age soils, and block streams to cause aggradation. Little substantial evidence of human activity over wide areas. |
| Major wet phase | < ?40,000 yr BP – Middle Stone Age | Major phase of weathering, decalcification, and stability of dunes. Through flowing rivers with coarse debris loads. Substantial human occupation. |
| Major dry phase | pre-Middle Stone Age | Major sheets of aeolian sand with some slope-wash material containing $CaCO_3$ nodules and rolled lower-Palaeolithic artifacts. |

## SLOPE FAILURE IN THE WESTERN HIMALAYA

Since the collision of the Indo–Pakistani and Eurasian plates produced such intense igneous and metamorphic activity, the resulting massive crystalline rocks are resistant to slope failure. However, the immense relief, strong fracturing and high seismicity guarantee major failure in some places. Where slopes are undercut by the strongly erosive rivers and where pressure release occurs in over-steepened rock slopes following retreat of glacial ice, large falls are pronounced. Even at the bottom of deep mountain valleys and in the foothills, failure of residual soils, valley-fill gravels and deeply weathered saprolites is common wherever gradients are sufficient (Shroder, 1989b). The Karakoram Highway is constantly cut by large and small landslips. In fact, as the exponentially increasing population pressure pushes ever further into these fragile environments, the deforestation, overgrazing and construction contribute significantly to further slope-failure hazards.

Many large slope failures have been recognized in the middle and upper Indus,

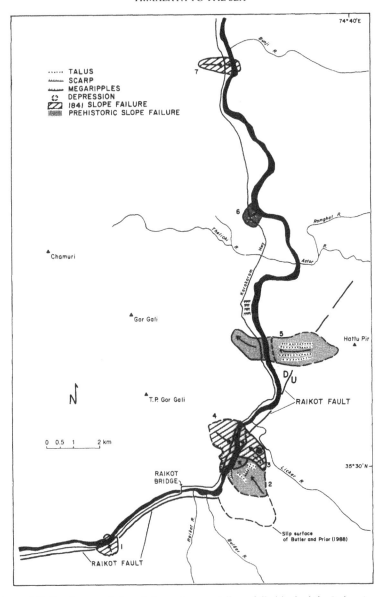

*Figure 1.7* Sketch map of slope failures that partially or fully blocked the Indus river:

1 Tatta Pani debris falls and slides (probably 1841 and recently);

2 Prehistoric Lichar rockslide;

3 Lichar rockslide of 1841;

4 Gor Gali debris slide of 1841;

5 Prehistoric Hattu Pir rockslide;

6 Prehistoric Thelichi debris slide;

7 Bunji rockslide of 1841.

Gilgit and Hunza river valleys (Burgisser *et al.*, 1982; Goudie *et al.*, 1984a), but only a few examples are mentioned here (Figure 1.5). Along the Raikot fault zone, 10 km south of Sassi village, steep dips on foliation planes into the Indus gorge have produced a massive block glide and rock slide > 1.5 m wide. Sumari village sits in a till-filled graben on another large failure above Sassi. Across the gorge on the ridge cut by the Raikot fault between Haramosh and Nanga Parbat, a large number of antislope scarps show characteristics of massive failure. Directly on top of the narrow ridge, 1600 m above the Indus river, Sarkun lake shows the *sackung* (literally 'sagging') style of failure in which a mountain fractures and moves down (Bovis, 1982). An alternative hypothesis to the *sackung* idea would involve disruptions along the Raikot fault to explain the observed phenomena without gravitative settling. In any case, the 1600 m of relief in a narrow ridge only 7 km wide at the base, however, indicates that *sackung* or other failure along the Raikot fault could be expected to disrupt the course of the Indus at sometime in the near future.

Two major rock slides blocked rivers in the western Himalaya in the nineteenth century. A slide in 1858 at Pungurh Serat in the Hunza river gorge caused significant flooding downstream when the dam broke. Large-scale failures have continued on both sides of the valley there ever since because of the steeply dipping bedding planes (Goudie *et al.*, 1984a). The most famous slope failure, however, was that in 1841 when an earthquake brought down part of the Lichar spur of Nanga Parbat into the Indus gorge. Relocation of the original site has been fraught with controversy (Code and Sirhindi, 1986; Butler and Prior, 1988a; Owen, 1989a; Shroder, 1989b; Shroder *et al.*, 1989), but our 1991 fieldwork was designed to resolve the problem (Figure 1.7).

Seven large slope-failure zones occur in the region and their overlapping spatial and temporal relations have caused variable interpretations. The prehistoric Thelichi, Hattu Pir and Lichar failures attest to predominant instability in antiquity. The prehistoric Lichar rockslide is a special case resulting from bedrock thrust over sediments along the active Raikot fault. Following its failure in antiquity, thick talus and debris flow sediments accumulated above it for centuries. A major seismic event in the winter of 1840–1 then caused the collapse of the Gor Gali talus and alluvial fan slope directly across the valley, which was partly overridden by another part of the collapsing Lichar ridge, as well as some remobilization of the prehistoric Lichar failure. The result was a major dam across the Indus river. The Tatta Pani area, 6 km downstream, appears to have failed at the same time and blocked the Indus as well. The dam at Lichar backed up water > 150 m deep and perhaps as much as 30 km upstream into the Gilgit and Indus rivers. As the impounded Indus backed up to its full height on the terrace plateau by Bunji fort, the opposite rock slope failed as well and made a giant wave in the lake (Drew, 1875). Failure of the main dam on the Indus six months later generated a flood near Tarbela that destroyed an army of Sikhs encamped near there. Subsequently the Indus was diverted around the Bunji rock slide, undercut the Bunji terrace plateau, and part of it failed as well.

The uplifted foothill regions of the Himalaya have other kinds of failure. On the Chattar plain, for example, the deeply weathered, clay-rich saprolites slide and flow slowly downhill (Lawrence and Shroder, 1985). The Karakoram Highway in this region is cut repeatedly by such slow failures. Large and small debris falls and slides occur in the valleys where the thick valley-fill gravels are undercut by rivers and fault scarps. Talus slopes, however, are probably the most conspicuous and dramatic element in the mountain landscape. They are created by the constant rock falls, dry grain flows and wet debris flows that form simple small cones up to huge compound accumulations several km wide and up to 1 km long (Brunsen et al., 1984).

Probably the most common catastrophic mass-movement mechanism in the Himalaya is the debris flow in which surface debris is mobilized by rain or snowmelt. Waves or walls of this slurry rush down gullies, cross fans and devastate bridges and fields. Small storms on the peaks can generate many flows that rush in the arid lower valleys in the mountain rainshadows. Larger, more destructive debris flows have blocked some of the larger rivers and overwhelmed major structures as a result of intense storms or breakout floods from ice-blocked water bodies in ice and rock glaciers (Goudie et al., 1984a).

## EVOLUTION OF THE PUNJAB AND LOWER INDUS DRAINAGE SYSTEM

The course of the Indus river from the Salt range down the Punjab–Indus plain to the sea is a gentle one, only $\sim 4.8$ cm km$^{-1}$. The Indus river is a braided stream from the Salt range south, but establishes a well defined meander belt in the lower reaches. Sediments of the meander belt consist of river bar deposits and natural levees on a landscape marked by abandoned channels and crescentic meander scars (Holmes, 1968). The past and present courses of the lower Indus river are largely covered by Jorgensen et al. and Flam in this volume and will not be discussed further in any detail here. Nevertheless, the general geomorphology of the riverine plains of the Punjab, the Indus plain and the Indus delta does need further exposition and is treated herein (Figure 1.8).

'Punjab' means the Land of Five Rivers, but the Indus was not counted as one of them because it was regarded as the western edge of the region. Instead the Punjab is defined and characterized by the Indus tributaries and interfluvial plains that converge on the Indus river from the northeast. From northwest to southwest, the five rivers are the Jhelum, Chenab, Ravi, Beas and Sutlej (Figures 1.1 and 1.2). All of these rivers join together in different places until finally only the Chenab and Sutlej meet in the Panjnad before flowing into the Indus. Two additional small rivers of the region also occur and are important in the literature because of their possible involvement in capture history in late Cenozoic: the Soan river, between the Indus and Jhelum on the Potwar plateau, and the easternmost Ghaggar (and Hakra) and its tributaries to the southeast in India, which disappears into the wastes of the Thar desert.

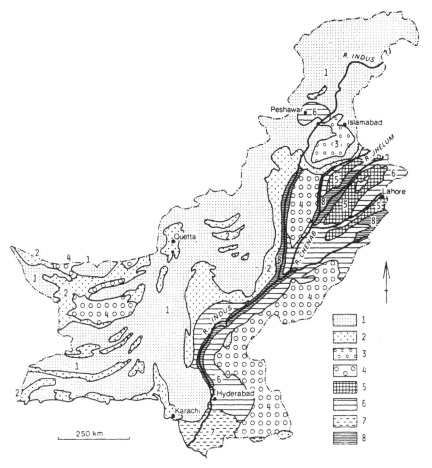

*Figure 1.8* Generalized landforms of Pakistan with soil parent materials. Only plains, plateaux and lowland areas are treated in any detail because of the complexity of the plentiful mountain areas (after Higgins *et al.*, 1973).

1 Mountains: rock, residual material and local colluvium;
2 Piedmont plains and terrace remnants: subrecent and some old gravelly and loamy to clayey piedmont alluvium;
3 Dissected old loess and alluvial terraces (Potwar uplands): loess, residual material, old river alluvium and subrecent outwash;
4 Rolling sand plains: old and subrecent wind-reworked sands;
5 Old river terraces (bar uplands): old mainly silty river alluvium;
6 Subrecent river plains: subrecent silty and clayey river alluvium;
7 Indus delta: subrecent and some recent silty and clayey estuary alluvium with some clayey coastal alluvium;
8 Active and recent river plains: recent and some subrecent silty and sandy river alluvium.

The geomorphic evolution of the Punjab is problematic because little modern work has been done there. Some of the older ideas are probably valid but most are rather too speculative. Greatly needed are modern studies using acceptable dating techniques. In any case, as previously mentioned, Raynolds (1981) used palaeocurrent directions of crossbeds in the Siwalik sediments to show that an east–northeast flowing river of the eastern Potwar plateau was replaced by a southward flowing drainage. Whether or not this was actually an ancestral Jhelum or the upper headwaters of a once larger Soan is open to question. Pascoe (1920a,b) noted that the upper Jhelum has the appearance of having been headwaters to the Soan and having been subsequently captured by the lower Jhelum. He felt that the gravels and alluvial deposits of the Potwar plateau through which the Soan flows are thicker and more extensive than can be attributed to the underfit Soan of today. A glance at the map (Figure 1.2) also shows that the upper Jhelum, after passing from the Kashmir basin through the Pir Panjal range, is trained and deflected south along the left lateral strike-slip fault on the west side of the Hazara syntaxis, before finally passing around and being further pushed south by the frontal thrust of the Salt range.

Wilhelmy (1969) assessed all of the prior work (Figure 1.9), including the texts of antiquity, and showed that two independent major river systems existed in the Indus lowlands ∼ 2000 BC ( ∼ 4000 yr BP): the frequently changing Hakra–Nara courses and the Indus river. At that time the watershed divide between the Indus and the Ganges was determined by the Sutlej and the Ur-Jumna (Chautang), both passing into the Indus basin. The mythical Saraswati river (also Sarasvati or Ghaggar) of the Rig Veda was tributary to these two rivers as well (Figure 1.9A). Ghose et al. (1979) used Landsat imagery to define the prior courses of the Saraswati, and suggested that it once flowed several hundred km further east of the Indus river of today.

From ∼ −1000 BC ( ∼ 3500–3000 yr BP) the Jumna was captured and switched into the Ganges drainage (Figure 1.9B). The Saraswati continued as before, but with some of the Sutlej changing position through avulsion to move northwest to become the Saraswati–Sutlej and to extend its lower channel. Up to ∼ 600 BC (2600 yr BP), a later Sutlej course (Hakra–Sutlej) provided the water supply for a primary river valley at the eastern margin of the Indus plain throughout the year. Aside from their different upper courses, the first Jumna and the later Sutlej, one after the other, used the same Hakra river bed. Sometime prior to ∼ 600 BC the Sutlej avulsed to the northwest again to become the Hakra–Sutlej (Figure 1.9C), and the Hakra–Nara system began to become underfit.

Between 500 BC and AD 1100 the Sutlej was captured or avulsed northwest once again and was confluent with the Beas (Beas–Sutlej). The old Sarasvati-Ghaggar was greatly reduced in discharge and died out in the desert, while all of the old Sutlej channels had dried out (Figure 1.9D). These many avulsions and cross-divide captures between the Indus and Ganges tributaries explain the similarity in river faunas commonly cited in the older literature as evidence for the west flowing 'Indobrahm' river and subsequent drainage reversal.

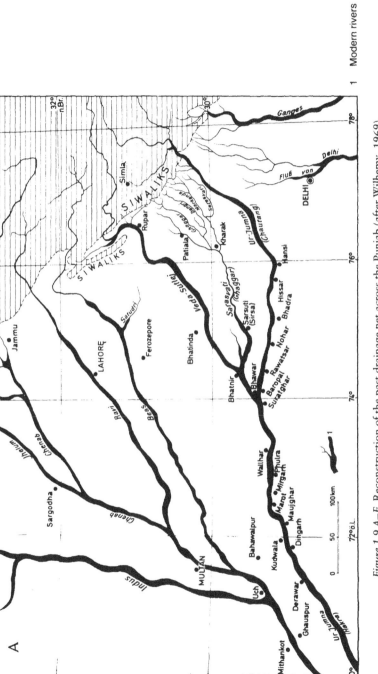

*Figure 1.9 A–F* Reconstruction of the past drainage net across the Punjab (after Wilhemy, 1969).

A ~ 2000 BC;

Comparison of these different courses is facilitated by making photocopies which can then be overlain on a light table to show progressive course changes through time.

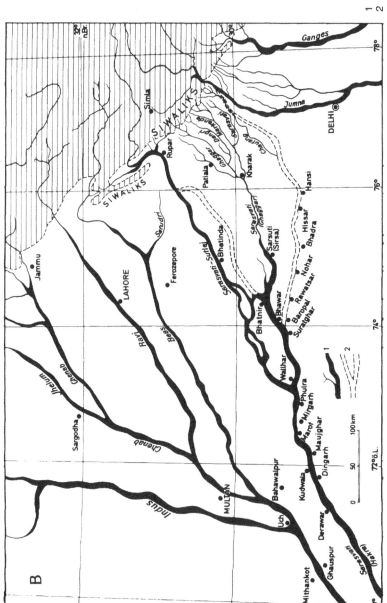

B 1500–1000 BC;

1 — Modern rivers
2 — River channel straths

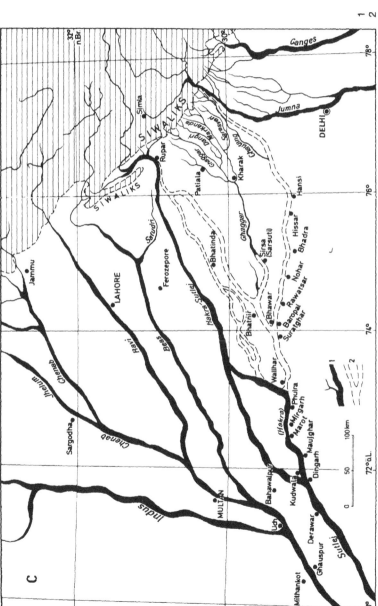

C ~ 600 BC;

1 Modern rivers
2 River channel straths

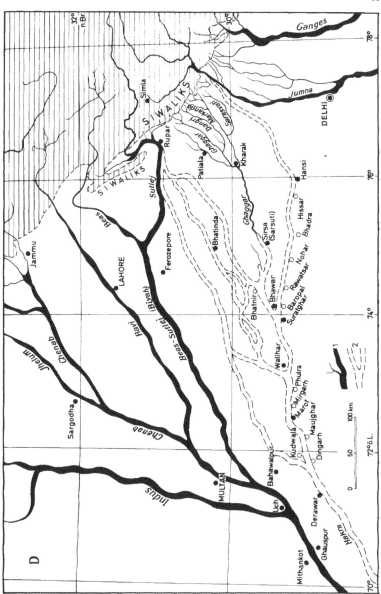

D 500 BC – AD 1100;

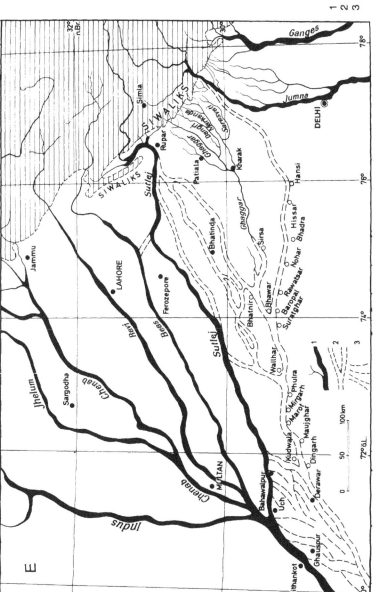

E AD 1300–1500;

1   Modern rivers
2   River channel straths
3   Inland delta near
     Derawar

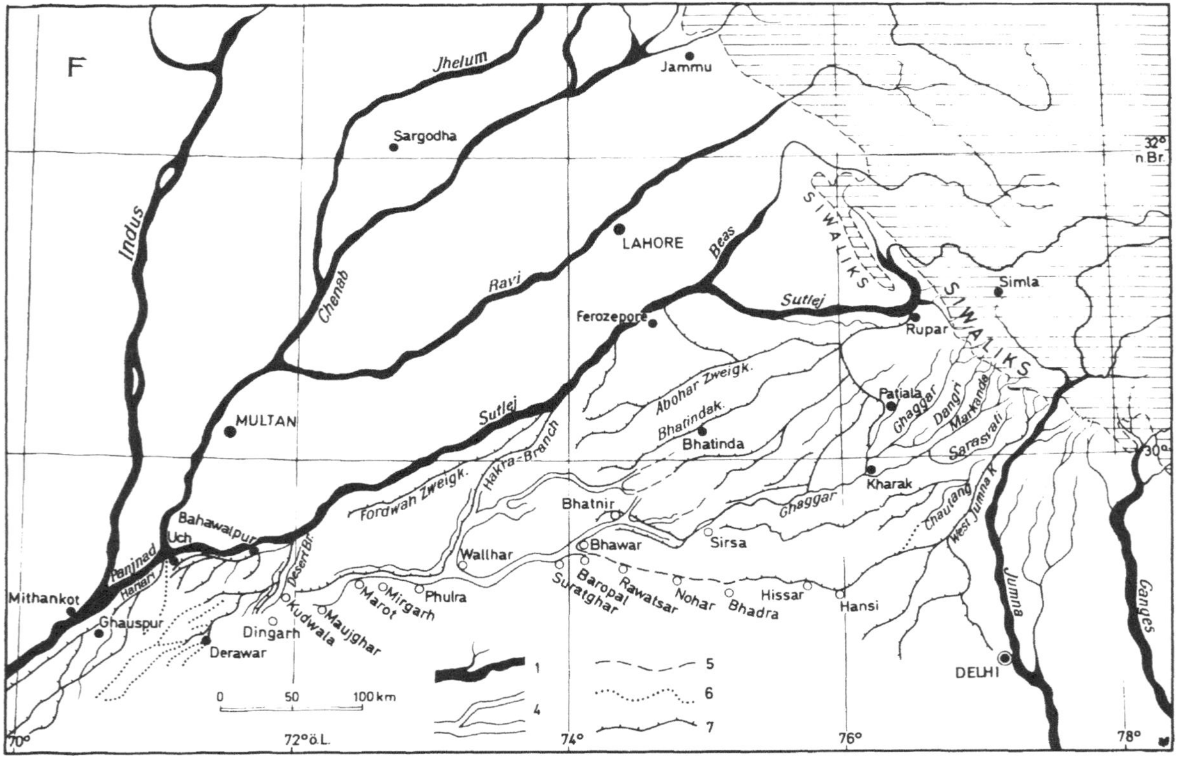

F ~ 1940.

1 Modern rivers
2 River channel straths
3 Inland delta near
  Derawar
4 Preserved old channels
5 Intermittant meadows
  along old river line
6 Temporarily employed
  drainage way
7 Irrigation canal

From AD 1300 to 1500 the Sutlej avulsed southeast away from the Beas, back closer to its original channel. The Ravi, Chenab and Jhelum had a confluence to the northeast of Multan about 1245. By the time of Timur's invasion in 1398, the Chenab had again changed its course and was flowing to the west of Multan. The Ravi continued to flow for a while to the southeast of Multan, until it too changed its course to the present one. In times of high flood today it still reaches Multan by the older channel, which dates back to at least AD 800 (Seth, 1978). Also during this time the confluence of the Jhelum and Chenab had shifted ~ 125 km downstream to the southwest as the Jhelum shifted west and the Chenab slightly east. The confluence of the Chenab with the Indus, however, had shifted ~ 48 km upstream. Numerous spill channels and course changes were consequently developed on the left bank of the Indus just downstream from the confluence of the five rivers of the Punjab. The Indus itself had shifted ~ 50 km westward as well (Figure 1.9E).

The confluence of the Sutlej and the Beas continued to change position with time. Burnes (1835), for example, saw that the old channel of the Sutlej where it joined the Beas had been at Ferozepur rather than its later location 25–30 km to the northeast. By 1940 both the Indus and the Chenab had shifted more to the west. The headwaters of the Beas, however, was entirely diverted into the Sutlej, leaving only an abandoned channel. Seth (1978) also showed that the Ravi had shifted its confluence with the Chenab north by this time as well (Figure 1.9F).

In general the plains of the Punjab and the Indus are virtually featureless (Figure 1.8), but elements of microrelief assume considerable importance because of their relationship to flooding, and probably also because of their Quaternary history, although little is known of this. Only a few areas of hills of circumalluviation occur throughout this extensive region. A number of these inselberg-like hills occur between the Jhelum and Chenab, between the Chenab and Ravi rivers and in the extreme southeast along the Indian border. These bedrock hills are projections of basement rock from the top of the buried Sargoda–Shahpur and the Nagar Parkar structures. A few other hills of Eocene limestone occur along the lower Indus as well. Five distinct microrelief landforms occur throughout the entire region: active floodplains, meander floodplains, cover floodplains, scalloped interfluves or bar uplands (Kureshy, 1977), as well as the deltaic plain and tidal delta that are treated separately.

The active floodplains lie directly adjacent to the rivers and are inundated almost every rainy season or when the snows melt in the high mountain areas. During the low water season considerable areas of coarse-textured sand and silt are exposed, and the surfaces are scarred by numerous active or abandoned channels. Active floodplains occur along all of the rivers except the lower half of the Ravi. This belt is quite wide from Kalabagh to the Indus delta.

The meander floodplains usually adjoin the active floodplain but are somewhat higher. They also occur away from the present courses of the rivers on the site of the old channels. Bars, meanders, levees and oxbow lakes are common. Relief is generally not more than a metre and the soils differ because of the diversity of materials deposited. Meander floodplains are widespread along the Jhelum, the Chenab and the upper reaches of the Ravi. Along the Indus it is absent in the upper reaches but widespread in Sindh.

Cover floodplains consist of recent alluvium spread over former riverine features as a result of floods. Because of the varying speeds of the flood waters at different locations and different times of deposition, frequent changes of soil texture occur. The boundary between the meander floodplains and the cover floodplains is indistinct as the two commonly merge together. Cover floodplains are the most extensive of the plain areas in Sind, the Bahawalpur plain between the Sutlej river and the Thar desert, the Ganji Bar area between the Ravi and the old bed of the Beas, and the Rechna doab area between the Chenab and the Ravi rivers.

Scalloped interfluves or bar uplands occur in the central and higher parts of the Chaj, Rechna, Bari and Sind Sagar doabs. Their boundaries are mostly formed by river-cut scarps, commonly $> 5$ m high. Low sand or silt dunes occur at their southern ends in some places. The Thal desert area on the Sind Sagar doab is the interfluve south of the Salt range between the Indus and the five rivers of the Punjab. It is swept by multidirectional winds that have produced several different dune formations (Higgins *et al.*, 1974; Figure 1.10). Longitudinal sand ridges occur primarily in the north portion and are aligned parallel to the dominant winds that shift direction $< 120°$ between summer and winter. Transverse sand ridges are at right angles to opposing winds of similar strength, and angles between dominant summer and winter winds $> 120°$. Transitional forms between longitudinal and transverse dunes occur as alveolar sands, where one of the opposing winds is slightly predominant, and ridgy alveolar sands, where one of the winds is markedly predominant. Barchan dunes occur under complementary wind regimes in severely devegetated areas and occupy less than 5 per cent of the region.

The Thar desert of southeastern Pakistan and northwest India extends west from the vegetated floodplain of the Indus river to the Aravalii range in India, and from the Rann of Kutch northeastward into the Punjab. The region consists of low hills and sand dunes lying on sandy alluvium (Breed *et al.*, 1979). Former river courses cross the region, with the dry bed of the Ghaggar–Saraswati being the best known (Oldham, 1893; Wilhelmy, 1969; Verstappen, 1970; Allchin *et al.*, 1978).

Average annual rainfall ranges from $< 100$ mm in the northwest to $> 400$ mm in the southeast. Monsoon-driven, effective surface wind flow is from southwest in summer and northwest in winter. Controversy over the origin and timing of climate changes in the Thar region has raged for some time (Verstappen, 1970). Overall, however, one can infer that substantial geomorphic changes have occurred in the area during late Pleistocene and Holocene (Allchin and Goudie, 1978; Allchin *et al.*, 1978). Conditions have been both wetter and drier than they are at present, but in general the Holocene seems to have been somewhat moister than the late Pleistocene, so the old concept of glacial pluvials

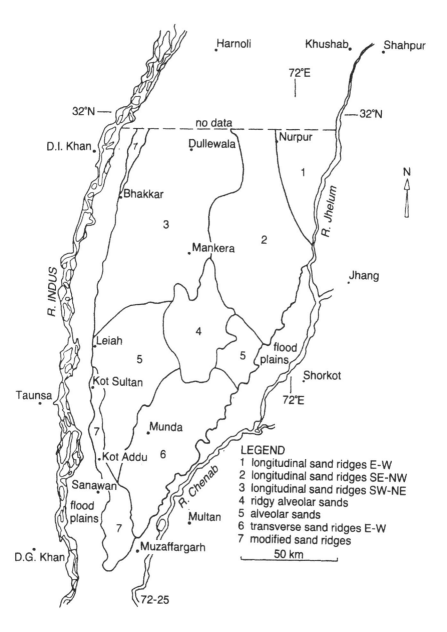

*Figure 1.10* Generalized map of sand ridge landforms in the Thal desert interfluvial area between the Indus river and the Jhelum and Chenab rivers (after Higgins *et al.*, 1974).

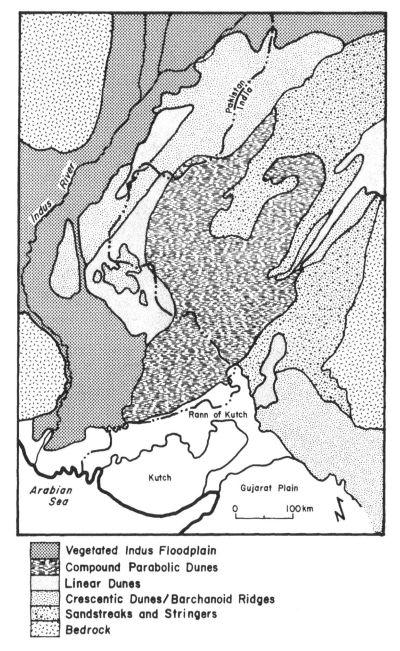

**Vegetated Indus Floodplain**
**Compound Parabolic Dunes**
**Linear Dunes**
**Crescentic Dunes/Barchanoid Ridges**
**Sandstreaks and Stringers**
**Bedrock**

*Figure 1.11* Distribution and morphology of aeolian sand in the Thar desert of India and Pakistan. Map based on Landsat (ERTS) imagery (after Breed *et al.*, 1979 and Allchin *et al.*, 1978). Kar (1990) remapped this area somewhat differently but the overall plan is similar.

and post-glacial desiccation may not apply here (Table 1.3).

Vegetated compound parabolic dunes of a distinctive rake-like shape are the most widespread type of aeolian landform in the Thar desert (Breed *et al.*, 1979; Figure 1.11), although Kar (1990) considered many of them to be mega-barchanoids. Compound crescentic dunes are another common type that occur in a broad band along the Pakistani–Indian border on the northwest side of the Thar desert. Parallel linear dunes are also an important dune type along the frontier.

Ancient courses of the Indus river in its lower reaches have been documented but not well dated as yet (Seth, 1978; Kazmi, 1984). Geophysical and borehole data have revealed a number of buried channels, presumably entrenched during lower sea levels in Pleistocene, and filled to the present time (Figure 1.12). One buried channel occurs about coincident with the modern Indus upstream from Hyderabad and the other is beneath the old Nara channel to the east alongside the Thar desert. The two buried channels are confluent northeast of Hyderabad and pass as one channel due south from there and east away from the present course of the Indus to the coast. Since the time that these channels were emplaced and filled, throughout the Holocene the Indus has shifted west through a series of course changes or delta switches to its present position. Seth (1978) noted that one of these relatively recent course changes occurred as a result of the 1819 earthquake, when 5120 km$^2$ was submerged for two years through the elevation of the so-called 'Allah bund' which extends over 32–80 km and rises 3–8 m above the surrounding land. The mouth of the Khori Creek was also depressed (Figure 1.12). In 1826 the Indus burst all dams on its course and eroded a passage through its old channel to discharge into Khori Creek close to the Rann of Kutch. On the way it overspread the Sind desert and cut its way through the Allah bund (Sahni, 1956).

The modern Indus delta covers an area of about 1600 m$^2$ between Karachi and Cape Monze to the Rann of Kutch in India. From inland extending towards the sea, the Indus delta is comprised of deltaic floodplain deposits with an intervening meander belt of deposits from the distributaries, an arcuate zone of older tidal deltaic deposits, followed by more recent deposits of the tidal delta and coastal sand dunes (Kazmi, 1984). The delta itself is thus a product of energetic interaction between fluvial and marine processes (Wells and Coleman, 1984). Historically the delta formed in an arid climate with a high river discharge (400 × 10$^6$ tonnes of sediment yr$^{-1}$), moderate tidal range (2.6 m), extremely high wave energy (14 × 10$^7$ ergs sec$^{-1}$), and strong monsoon winds from the southwest in summer and northeast in winter. The resulting barren, sandy, lobate delta is dissected by numerous tidal channels and has prograded seaward during the last 5000 yr at ∼ 30 m yr$^{-1}$. Delta morphology is midway between that of fluvial domination (elongate, protruding distributaries) and high energy, wave-dominated (beach, beach ridge and long shore drift deposits).

Coarse sediments from the Indus river generally remain in the delta and inner self or undergo transport to deeper water via the Indus submarine canyon. Little

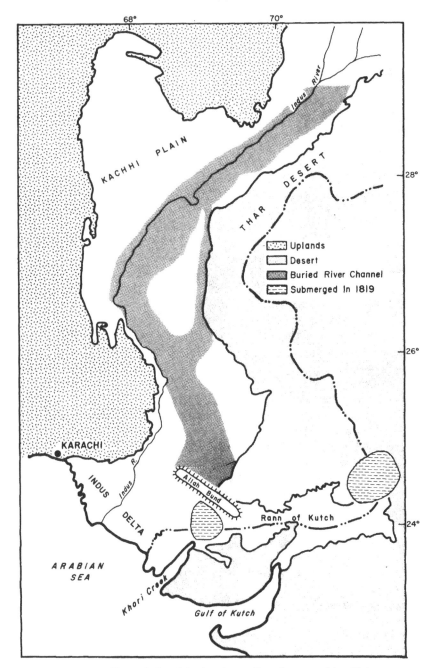

*Figure 1.12* Map of lower Indus plain showing outline of presumed late Pleistocene entrenched channel of the Indus. The Allah Bund was raised in the earthquake of 1819 and several other areas were depressed (after Seth, 1978; Kazmi, 1984).

fine-grained sediment stays in the delta because maximum discharge occurs during the southwest monsoons, resulting in transport of muds southeast into the Rann of Kutch. Extensive irrigation and flood protection works have reduced discharge to about one-fifth, sediment loan to $100 \times 106$ tonnes $yr^{-1}$, and may continue to reduce the load to zero in a few decades (Milliman *et al.*, 1984). As a result, the active delta has shrunk to ~ 160 km², one-tenth of its former size. Decreased load and high wave energy will cause erosion and marine transgression of the delta so that the end product will be a wave-dominated feature with a transgressive sand body capped by extensive aeolian dunes.

## EVOLUTION OF THE SOUTHWESTERN BASINS AND RANGES

Only minor geomorphologic research has been done in the basins and ranges south and west of the Salt range in the Northwest Frontier Province, Sindh and Baluchistan. This is partly the result of physical or politically motivated inaccessibility and partly because the 'plums' of research problems have been perceived as the landforms, processes and sediment records located close to or in the Himalaya. Perhaps the most important reason, however, is that much of the geomorphology in this barren desert landscape is largely structurally based and not process related. Structural geomorphology is of great interest to tectonicists but not so much to process geomorphologists or to many scientists of the Quaternary. Only the geomorphology close to the active Chaman transform fault or that associated with active or potential hydrocarbon or mineral extraction zones has received much attention to date.

The basic structure and landforms of the region, as with so much of the rest of Pakistan, are a direct result of sedimentation, deformation and erosion of rocks produced by the collision of the Indo–Pakistani plate with the Afghanistan microplate and the main Eurasian plate. The Chaman transform fault and the Makran subduction zone (Figure 1.1) are the chief controlling factors in the direction and character of the structural and earth-surface processes involved in the area. The collision process incorporated in an unusual fashion in Las Bela valley northwest of Karachi (the piedmont plains and terrace remnants of Figure 1.8) a triple junction on land, formed by the intersection of the Makran subduction zone with the Owens fracture zone–Murray Ridge zone of the Indian Ocean, and the landward extension as the Ornach Nal–Chaman–Panjsher fault system. The Chaman transform fault zone connects the Makran convergent margin in the southwest with the Himalayan convergence zone in the northeast (Farah *et al.*, 1984a,b). The whole zone is nearly 900 km long and is a large and complex group of faults with a range in width from a few tens of km to several hundred. Offset along the fault system is no less than 200 km and probably over 450 km since Early Miocene ( ~ 20 myr ago; Farah *et al.*, 1984a,b). These faults seem to continue to function as a transform zone along which rapid neotectonism occurs in Pakistan and Afghanistan (Abdel-Gawad, 1971; Jacob and Quittmeyer, 1979; Farhoudi and Karig, 1977).

Directly east of the Chaman zone occurs the internal convergence zone of Quetta–Loralai–Barkhan (Farah *et al.*, 1984a,b), which is the fold-belt ranges of Suleiman and Kirthar (Figure 1.1 and Table 1.1 [I: PD; V: PK]). This region of linear and zig-zag ridges is a complex of large upright fold 'festoons' that are the result of a combined northward and westward convergence and counterclock-wise rotation of the Indo–Pakistan subcontinent with the Eurasian plate.

The Chagai–Makran region west of the Chaman fault in southern Afghan-istan and southwest Pakistan is a subduction–convergence zone with an unusually wide and exposed volcanic island arc and trench gap which is on land. The general geologic setting appears to involve an Andean-type of andesitic volcanic arc in the Chagai hills, Ras Koh and Saindak areas. Marginally active volcanoes include Koh-i-Sultan and others. South of this arc belt is a wide area of Quaternary basin fill with rolling sand plains (Figure 1.8) in the Hamun-i-Mashkel, and from there to the coast are the Makran ranges where arc–trench gap materials are widely exposed.

The Chagai hills on the border with southern Afghanistan began as a volcanic island arc in Cretaceous which was subjected to repeated uplift through the Tertiary. Two important phases of volcanism and folding tectonism occurred around the Mio–Pliocene boundary ( ~ 5–6 myr ago) and in the early Quater-nary. The earlier Mio–Pliocene events were followed by erosional planation in Pliocene. Subsequently, alluvial fan deposition occurred around Chagai in early Quaternary, which marked renewed uplift of the arc. This uplift directly preceded the extrusion of the Quaternary volcanics of the Koh-i-Sultan area, members of which interbed with and cap this early Quaternary alluvium (Farah *et al.*, 1984a,b). Landforms of the region are rugged, wind-blasted hills of rock with intervening desert basins, all with extensive piedmont plains with terrace remnants, alluvial fans, dune sands and dry lake basins, some with salt crusts (Figure 1.8).

The Makran margin was formed as an accretionary wedge of sediments that developed between a buried offshore trench above a subduction zone, and the volcanoes of the Chagai area and elsewhere. The structural landforms of the Makran region are confined between two strike-slip fault systems to the west and east: to the west in Iran, the Neh faults of the Hari Rud zone of the eastern boundary of the Lut Block; and to the east in Pakistan, the Chaman transform zone.

Regional interpretations of the Makran ranges of southern Baluchistan suggest that by middle Miocene ( ~ 15 myr ago) an extremely large submarine fan had built westward in a moderately deep oceanic setting (Harms *et al.*, 1984). The quite sandy, thick-bedded nature of the deposits, which stretched at least 400 km west of potential source areas at the Indian plate boundary, indicate a fan of enormous proportions and energetic character. The fan from a presumed proto-Indus developed in the Sindh estuary, which was marked on its east side by a major transform fault (now the Chaman system) and on its west by a subduction zone. The sand-dominated event on the fan corresponds with the deposition of the Lower Siwalik Group in northern Pakistan (G.D. Johnson *et al.*, 1979), and thus is possibly linked to orogeny in that collision zone. Subsequently in the later Cenozoic the marine sediments shifted to mainly slope and shallow shelf facies, with heavy mineral and clay mineral assemblages unlike Indus drainage characteristics and similar to those of the early Tertiary beds in Baluchistan. This must record the movement of the lower Indus river plain away to the east from this region, and may indicate the establishment of drainage from north to south across at least part of the Makran.

Leggett and Platt (1984) mapped a variety of structures in the Makran, all with surface landform expression. Reverse faults striking parallel to bedding form linear ridges. Tight folds, which are mostly small synclines with half wavelengths of ~ 1–3 km, and large open synclines with half wavelengths of ~ 20–30 km form valleys and basins in linear zones. Conjugate wrench (small strike-slip) faults of minor displacement ( < 1 km) have offset other ridges, and large terrane bounding faults of unknown displacement set apart larger regions. In a few places occur large anticlines which are transverse to the regional structural grain of the country. In general, many of the structural landforms are relatively uneroded or unbreached, although cuestas (dip < 10°), monoclinal ridges (dip 10°–45°) and hogbacks (dip > 45°) are common landforms where erosion has exposed dipping beds. The curvilinear structures associated with the plentiful plunging folds produce zig-zag ridges on the landscape (Harms *et al.*, 1984). Wherever mudstones are exposed, complexly dissected badlands have developed across structure.

Snead (1967, 1969) divided the Makran coast into three broad geomorphic types: rocky cliffs, wide sandy beaches, and low, flat delta plains. Active tectonism associated with subduction is producing numerous raised beaches, as well as entrenching of pre-existing drainage. Mud volcanoes occur at several places along the coast. They are conical hills with reliefs of up to ~ 100 m that reflect escaping natural gas in the high pressure areas of a subduction zone. Several large tombolos occur, particularly where offshore fault-block horsts have been tied to the coast by spits up to 15 km long. Two small rocky islands also formed off the coast in an earthquake in 1945. The Sutkagen-dor archaeological site on the west coast near the Iranian border (in the extreme southwestern piedmont plains area of Figure 1.8) was thought to be an Harappan seaport of ~ 2000 BC ( ~ 4000 yr BP). Today it is ~ 55 km from the coast and 41 m above sea level. Both silting of the Dasht river and uplift are probably responsible.

## CONCLUSION

Study of the regional geomorphology and Quaternary history of Pakistan has received increasing attention over the past decade and this paper has attempted to address a general overview of some of the newer ideas. Modern methodology has made only small inroads into the understanding of some of the high magnitude, high frequency landform-shaping events that have occurred in this active region. Inasmuch as parts of the Himalaya are rising at such considerable rates and are being degraded equally as rapidly, a clear exposition of the course of these events in Quaternary times should be the goal of future research. We know a good deal now but a far greater amount remains to be discovered.

# 2

# MIOCENE TO HOLOCENE UPLIFT AND SEDIMENTATION IN THE NORTHWESTERN HIMALAYA AND ADJACENT AREAS

*Michael E. Brookfield*

## ABSTRACT

Sedimentary basins marginal to the Himalaya and adjacent ranges subsided markedly and continuously from Miocene times onwards, accumulating progressively thicker continental clastics. In contrast, recent radiometric and fission-track dates indicate that the northwestern Himalaya, Karakoram, Pamir and adjacent ranges underwent major phases of erosion (and presumably uplift) starting around 15 myr in the mid-Miocene. These cooling dates seem remarkably consistent in the various ranges north of the Main Central Thrust in India and north of the Northern Suture in Pakistan. In conjunction with the late Tertiary deformation of the northwestern Himalaya and Pamir, this suggests that the late Cenozoic uplift of the Himalaya, Pamir and Tibet is related to the start of intracontinental thrusting along the Main Central Thrust and Northern Suture, after earlier Tertiary telescoping of the northern Indian passive margin and crustal thickening by shortening in Tibet. Sediment budget studies indicate progressively increasing sediment volume delivered to the Indian Ocean and to areas north and west of the mountains in post-Oligocene times by erosion of the rising mountains. These increasing rates of erosion indicated by radiometric and sediment budget studies are confirmed by evidence of rapidly increasing rates of net uplift derived from palaeontological and palaeoclimatic evidence and from river gradient studies. Most of the uplift of the Himalaya, Karakoram, Pamir, Kun Lun and Hindu Kush has taken place in the last 15 myr and cannot be directly related to continental collision. A more plausible mechanism involves changes in upper mantle structure as a consequence of the end of subduction along the Pamir and Himalayan arc.

## INTRODUCTION

The Himalaya has been considered the best example of a collision mountain belt because mountain building is still active, forming the highest range and plateau in the world; the architecture is relatively simple, with comparable structures extending along strike for over 2000 km between the western and eastern syntaxes; mountain building is clearly related in space and time to the collision of India and Asia; and the history of Indian plate movement is available from the magnetic anomalies of the Indian Ocean (Mattauer, 1986). Nevertheless, the mechanics of the collision are disputed. Did the Indian shield underthrust the entire Tibetan plateau (Powell, 1986)? Was limited underthrusting followed by indenting of India into Asia, with Asian blocks being moved northwards and eastwards along major faults, particularly at the western and eastern syntaxes (Molnar and Tapponnier, 1975)?

A particular problem is the time-lag between initial late Eocene collision of India and Asia and the post-Miocene major uplifts of the Himalaya and adjacent ranges and Tibet (Tapponnier *et al.*, 1986).

In this paper I wish to evaluate various lines of evidence for uplift of the northwestern Himalaya and adjacent areas which indicate that major uplift did not start until the early Miocene and progressively accelerated through the late Cenozoic. This evidence includes: foredeep and intermontane basin stratigraphy and sedimentology; radiometric and fission-track dating; sediment budgets of erosion of the mountains and deposition in foredeeps north and south of them; palaeobiological evidence of climatic change and uplift; and changing drainage patterns.

Before discussing the general features of the various late Cenozoic sedimentary basins and mountain areas, the components of uplift and subsidence must be defined to establish a common ground for discussion. Uplift is positive vertical elevation with respect to the geoid (mean sea level). Similarly, subsidence is negative vertical elevation. Uplift normally refers to *net* uplift, combining the effects of *gross* (or total) uplift minus the effects of erosion. Thus:

Net uplift = Gross uplift − Erosion.

Net uplift can be obtained by geodetic measurements which give present values of about 1 mm yr$^{-1}$ for the High Himalaya and Tibet (Chi-yen Wang and Yaolin Shi, 1982).

Gross uplift can be determined from sedimentary and fossil evidence, if erosion has not removed it. Thus, sediments and fossils characteristic of Miocene lowlands now occur on the Tibetan plateau. If erosion has stripped this material, then erosion rates can sometimes be determined with radiometric and fission-track dating, as well as by working out the sedimentary budgets of erosion/deposition of the mountain belts, although a number of assumptions must be made.

Net uplift rates in the Himalaya and Tibet have often been inferred from radiometric and fission-track dating. In fact, radiometric and fission-track dates give only cooling temperatures, and these can be related to depth from the contemporary land surface only by assuming an average geothermal gradient. Such average gradients are unlikely to be correct, particularly in mountain belts

44

undergoing deformation magmatism and metamorphism. Nevertheless, without geothermobarometric data average gradients must be used.

Thus, erosion rates can be calculated from the depth of rock removed between successive closure temperatures of the systems. On the other hand, such erosion rates do not necessarily relate to uplift; the land surface could stay almost at sea level as successive layers are eroded from the land surface (see England and Molnar, 1990, for a good discussion of this). In addition, geothermal gradients vary depending on heat flow, on thermal conductivity of rocks and on topography and an average gradient rarely approaches natural conditions. The effect of topography in particular provides a way of reconstructing ancient landscapes.

Nevertheless, in general terms there is an increasing logarithmic relationship between erosion and height in mountains (Schumm, 1963). If the volume of rock removed from a mountain belt can be calculated (for example, by adding up the amount of solid and dissolved material delivered to adjacent sediment traps), then some idea of the total erosion of the belt major can be obtained. This does not necessarily hold for plateaux, where sediments can be uplifted several km without significant erosion, as in Tibet. Such uneroded sediments give values for gross uplift.

This paper will show that: (a) erosion rates, which are derived from radiometric and fission-track data, coupled with sediment budget studies, can give a consistent view of the accelerating late Tertiary uplift and erosion of the Himalaya and adjacent areas; and (b) gross rates of uplift, which are derived from uplifted marine deposits, river gradients and palaeoclimatic data, are consistent with the erosion rates from the Himalaya and with a geodynamic model involving early Tertiary crustal thickening in Tibet followed by southward thrusting of the Himalaya and rapid uplift of the Himalaya, Pamir and Tibet. These concepts are also applicable to other mountain belts where crustal shortening has preceded the main topographic uplifts.

First, the general features of the various late Cenozoic sedimentary basins and mountain areas are noted. The time scales used are of Berggren *et al.* (1985b) and Repenning (1984).

## NORTHERN FOREDEEPS

The bulldozing of the Pamir into Central Asia has formed a narrow trough, the Surkhab basin, flanked by larger foredeep basins, the Tadzhik basin on the west and the southwestern Tarim basin (Figure 2.1). These northern foredeeps are outlined in some detail, since data on them is difficult to obtain and has not yet been collated in easily accessible summaries.

### Surkhab basin

The Surkhab basin is ~ 300 km long and 30 km wide and contains a thick (3.5 km) Upper Oligocene–Pleistocene section (Krestnikov, 1963). Little

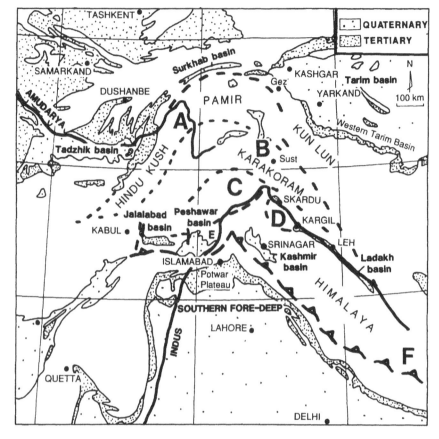

*Figure 2.1* Location map for the northwestern Himalaya syntaxis and adjacent areas. Dashed lines are main faults bounding tectonic units. Letters refer to units used in radiometric/fission-track uplift plots:

A Pamir;
B Karakoram;
C Kohistan;
D Ladakh;
E Peshawar;
F Himalaya.

information is available. The basin is being overthrust from both the north and south. The Mesozoic clastics of the North Pamir are being bulldozed northwards against the southerly thrusting crystalline rocks of the southern Tien Shan (pers. obs., 1990).

Geodetic measurements show that the basin is presently subsiding in places by ~ 0.3 mm yr$^{-1}$: Holocene fault escarpments reach a height of 13 m. South of the Surkhab fault, the land is rising at > 10 mm yr$^{-1}$ (Finko and Enman, 1971). The

Surkhab basin is likely to end up as a most peculiar crush belt between the North Pamir and Tien Shan crystalline massifs.

## Tadzhik basin

The Tadzhik basin is a depression filled with 10–14 km of Mesozoic–Cenozoic deposits. It lies between the late Paleozoic Gissar–Alai range on the north and west, and the Pamir and Hindu Kush on the east and south (Loziyev, 1976; Figure 2.1). The Hindu Kush and Pamir are actively overthrusting the Tadzhik depression and the basin is closing rather than being displaced to the west (Abers *et al.*, 1988). The depression is thus the site of active crustal shortening by folding and faulting as well as tectonic and sedimentary loading of a subsiding foredeep.

Northwesterly directed folding and thrusting commenced after deposition of Palaeogene deposits. The Mesozoic–Palaeogene deposits are ~ 4–5 km thick and may form a subsiding passive margin sequence (Leith, 1985). On the other hand, these deposits could equally well form a post-orogenic successor basin sequence, deposited after Mesozoic collision orogeny (pers. obs., 1990). The last marine deposits in the basin are red lagoonal gypsiferous clastics of the Shurysay formation of Oligocene age (Glikman and Ishchenko, 1967). This forms a sea-level datum for subsidence measurements. At Ariktau, south of Dushanbe, the thin Oligocene is now 2 km *below* sea-level.

The succeeding Neogene non-marine clastics (Abdunazarov *et al.*, 1984) are ~ 3–7 km thick and mark the deformation of the Tadhzik depression and the uplift of the Pamir range to the east (Leith, 1985).

The Boldshuan formation of Miocene age is up to 1200 m thick. It consists of clays, red sandstones and siltstones, interbedded with polymict conglomerates. The Khingouss formation is 400–1700 m thick. It consists of purple-red polymict conglomerates with rare interbeds of siltstone, sandstone and red-brown clays. This formation marks a major uplift of the Pamir to the east. The Tavildarya formation is ~ 1700 m thick. It consists of alternating red and grey siltstone, sandstone and conglomerate. Pollen, vertebrates and palaeomagnetism indicate a late Miocene age. The central part of the basin consists of 300–1600 m of grey sandstones with rare siltstone beds, the Kafirnigan formation, which is mid- to upper Miocene on vertebrate evidence. Thus, it is probably equivalent in age to parts or all of the Khingouss and Tavildarya formations.

The Pliocene overlies the Miocene with an angular unconformity. The Karanak formation is up to 1800 m thick and consists of coarse to medium grained conglomerates with lenses of siltstone and rarely clay. This coarse braided alluvium may pass laterally into overbank siltstones. The Polizak formation is up to 1280 m thick and fills synclinal depressions. It thus postdates an important phase of folding which is reflected in its lithology. This formation consists of coarse polymict conglomerates and grey sandstones which rest conformably on the Karanak formation where present. The Kuruksai formation rests with angular unconformity on underlying Miocene to mid-Pliocene deposits. It is up to 500 m thick and consists of coarse boulder conglomerates interbedded with sandstone and siltstone. It contains a diverse late Pliocene (Villafranchian) mammal fauna with savannah and open woodland forms such as hyaena, elephant, camel, gazelle and others (Pen'kov et al., 1977). Palaeomagnetic data suggest that the formation includes the Gauss and lower Matuyama episodes.

The Kairubak formation is up to 300 m thick. It consists of pebble beds alternating with siltstones and sandstones and includes a lot of loess and red-brown loess soils. This semi-desert deposit is usually considered to span the Plio–Pleistocene boundary. Its upper boundary seems to correspond to the Brunhes–Matuyama inversion (Pen'kov et al., 1977).

The Kyzylsui formation consists mostly of loess and fine alluvium. Ten terrace–loess–soil complexes are recognized. This formation corresponds to the main glacial/interglacial phases of the late Pleistocene when extensive glaciers existed in the Pamir (Velichko and Lebedeva, 1973).

From these deposits we can conclude that rapid and exponentially increasing sedimentation commenced in the early Miocene. Two pulses occurred in early and mid-Miocene, followed in the Pliocene by repeated and increasing pulses of sedimentation, alternating with erosion (Figure 2.2). Rates commenced at 0.4 mm yr$^{-1}$ and reached 1.4 mm yr$^{-1}$ in the Pliocene. These sedimentation rates differ little across the Tadzhik depression, suggesting that the rates are a general feature of the basin and not simply a result of local fault-block subsidence (Melamed, 1966). Continued folding and thrusting is shown by the angular unconformities within the Pliocene as far from the mountain front as Dushanbe. Many Plio–Pleistocene beds are intensely deformed.

In the Dushanbe area, two major pulses of coarse conglomeratic sedimentation can be dated palaeomagnetically as late Gilbert (4 myr) and Olduvai (2 myr). These Olduvai sediments contain a rich vertebrate fauna, including elephant, rhinoceros, hyaena, gazelles and other savanna animals (Pen'kov et al., 1977), and pollen of grasses (Abdunazarov et al., 1984). The general trend in pollen is from early Pliocene temperate dry coniferous forest through early Pleistocene grassland to the present desert environment.

The area is now undergoing erosion with the incising of the Amu darya river. The drainage system of the upper Amu darya changed from longitudinal to transverse in the early mid-Quaternary in the Darvaz range (Nikonov, 1970). In the upper Amu darya, there was a phase in which valleys formed in earlier Pliocene times were filled with as much as several hundred m of upper Pliocene sediments of the Kilimbinskaya and Kuruksay formations, which are dated by plant, molluscan and vertebrate fossils. Over much of the area sedimentation continued into early Quaternary times. This phase is well represented by reaches of old, flat, greatly uplifted (to 3000 m) longitudinal valleys in the Darvaz range.

All investigators agree that in mid-Quaternary time, there was intense valley downcutting, to depths of 1500–2000 m in the mountains and several hundred m in the plains. Virtually all valleys in the upper Amu darya basin expose sediments that are not directly associated with their present terraces. It is signifi-

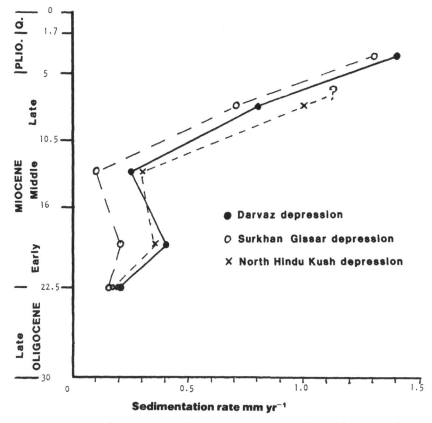

*Figure 2.2* Rates of subsidence (really sedimentation rates) in the Tadzhik basin (after Melamed, 1966).

cant that the rivers flow not over bedrock but over Quaternary sediments, even in alpine areas, except in the upper reaches of side tributaries. For example, in the Pyandzh river virtually no bedrock is exposed throughout the Darvaz range and Badakhstan. Depth to bedrock in the river ranges from 40 m in narrow areas to 150 m in wider areas of the valley (Nikonov, 1970). Throughout the enormous Amudarya basin, there are only isolated reaches on which rivers are incised into bedrock. Successive late Quaternary and Holocene terraces lie along the valley sides.

The primary downcutting of the rivers to bedrock must predate the late Quaternary. The aggradation level of the subsequent late Quaternary filling reaches 50–100 m in the Amu darya basin, 200–300 m in the mountains and foothills and 700–800 m in the upper Amu darya valley. Fluctuations related to local glaciation can be see in the subsequent downcutting and terracing of the valleys (Nikonov, 1970). At least for the Quaternary, these events in the plains can be related to tectonic movements in the adjacent Pamir range (Figure 2.3).

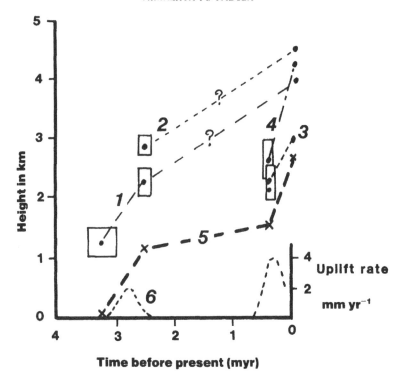

*Figure 2.3* Uplift of the Pamir in the last 3 myr. Greatly modified to emphasize uncertainties (after Pakhomov and Nikonov, 1983). Boxes indicate confidence limits of estimates (1–4: changes in elevation based on palynology).

1  Southern Pamir valley bottoms;
2  Southern Pamir, northern slopes of South Alichur range;
3  Southeastern Pamir;
4  Southwestern Pamir, Pyandzh valley;
5  Combined data showing inferred *amount* of uplift of the valleys of southern Pamir;
6  Rates of uplift at different periods in the southern Pamir.

### Southwestern Tarim basin

The southwestern Tarim basin lies between a buried basement high and the Kun Lun range to the south (Figure 2.1). Mesozoic isostatic subsidence allowed the accumulation of more than 10 km of sediment in the basin centre. The basin was essentially filled by the Cretaceous, after which thin marginal marine and terrestrial red beds were deposited during the Cretaceous and Palaoegene. The basin was reactivated during Neogene times, when very thick continental clastics were laid down in the marginal troughs (Hsu, 1988). Over 12,000 m of sediments have accumulated in the deepest parts of this trough since the Miocene; 6400 m since the start of the Pliocene (Hsu, 1988). At the same time, the western Kun Lun have been rising at equally remarkable rates. Unlike the Tadzhik basins, much of the deformation away from the Kun Lun seems to be caused by faulting, not by folding. Long fault-bounded blocks parallel to the Kun Lun are now rising in the Tarim basin. This style of deformation is in keeping with the contrast between the western and eastern sides of the Pamir indenter (Molnar and Tapponnier, 1975). On the west, the Tadzhik depression is being squeezed between two large continental masses. On the east, the various blocks of southest Asia are moving eastward over the Pacific and eastern Indian oceans (Tapponnier et al., 1986). As Hsu (1988) notes, it is likely that at least parts of the Tarim basin are now underlain by oceanic crust.

Between Kashgar and Yarkand and beyond, Tertiary deposits flank the Kun Lun range and pass beneath the sands of the Taklimakan desert. Very little is known of these deposits, which have been noted in de Terra (1932), Norin (1946), Belyayevskiy (1966) and Kaz'min and Faradzhev (1963). When completed, current Chinese work will give a good stratigraphic framework for the first time.

Near Yarkand, coarse red sandstones and conglomerates overlie upper Eocene marine beds (de Terra, 1932; Norin, 1946). De Terra (1932) described 400–500 m of reddish-yellow fine-grained sandstones, red claystones and red gypsiferous marl conformably overlying the marine Eocene. These fine-grained clastics are probably of Oligocene age (Tang Tianfu et al., 1984). Near Kashgar, the Eocene is overlain by conglomeratic beds with interbedded gypsiferous and saline clays (de Terra, 1932). These gypsiferous sediments are in turn overlain by coarse conglomeratic clastics. South of Yarkand, in the Kokjar basin, are 1400 m of coarse boulder conglomerates and sandstones. Above are massive coarse conglomerates with unrounded boulders with a greater number of crystalline pebbles than the beds below (de Terra, 1932). All of these beds are folded near the Kun Lun and are overlain, often with angular unconformities, by Quaternary(?) fluvial, glacial and aeolian deposits. The lower Quaternary erosion surface has been uplifted to over 3000 m, with an actual height change (based on river gradients) of 800 m (Kaz'min and Faradzhev, 1963).

In Neogene times, the foredeep north of the Kun Lun subsided greatly. A narrow trough accumulated up to 3500 m of coarse clastic sediment, while to the south, in the Kun Lun, marine Eocene was uplifted to 5000 m (Kaz'min and Faradzhev, 1963). In the early Miocene, uplift of the Kun Lun caused the formation of a large lake into which coarse alluvial fans fed from the south. Nevertheless, the geomorphological contrast was relatively weak. The lake was shallow and the marginal fans relatively fine-grained. In mid- to late Miocene, the Kun Lun increased its rate of uplift. A deep lake was fed by coarse fan deltas building directly into deep water (Giu Dongzhou, 1987).

Much of the Neogene tectonic movement took place along the Karakoram fault zone, a dextral strike-slip fault with up to 270 km lateral displacement (Trifonov, 1978). Block uplifts along this fault zone have formed the huge 6–7.5 km high Kongur, Muztagh Ata and Kung Ata Tagh massifs southwest of

Kashgar. Uplift is also still taking place in the foothills and plains. Norin (1932) noted a post-glacial lake shoreline which had been tilted considerably to the east in the last few thousand years.

## INTERMONTANE BASINS

Inermontane basins are those formed *within* mountain ranges by faulting, erosion and subsidence. Most of them are formed by a combination of processes. In the Himalaya–Pamir, most of the basins are late Quaternary. Erosion during the great late Cenozoic mountain uplifts have in most cases removed any earlier basins. Thus, the intermontane basins of the Pamir, Karakoram and Himalaya are almost entirely Quaternary in age, are sites of only temporary storage of sediment and are mostly now being eroded (Figure 2.4). Only along the Indus Suture Zone has Quaternary backthrusting partially preserved a Miocene basin.

A convenient distinction in the Himalaya–Pamir is between 'true' inter-montane basins entirely within the mountains, and 'piggyback' basins formed by late Cenozoic thrusting. The latter are now being carried southward on the frontal nappes of the Himalaya.

### 'True' intermontane basins (and valleys)

#### Pamir

Marine deposits were laid in the northern Pamir (Figure 2.1) until the late Eocene (Moralev *et al.*, 1966). The entire Pamir were involved in uplift, starting in the Oligocene (Krestnikov, 1963). At that time they were gentle rolling hills. The highest elevations were around 1500–2000 m in the northern and western Pamir (Belousov, 1976).

Neogene uplift increased erosion throughout the Pamir with the result that few Neogene deposits are preserved, apart from the Quaternary deposits of recent intermontane valleys and depressions (Figure 2.1). The most intense uplift occurred in the northern and western Pamir. Here, uplift and deep erosion are responsible for the development of steep-sided valleys and sharp-crested watersheds. Uplift of the eastern Pamir was slower. Here, broader valleys and watersheds occurred.

The end of the Pliocene and beginning of the Pleistocene were tectonically quiet. The northern and western Pamir continued to rise but at a slower rate, resulting in greater lateral and less vertical erosion of valleys. In the Tadzhik depression, marginal to the range, valley filling occurred. The valley floors apparently reached altitudes of 700–1700 m and the mountains stood as high as 3000 m, reaching 4000 m in places (Belousov, 1976; but no basis was given for this evaluation). The eastern Pamir were also around 3000 m but with broad alluvial valley floors at 2000–2800 m. This suggests greater uplift, but with less

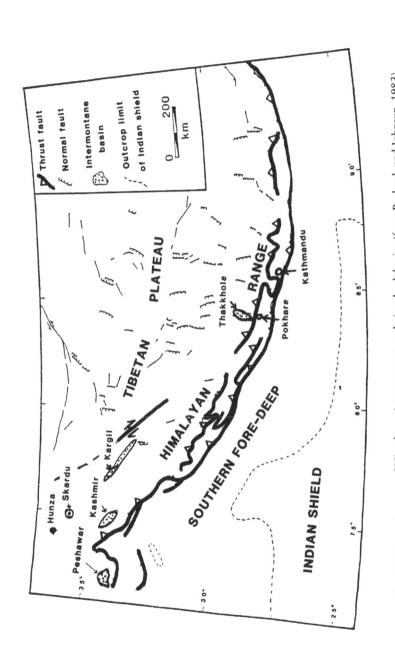

*Figure 2.4* Location map of Himalayan intermontane and piggyback basins (from Burbank and Johnson, 1983).

erosion in the eastern Pamir. At any rate, by the Pleistocene, the Pamir as a whole had been uplifted 1500–2000 m since the end of the Oligocene. This averages to a rate of 0.12 mm yr$^{-1}$.

The late Pliocene decline in uplift rate is also indicated by palynological data from the eastern Pamir (Figure 2.3). Here, uplifts of 1000 m and almost 2000 m occurred before the end of the Pliocene ( ~ 3 myr) and in the mid- to late Pleistocene ( ~ 400,000 yr BP) (Pakhomov and Nikonov, 1983).

The increasingly rapid uplifts of the mid- to late Pleistocene mark the stage that has determined the present structure and relief of the Pamir (Belousov, 1976). In the western Pamir, the mountains rose at increasing rates of between 2.6 and 4.0 mm yr$^{-1}$ for the mid-Pleistocene ( ~ 300,000 yr BP) and between 4.8 and 5.6 mm yr$^{-1}$ for the late Pleistocene (100,000 yr BP). Rates substantially decreased, then rose to incredible average rates of 15–20 mm yr$^{-1}$ in the Holocene (Belousov, 1976). In the eastern Pamir rates were slightly lower at 2–3.4 mm yr$^{-1}$ for the mid-Pleistocene (Belousov, 1976). Also in the eastern Pamir, at a height of 4300 m, lake deposits below a mid-Quaternary moraine contain 10–30 per cent tree pollen (pine, Himalayan cedar, alder, etc.) generally living at much less than 4000 m (Pen'kov et al., 1976). Pleistocene and Holocene movements in the Pamir were thus consistently upwards. The total uplift for the region as a whole was 1500–2700 m for the period.

Calculations indicate that the amount of purely tectonic uplift must have been 4 km since the early Miocene (Artem'yev and Belousov, 1979). This involves a thickening of the crust by ~ 20–40 km. The cause of this is the underthrusting of the entire Pamir by Asian crust since the early Miocene start of continental collision in the area (Leith, 1985). Mass redistribution near the surface, owing to faulting and folding adjustments to tangential stress, resulted in very rapid vertical block movements and had virtually no effect on isostatic anomalies. However, these local movements together with erosion can explain the differences between the northern and eastern Pamir. In the northern Pamir, the mountains are 0.8 km higher because the valleys are more incised. Though the average elevations of the northern and eastern Pamir are the same, the mountains are higher and the valleys deeper in the northern Pamir because of isostatic compensation.

The limited radiometric dating in the Pamir suggests Neogene intrusions in the late Oligocene/early Miocene followed by steady cooling until Holocene (Figure 2.5). The meagre data simply confirm the uplift since the Miocene. Displacements on Holocene faults in the Pamir have been estimated from radiocarbon dating of offsets. Rates of displacement for three faults in the last 10,000 yr are 1–3 mm yr$^{-1}$ (Nikonov, 1981).

In general terms, the Pamir closely follow the Neogene histories recorded in the High Himalaya, Tibetan plateau and Karakoram. Progressively increasing uplift is associated with continental subduction and formation of anatectic granite magmas from about 20 myr onwards (Le Fort, 1986).

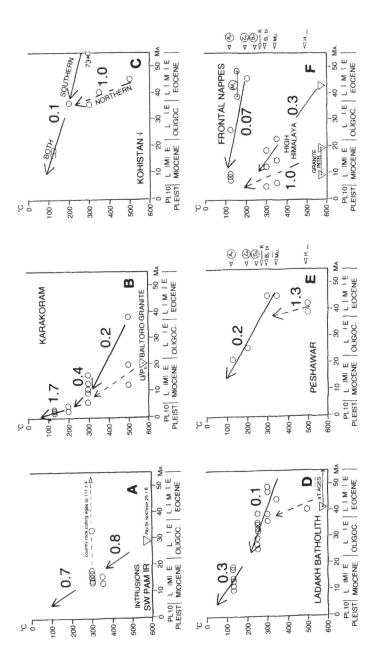

*Figure 2.5* Cooling curves for various tectonic units of Figure 2.1. Arrows show resultant cooling curves for data. Minerals used at right. Circled are fission-track minerals: upper case are K/Ar; lower case are Rb/Sr. Crystallization ages are shown at base where available. Numbers on arrows are erosion rates in mm yr$^{-1}$ based on average geothermal gradient of 30 °C km$^{-1}$.

## Karakoram

The Quaternary of the southern Karakoram and Himalaya has been summarized by Shroder et al. (1989). Here, I wish to consider two reasonably well preserved examples: the Indus valley at Skardu and the Hunza valley north of Gilgit (Figures 2.1, 2.4). Neither contains any pre-Quaternary deposits.

At Skardu, a thick clastic sequence with glacial deposits has been preserved on the northern side of the present wide valley. This Bunthang sequence has been described in some detail by Cronin and Johnson (1989), and elsewhere in this volume.

The section starts with a coarse diamictite. Above, the lower Bunthang sequence consists of a 50 m micaceous sand passing up into 300 m of silty clays. The mineralogy of the silts is very immature and they thus represent glacially derived unweathered particles. Identical silts are present in modern glacially derived streams and lakes. The silts also contain dropstones, rhythmites and possible varves. They are glacial lake deposits. The middle Bunthang sequence consists of boulder conglomerates and interbedded sandstones up to 190 m thick. The conglomerates thicken towards the mountain front to the northeast, while the sandstones thicken toward the basin. This sequence was formed by alluvial fans interfingering with valley-floor stream deposits. The upper Bunthang sequence onlaps underlying inclined fanglomerates. It consists of up to 70 m of mixed sand, slit and clay. It was deposited by longitudinal streams and associated lakes. The sequence predates the last ice advance into the Skardu valley, since the deposits were eroded and sculptured by it.

No fossils have been found in the Bunthang sequence, but the lower and upper sequence are reversely magnetized throughout (most of the middle sequence is too coarse for palaeomagnetic analysis). These thick bodies of reversely magnetized sediment may have been deposited during the Matuyama reversed chron (between 2.48 and 0.73 myr). If this date is provisionally accepted, then the sequence was deposited in the latest Pliocene to mid-Pleistocene (Repenning, 1984). However, exactly when during this interval the sequence was deposited is unknown. Cronin and Johnson (1989) calculated that the entire sequence could have been deposited within 500,000 yr. At any rate, the whole sequence now towers up to 500 m above the present level of the Indus, which was actively aggrading until recently in the Skardu basin.

The deposition of the thick lower Bunthang lake sequence could be caused by blocking of the Indus valley by a glacier, by glacial moraines or landslides, or by glacial erosion of a bedrock basin. I prefer the bedrock basin model, since I find it hard to believe that a lake fed by the Indus river could last long enough to form the sequence unless it were overflowing through bedrock. In that case the present altitude of the Bunthang would result from uplift, since the Indus gradient is, even now, greatly oversteepened.

In the Hunza valley, a complex of eight glacial deposits is mostly younger than 150,000 yr BP, and no older deposits are preserved (Derbyshire et al., 1984). The earliest widespread phase, attributed to the early Pleistocene, left isolated and deeply weathered erratics on summit surfaces above 4150 m. The succeeding phases occur at 2500 m (older than 139,000 yr BP) and lower. These lower elevations reflect valley incision during the Pleistocene and are related to uplift of the Karakoram. If we assume that the earlier glaciers reached around the same height, then early Pleistocene uplift of over 1500 m is indicated.

## Kargil basin

The Kargil basin forms the western end of the long elongate intermontane late Tertiary basin along the Indus Suture Zone. It contains a Miocene–Pliocene late orogenic clastic sequence (Brookfield and Andrews-Speed, 1984a,b). Only the Kargil section has been studied sufficiently to determine rough ages and environments for the intermontane clastics.

The sequence consists of coarse clastic alluvial fan deposits of upper Miocene age passing upward into braided and meandering stream deposits and finally into lake sediments, which are overlain by coarse alluvial fan deposits. The basin was deformed and overthrust from the south by Indus Suture units in probable Pliocene times and terraces related to the present Indus then formed (Fort, 1982). This ubiquitous late Tertiary backthrusting north of the High Himalaya has destroyed much of the late Tertiary intermontane lake record along the Indus Suture Zone (Searle, 1983).

In the upper Miocene section, angiosperm pollen suggest a temperate climate (Bhandari *et al.*, 1977) and the gastropod *Subzebrinus* a high altitude of at least 2000 m (Tewari and Dixit, 1972). The high altitude of the basin by the upper Miocene may be related to accelerating uplift. A lake developed after this, possibly owing to blocking of the ancestral Indus by large landslides; downcutting by the river would undoubtedly be capable of keeping pace with any gradual uplift. The change to semi-arid climates in the topmost units is possibly related to breaching of the lake with consequent return to semi-arid conditions typical of the later Tertiary. There is no precise fossil or palaeomagnetic dating of the deposits and their history cannot therefore be used to confirm ideas on mountain uplift.

## Thakkhola basin

This basin of > 850 m of sediment lies at 3000–4000 m altitude and developed as a graben at right angles to the Himalayan trend in late Tertiary time (Fort *et al.*, 1982a,b; Figure 2.4). The sediments of the basin are divided into the Tetang and Thakkhola formations. The Tetang outcrops only in the southern part of the basin. It consists of two gravel/sand fluvial deposits with a lake sequence, including carbonates in between. The sediments were derived from the north and east. This formation is probably Pliocene. It contains dominant conifer pollen comparable to the Middle Siwalik floral zones II and III, and Pliocene ostracods.

Yoshida *et al.* (1984) reported pollen assemblages now characteristic of 900–1800 m, i.e. over 2000 m lower than the present altitude. The Thakkhola formation lies unconformably on the underlying beds and consists of basal fluvial boulder conglomerates overlain by mixed fluvial/lacustrine beds. These sediments were derived from the south and west (Yoshida *et al.*, 1984). Since these units were deposited a river gorge has been cut, filled with sediment (Sammargaon formation), and cut again (Fort *et al.*, 1982a). There is no good fossil dating of the Thakkhola basin sediments. Yoshida *et al.* (1984) made palaeomagnetic studies of the sediments.

Cross-correlation of the palaeomagnetic reversals in the combined sections of the Tetang formation with the standard scale gives Matuyama reversed epoch to epoch 5 (i.e. early Pleistocene to late Miocene; Yoshida *et al.*, 1984). However, this is unlikely, especially since Yoshida *et al.* (1984) take no account of facies changes in their sections, and reinterpreting their data indicates the Gauss to early Matuyama epochs for the Tetang formation (i.e. 3.4–2 myr). The Thakkhola formation was deposited during a mainly normal epoch, probably the Brunhes ( < 0.75 myr), and the younger gorge sediments of the Sammargaon formation in relatively recent times.

Although the dating of events in the Thakkhola basin is still rather imprecise, in general terms it seems to fit the Kashmir chronology. Like the Kashmir basin, it probably started its development in the late Pliocene. It also shows reversal of drainage and uplift sometime in Pleistocene times.

## Piggyback basins

These are basins carried passively on the backs of thrust sheets (Ori and Friend, 1984). They usually involve ponding of the former drainage and the formation of temporary lakes which are then breached by overflow and downcutting. The Peshawar, Kashmir and Katmandu basins are examples, all now breached, which were carried southward and upward during development of the Main Boundary Thrust during the late Cenozoic (Burbank and Raynolds, 1984) and consequent uplift of the frontal ranges of the Himalaya. The present Potwar plateau is an incipient example, which emerged in Plio–Pleistocene times as the Salt range developed (Burbank and Raynolds, 1984).

### *Katmandu basin*

The Katmandu basin, like the Kashmir basin in India, evolved in response to tectonic movements associated with southward thrusting on the Main Boundary Thrust and with climatic changes owing to Quaternary glaciation. It has been described by Dongol (1985, 1987). The maximum exposed thickness of sediment is 280 m and the top of the sequence lies at only 1500 m altitude.

The oldest exposed deposits are up to 60 m of alluvial fan stream flood and debris flow breccias, derived from the south (Mahabharat granite clasts). These deposits initiated sedimentation in the Katmandu basin as uplift of the Mahabharat range occurred to the south.

Disconformably above the basal gravels is a 50 m thick deltaic–lacustrine sequence of alternating sandstones, siltstones and mudstones with interbedded thin lignites. The lignites contain the only Quaternary vertebrate fauna recorded from the Katmandu basin: a Plio–Pleistocene fauna of water buffalo, swamp deer, river hog, bush pig, elephant and rhinoceros. These indicate environments varying from reed swamp, humid jungle, light forest around the lake, open woodland and grass jungle (Dongol, 1985).

Above the lignites are 40 m of massive to laminated mudstones with occasional leaf impressions. They represent a more offshore, deeper euxinic lake environment. Northern exposures show more variable and more shallow water conditions including rapid deposition from flooding streams. Pollen assemblages suggest an initial subtropical moist climate changing to a much cooler and drier one at the top (Yoshida and Igarashi, 1984).

Overlying coarse gravels consist of a variable, < 50 m thickness of massive stream flood alluvial fan gravels interbedded with coarse felspathic sandstone. The pebbles are all derived from the south, indicating uplift rejuvenation of the Mahabharat range.

The last and youngest deposit of the basin is a dark organic clay. It formed in a residual lake dammed behind a bedrock ridge, the Chovar Barrier, in the northern part of the basin.

Thus, the Katmandu basin shows two phases of uplift of the southern Mahabharat range, owing to thrust movements on the Main Boundary Thrust. The first ponded back a large lowland Pliocene lake in which the deltaic–lacustrine and euxinic lake deposits accumulated. The upper part of this sequence is entirely normally magnetized (Yoshida and Igarashi, 1984). They correlated it with either the late Gauss (3–2.5 myr) or Olduvai ( ~ 1.8 myr) chrons. The mammal faunas suggest the earlier late Gauss interval. At any rate, the climate cooled during accumulation of these lake deposits.

This lake drained, probably to the east (Dongol, 1987), and was followed by a long period of non-deposition lasting the entire lower and middle Pleistocene (Yoshida and Igarashi, 1984). Then, within the Brunhes normal chron ( < 0.74 myr) a second uplift led to rapid deposition of coarse gravels in a higher intermontane basin, followed by local tectonic ponding and deposition of the dark organic clays.

Gross uplift of the Katmandu basin has been relatively small, certainly not exceeding 1500 m (its present height) since the Pliocene.

### Kashmir basin

This basin developed as a piggyback lacustrine basin above the developing Pir Panjal Thrust sheet. Its late Cenozoic history has been summarized by Burbank and Johnson (1983) from which most of the following is taken.

The basin developed ~ 3–4 myr, based on stratigraphic sequences in the basin and changes in Siwalik sedimentation south of the Pir Panjal. The stratigraphic sequence, well dated palaeomagnetically, shows initial ponding of a lake, followed by rapid uplift of the northeastern margin ~ 3.5 myr. Slower uplifts between 3.5 and 0.4 myr were followed by rapid uplift of the Pir Panjal and northward tilting of the lake deposits. This last event involves a minimum of 1400–1700 m differential uplift between the southern and northern margins of the basin: the minimum uplift rate is thus about 4 mm yr$^{-1}$. The actual total uplift is probably much greater. Thus, the Kashmir basin fits the late Pliocene and late Quaternary uplift sequence of the Pamir, Karakoram and northwest Himalaya. From 4–1.8 myr the conglomerates were shed mostly from the northeastern boundary faults. After 1.8 myr the conglomerates were shed from the south and southwestern Pir Panjal margin.

### Peshawar basin

In the Peshawar basin, sedimentation began ~ 3 myr. Before this, movement on the Attock thrust had caused uplift of the ancestral Attock range and generated a tectonic depression to the north. This ponded the fluvial systems which had previously drained across the area and localized sedimentation in the newly formed basin (Burbank and Raynolds, 1984). Intermontane basin sedimentation was terminated by accelerated uplift of the Attock range after ~ 0.6 myr. Subsequently, there was catastrophic flooding probably caused by landslide and glacial dams in the main Himalayan valleys to the north (Burbank and Tahirkheli, 1985).

The development of the Peshawar and Kashmir basins are not synchronous. The Kashmir basin seems to have formed at least 1 myr earlier than the Peshawar basin (Burbank and Tahirkheli, 1985). Phases of active faulting on either side of the western syntaxis do not correlate.

### Jalalabad basin

The Jalalabad and adjacent Lagman basins were formed by block faulting in the late Cenozoic and filled with continental clastics up to 650 m thick. Raufi and Sickenberg (1973) described a unit consisting of alternating conglomerates and sandstones in several cyclical sequences, and containing upper Pliocene mammals and plants. The mammals are a low diversity fauna of rodent, camel and ox species, suggesting dryish savannah conditions. These beds are disconformably overlain by conglomerates which thicken northward: these are probably of Pleistocene age. Pebble orientations show southerly and southwesterly current flow from the mountains to the north. The entire sequence was then folded prior to the development of the present Kabul river system, with its three terraces. No palaeomagnetic work has yet been done.

## SOUTHERN FOREDEEP

The foredeep in front of the rising Himalaya developed progressively from late Eocene times onward as India collided with Asia. The early stages involved progressive telescoping of a northern Indian passive margin with sediments being deposited primarily in a marine foredeep in front of the progressively southward moving Asian margin. The sedimentary sequences show that internal deformation of the Indian shield and the start of uplift of the main Himalaya started in the lower Miocene. The sediments of the Kasauli, Murree and Dharamsala Groups mark the start of continental sedimentation derived from the Himalaya. The change from marine to continental sedimentation, though naturally diachronous in different areas, occurred mostly between the Aquitanian and Burdigalian (Sahni and Chandra, 1980), i.e. ~ 20 myr. The overlying Siwalik Group and related continental sediments record the following 20 myr of uplift and denudation of the adjacent Himalaya (G.D. Johnson et al., 1981).

This 20 myr datum is a convenient base, since it also marks the start of movement on the Main Central Thrust and intrusion of granite and metamorphism in the High Himalaya (Le Fort, 1986).

From early Miocene times onward, the Siwaliks generally increase in coarseness and in the grade of the source rocks. The variation in heavy mineral frequency distributions suggests increasing intensity of denudation, and hence uplift, with increasing age (Chaudhri, 1975). The coarse-grained Upper Siwaliks (Upper Pliocene–Pleistocene) have rock fragments of schist and gneiss occasionally exceeding 50 per cent of the sandstone modal compositions. They contain increasing amounts of unstable detrital heavy minerals, including sillimanite. Weathering and attrition were thus mainly by mechanical disintegration and distance (Chaudhri, 1975). The Upper Siwaliks mark a significant coarsening of the debris from the adjacent Himalaya and stretch the entire length of the range. If not markedly diachronous, they would reflect major uplift of the Himalaya.

In the northwestern Himalaya, the topmost Siwalik is marked by the Boulder Conglomerate, a coarse braided stream fanglomerate which sharply overlies meandering stream deposits. In the Punjab, the base of this is dated palaeomagnetically as the top of the Olduvai event ( ~ 1.7 myr) and sedimentation becomes coarsely conglomeratic at the Jaramillo event ( ~ 0.97 myr) (Azzaroli and Napoleone, 1982). In Jammu, two different areas show conglomeratic sedimentation starting at the same Olduvai and Jaramillo events. The earlier Olduvai event at Jammu is related to southward thrusting of the Pir Panjal (Ranga Rao et al., 1988). This is confirmed by the synchronous change from a northerly to a southerly source for the Kashmir basin conglomerates (Burbank and Johnson, 1983). However, the Pleistocene also marks a sharp climatic cooling and this may have contributed to increased physical denudation in the mountains.

An earlier ( ~ 9 myr) phase of uplift is recorded in the Potwar plateau, an incipient piggyback basin which rose on the subhorizontal detachment surface thrusting the Salt range over the Indian shield (Burbank and Raynolds, 1984). It contains Miocene–Holocene fluvial sediments. Southward thrusting commenced ~ 2.1 myr. Very rapid uplift is shown in the Soan syncline, where the 1.9 myr old Lei Conglomerate truncates earlier deposits. Here, at least 3000 m of structural relief was developed within 200,000 yr (Burbank and Raynolds, 1984). At around 9 myr, an earlier phase of uplift to the north is recorded in a doubling in the sedimentation rate from 0.22 mm yr$^{-1}$ to 0.49 mm yr$^{-1}$ (N.M. Johnson et al., 1982).

## RADIOMETRIC AND FISSION-TRACK DATES

We can determine the cooling history of a rock by using various dating methods on different minerals. This follows from the fact that each datable phase within each isotopic system can be assigned a characteristic closure temperature. Thus, by using the $^{40}Ar/^{39}Ar$ method on hornblende and mica, and the fission-track method on sphene, zircon and apatite, a rock's thermal history can be sampled at roughly 500°C, 350–300°C, 250°C, 200°C and 100°C respectively (Zeitler, 1985).

Cooling curves plotted for major tectonic units are based on closure temperatures of the various minerals used for dating: information is very poor for the Pamir, there are few Tertiary dates from the Hindu Kush, and dates are nonexistent for the western Kun Lun (a list of the numerous other data sources is available from the author). The steepness of the curves (Figure 2.5) is directly related to rates of cooling: these by inference are related to rates of erosion. Such rates of erosion can be roughly determined by assuming an average geothermal gradient of 30°C km$^{-1}$.

Interesting points about these plots are as follows. Recent rapid cooling, and perhaps erosion, are restricted to areas *north* of the Main Central Thrust and Northern Suture. This suggests, as do the facies distributions in Mesozoic–Tertiary deposits, that the Main Boundary Thrust is a late, minor feature. Most of the shortening along the northern Indian margin has taken place along the Main Central Thrust and Northern Suture, linked by the strike-slip Karakoram fault zone (Figure 2.1). The $^{40}Ar/^{39}Ar$ ages obtained by Maluski et al. (1988) indicate very strong tectonometamorphic and magmatic activity during the Miocene both in the upper part of the High Himalaya slab and farther north in the Tethyan cover. Similar events occurred in the Karakoram (Brookfield, 1981; Coward et al., 1986).

In Kohistan, Ladakh and Peshawar (Figure 2.5C,D,E), cooling, and perhaps erosion, appear to have been relatively constant after late Cretaceous ophiolite emplacement. Higher apparent erosion rates in northern Kohistan and in places (dotted lines in Figure 2.5) in Ladakh and Peshawar are related to the rapid cooling of high-level intrusions (Zeitler, 1985). The Ladakh curves can be related to the start of emplacement of the Trans-Himalaya batholiths of the Asian margin over the northern Indian margin during and after collision. The frontal nappes of the High Himalaya also show slow erosion rates (Figure 2.5F top).

In contrast, the Pamir and Karakoram (Figure 2.5A,B) show slow cooling and erosion, starting in the late Eocene in the Karakoram, but accelerating markedly after the Miocene intrusions associated with the start of thrusting on the Main Central Thrust and Northern Suture (Mattauer, 1986). These intrusions have been related to partial melting of continental crust associated with subduction of Indian continental crust (Le Fort, 1986). In fact, the High Himalaya, north of the Main Central Thrust, show rapid erosion of crystalline rocks starting only in the Miocene (Figure 2.5F bottom).

The conclusion is that the areas south of the Main Central Thrust and Northern Suture were slowly squeezed southwards after initial collision, but that areas to the north did not start major erosion related to uplift until the Miocene.

## SEDIMENT BUDGET STUDIES

Another approach is to add up the amount of sediment delivered within successive time intervals to marginal sedimentary basins during development of the Himalaya. These figures of sediment amount give volumes of material eroded from the Himalaya front during each time interval. High grade metamorphic rocks are exposed only in the High Himalaya. There has been almost no erosion of the Tibetan plateau in the late Tertiary and the volume of exposed Lesser Himalaya pre-Tertiary rocks is quite small. Thus, the sediment delivered to marginal sedimentary foredeeps by the Indus, Ganges, Brahmaputra and associated rivers and palaeorivers is almost entirely derived from the High Himalaya, Pamir and adjacent ranges.

Apart from a small amount delivered to the Himalayan foredeeps in the Oligocene (Figure 2.6A), the vast bulk of sediment has been transported into the Indus and Bengal fans with accelerating rates from mid-Miocene times onwards (Figures 2.6B, 2.7). Not enough is known about late Tertiary deposition from the Amu darya river and related drainage to make calculations for the northern areas.

Table 2.1 shows the volume of solid sediment deposited in adjacent basins for the three roughly equal periods, Oligocene (15 myr), early to mid-Miocene (10 myr) and late Miocene to Holocene (12 myr). The dissolved loads of the Indus, Ganges and Brahmaputra rivers are at least one-tenth of the weight of the solid load at present (Subramanian *et al.*, 1987). If this was true in the past then the volumes of rock eroded from the High Himalaya have to be increased as in Table 2.1. The correction gives total sediment volumes (in $10^6$ km$^3$) of 1.25 (Oligocene), 5.36 (early to mid-Miocene) and 7.4 (late Miocene to Holocene). The volume of rock eroded from the Himalaya can be determined with an average ratio of the densities of sedimentary (2.5) to crystalline rock (2.8). This gives an underestimate, since a significant part of the Himalaya is sedimentary rock. Volumes of crystalline rock eroded are (in $10^6$ km$^3$) respectively: 1.11, 4.77 and 6.59.

Because neither Tibet nor the Indo–Gangetic plain underwent significant erosion in the later Tertiary (Chang Chengfa *et al.*, 1986), all of this material

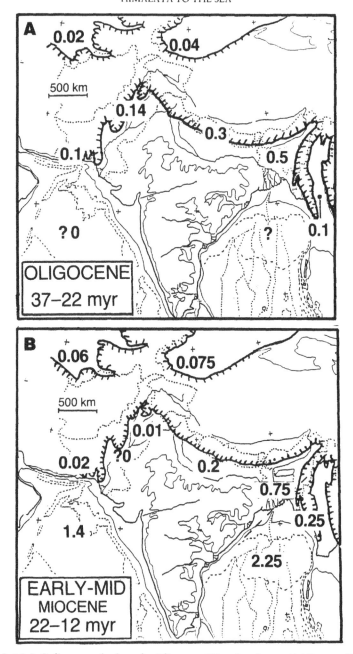

*Figure 2.6 A,B* Sedimentary budgets for Oligocene (A) and early to mid-Miocene (B). Time scale from Berggren *et al.* (1985b). Underlined values are in $10^6$ km$^3$ for individual basins. Value in box is total sediment *volume* preserved for the period.

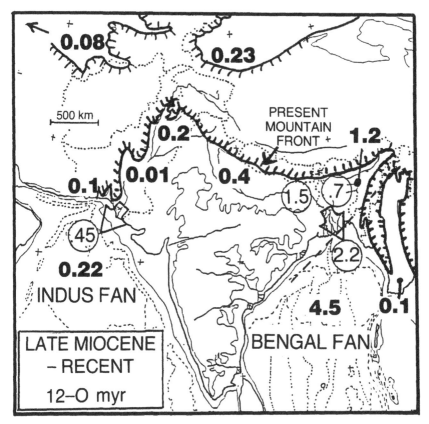

*Figure 2.7* Sediment budgets for the late Miocene to Holocene. Circled values are measured Holocene suspended sediment loads of the main rivers, in $10^6$ tonnes $yr^{-1}$. Note enormous value for Bengal fan volume and small values for northern foredeeps.

must be removed from the Himalaya. If we assume that this material came from erosion of a crystalline slab (the rising High Himalaya) and that this slab detached at the base of the Indian continental crust, then we can calculate the width necessary for a 35 km thick slab, 2500 km long, to supply these volumes. The calculations (Figure 2.8; Table 2.2) show minimum southward displacements across the Main Central Thrust for successive periods. It seems likely, however, that less than half of the thickness of the continental crust is involved in the southward thrusting. That is, the crust delaminates at mid-crustal levels rather than at the base of the Moho (Mattauer, 1986). Rocks exposed near the base of the High Himalaya slab, just above the Main Central Thrust, reached kyanite grade at temperatures of ~600°C (Le Fort, 1986). This corresponds to pressures of ~5Kb, i.e. 20 km depth (Thompson and England, 1984). In that case the given values must be multiplied by 1.75 (Table 2.2).

The end result, for a 20 km thick slab, is that the total southward displace-

Table 2.1 Sedimentary budgets for each basin (in $10^6$ km³).

| Basin | Late Miocene to Holocene (0–12 myr) | Early to mid Miocene (12–22 myr) | Oligocene (22–37 myr) |
|---|---|---|---|
| Indus fan | 0.22 | 1.4 | ?0 |
| Makran | 0.12 | 0.016 | ?0.1 |
| Indus valley | 0.01 | ?0 | ?0 |
| NW Himalaya | 0.19 | ?0.01 | 0.14 |
| Central–East Himalaya | 0.4 | 0.2 | 0.3 |
| Bengal basin | 1.2 | 0.75 | 0.5 |
| Bengal fan | 4.5 | 2.25 | ? |
| Nagaland | 0.1 | 0.25 | < 1 |

ments on the Main Central Thrust and Northern Suture are: about 20 km for the Oligocene, about 95 km for the early to mid-Miocene, and about 130 km for the late Miocene to Holocene. These values are in keeping with the increasing erosion rates derived from radiometric and fission-track studies. The total southward displacement on the thrusts is thus ~ 250 km for the Oligocene to Holocene and, more importantly, over 220 km since the early Holocene.

A consequence of these calculations is that most of the shortening in the Himalayan–Tibet area since the mid-Miocene has taken place by thrusting along the Main Central Thrust and Northern Suture. The calculated values are compatible with thrust displacements in other orogens (Wu-ling Zhao and Morgan, 1985) and with concepts of Miocene extensional tectonics in the Tibetan plateau (Royden and Burchfiel, 1987). They are incompatible with 1000 km underthrusting of continental crust below the Tibetan plateau (Powell, 1986). There is also no need to indent Asia in post-Miocene times, which is in keeping with the known small late Tertiary displacements along the marginal Tibetan plateau faults (Chang Chenfa et al., 1986). Nevertheless, some short-ening must have taken place in the Pamir, Kun Lun and Hindu Kush as deter-mined by the folding and thrusting now taking place in those areas. On the other hand, the amount of sediment supplied to the northern foredeeps is insignificant compared to the amount transported into the Bengal and Indus fans (Figures 2.6, 2.7). In addition, apart from the Karakoram, the ranges do not show such a deep level of erosion as the High Himalaya and mostly underwent smaller and less rapid uplift in the late Cenozoic. Therefore, much less underthrusting has occurred in the northwestern syntaxis than in the main High Himalaya.

Thus, southward thrusting along the Main Central Thrust and erosion of the Himalaya increased exponentially in the later Tertiary. The onset of accelerating uplift corresponds with the early to mid-Miocene anatectic intrusions of the High Himalaya and Karakoram and with the start of accelerating gross uplift of the Himalaya and Tibetan plateau as determined with palaeobiological evidence.

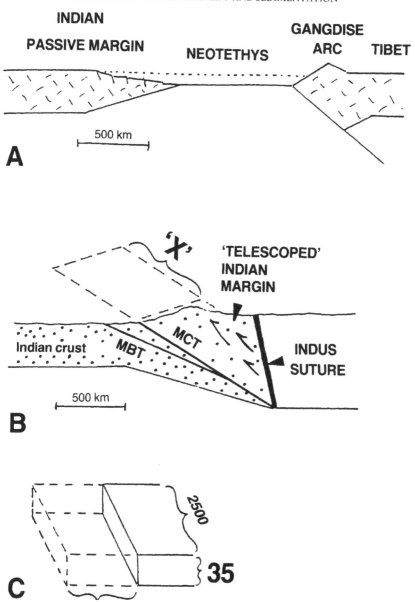

*Figure 2.8* Diagram showing method of calculating width (X) of block removed by erosion at different periods.

*Table 2.2* Width (X) of 2500 km long Himalayan slab required to account for sediment budgets at different thicknesses.

$$\text{Width (southward displacement) (X)} = \frac{\text{total sediment volume}}{\text{length (2500 km)} \times \text{thickness}}$$

$$\text{Rate of thrusting} = \frac{\text{southward displacement (X)}}{\text{time}}$$

| | Southward displacement (X km) thickness | | Rate (mm yr$^{-1}$) | |
|---|---|---|---|---|
| | A* @ 35 km | B† @ 20 km | A* | B† |
| late Miocene to Holocene | 75.3 | 131.8 | 6.2 | 11 |
| early to mid-Miocene | 54.5 | 95.4 | 5.45 | 9.5 |
| Oligocene | 12.7 | 22.2 | 0.85 | 1.5 |
| Total | 142.5 | 249.4 | | |

* A: total thickness of Indian continental crust overthrust.
† B: Indian crust delaminated at mid-crustal levels.

## PALAEOBIOLOGICAL EVIDENCE

Sediments and fossils, now at high altitude, can give values for *gross* uplift if the altitude at which they formed can be determined. Chinese and other workers (e.g. Xu Ren, 1982; Mathur, 1984) have provided abundant evidence for uplift from floras and faunas. These should, ideally, be calibrated against palaeolatitude and sea-level climate. For example, the sea-level temperatures for China have varied throughout the Tertiary and altitudes based on fossils should be corrected against a sea-level standard. Unfortunately, this is not yet possible and Figure 2.9A shows uncalibrated estimated heights for fossil floras and faunas in the Himalaya and Tibet. Note that there is practically no change in altitude for the Lesser Himalaya. From Figure 2.9A, it is obvious that the palaeobiological data corroborate the radiometric, fission-track and sediment budget studies. The start of rapid uplift on Xu Ren's summary uplift curve (Figure 2.9B) exactly corresponds with the other data.

## RIVER GRADIENTS

The average topographic profile across the Himalaya shows two distinct gradients. A relatively gentle slope across the Lesser Himalaya rises to 3 km at the junction with the High Himalaya. A steeper slope runs across the High Himalaya, levelling off at 5 km and extending across the Indus Suture into Tibet (Burrard and Hayden, 1907; Seeber and Gornitz, 1983). The High Himalaya

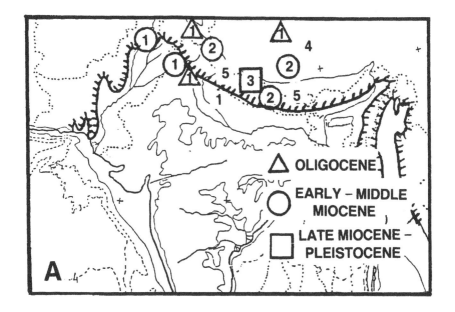

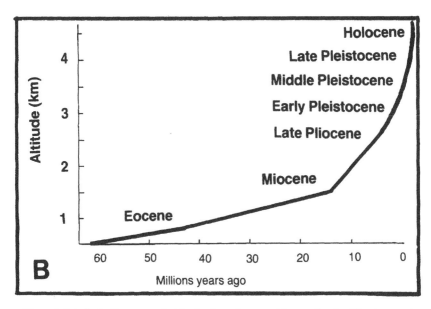

*Figure 2.9 A, B*  A  Uncorrected altitudes from floras and faunas for the Himalaya and southern Tibet. (Data from: Guo Shuang-Xing, 1982; Song Zhi-chen and Liu Geng-wu, 1982; Xu Ren, 1982; Mathur, 1984; Sharma, 1984).

B  Altitudes of the Tibet plateau based on floras and faunas (after Xu Ren, 1981).

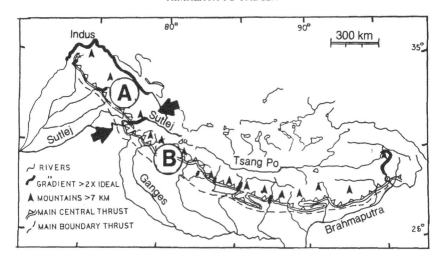

*Figure 2.10* River gradients (from Seber and Gornitz, 1983) and Himalayan biogeographic boundaries (from Mani, 1974).

A Western steppe fauna, derived from Ethiopian/Mediterranean area;
B Eastern humid tropical fauna, derived from the Indo–Chinese/Malaysian area.

Boundary is the Sutlej gorge.

topographic front is associated with a narrow belt of intermediate magnitude thrust earthquakes. The trace of the Main Central Thrust follows the topographic front. High river gradients are associated with this zone, suggesting continued major southward thrusting (Figure 2.10). No large inflections of gradient occur in the Lesser Himalaya, which is thus not a major tectonic boundary. Both the Indus and the Brahmaputra, where they cut the Main Himalayan range, have steep profiles. In contrast, the Sutlej shows a relatively even gradient (Burrard and Hayden, 1907). Since all three rivers have sources very close together, the Sutlej may have been the main drainage of the Tibetan plateau, until uplift or local ice-cap formation in the late Quaternary caused reversal of flow in both Indus and Brahmaputra rivers so that they now drain past the western and eastern syntaxes. Some feeder rivers to the Brahmaputra and Indus flow in opposite directions to the master streams (Figure 2.10). And these tributary streams join with acute angles pointing *downstream*, though in places they are now modifying their junctions.

The major biogeographic division in the Himalaya is between an Eastern humid tropical fauna and a Western steppe fauna (Mani, 1974). The boundary coincides precisely with the Sutlej gorge (Figure 2.10). Because endemism in the Tibetan plateau is so low, and since the biogeographic provinces have not yet been modified, these changes must have taken place very recently, i.e. in the late Quaternary.

## CONCLUSION

Uplift rates based on floras, on radiometric and fission-track data and on sediment budget studies show a remarkable correspondence for the late Cenozoic. This indicates that the exponentially increasing rates of uplift for the Himalaya and adjacent ranges are true, which places constraints on any models for collision mountain belts.

The delay between initial India–Asia collision in the late Eocene and the start of major uplift in the early Miocene may be related to two factors: the progressive telescoping of a thinned northern Indian passive margin back to an original continental thickness, and to thickening of the Tibetan plateau crust by compression.

The early Miocene start of exponential uplift can be related, in the Himalaya, to underthrusting of Indian shield crust below the Himalaya along the Main Central Thrust and to underthrusting of Asian crust below the Pamir and possibly Karakoram. However, since Tibet was not deformed, similar uplift of the plateau is more likely caused by changes in the upper mantle as a consequence of the start of continental underthrusting. Possible causes are the detachment of the Indian lithosphere and consequent thermal equilibration below Tibet.

But, on a cautionary note, subsidence and uplift are not strictly synchronous among and between basins and ranges. Each basin or range, though generally following the increasing rates of the late Cenozoic, has its own history. Thus, the Kashmir basin started its development before the Peshawar basin and the Pamir started a more gradual uplift before the Karakoram. The surface expression of deep underlying changes varies according to the position and orientation of the surface blocks, the local stresses acting on their edges, and the often unpredictable development of drainage patterns which differentially wear them down.

# 3

# PRESENT AND PAST PATTERNS OF GLACIATION IN THE NORTHWEST HIMALAYA: CLIMATIC, TECTONIC AND TOPOGRAPHIC CONTROLS

*Jonathan A. Holmes*

## ABSTRACT

This paper presents new findings regarding the present and past glaciation of Kashmir, in the northwest Himalaya. Equilibrium-line altitudes (ELAs) of modern glaciers in Kashmir basin rise 3900–4400 m in roughly a southwest-northeast direction. The glaciation threshold shows a similar trend, but lies 200–300 m above the ELA. Maximum depressions of equilibrium lines in Kashmir during the late Quaternary were 700–800 m. Reconstructed ELAs for former glaciers and cirque altitudes show an apparent reversal in trends of glacierization. Two possible explanations for this are discussed: a reversal of climatic gradients and differential tectonic uplift within the basin.

Previous studies have been carried out elsewhere in the northwest Himalaya, including Swat Kohistan, the Karakoram range, the Nanga Parbat massif, the Ladakh range and the Zanskar range. They show that modern ELAs range from 4100 m (Swat Kohistan) to more than 5500 m (northern side of the Karakoram range). The figures suggest that topography and precipitation control present patterns of glacierization.

Estimated maximum depressions of ELAs range from ~1100 m below present values (Swat Kohistan and the Hunza valley in the Karakoram range) to 600 m (southern side of the Zanskar range). The higher values for ELA depressions are similar to maximum depressions elsewhere in the world. Changes in precipitation gradients, tectonic uplift and topographic constraints on the past extent of glaciers have all been used to explain the ELA depressions of lesser magnitude.

## INTRODUCTION

Spatial variations in the size and distribution of glaciers are often studied in currently glacierized regions. Such information can provide valuable insights into

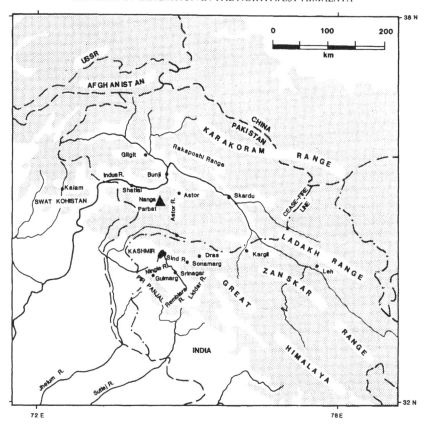

*Figure 3.1* Northwest Himalaya, showing the location of meteorological stations and other sites mentioned in the text. Shaded area is land over 3000 m altitude.

the nature of controls on glaciation, particularly if climatic data are available. If the extent of glaciers can be reconstructed for periods in the past, constraints can be placed on the magnitude of climatic change and any accompanying alterations in climatic gradient identified. Such information, therefore, makes an important contribution to the study of past climates and also provides inputs into climatic models (e.g. Broecker and Denton, 1990).

Northwest Himalaya (Figure 3.1), which for the purpose of this paper includes the Vale of Kashmir, Ladakh, Zanskar, the Nanga Parbat massif, the Karakoram range and Swat Kohistan, contains many glaciers of widely varying size. Locally, such as in the Karakoram range, the mountains are heavily glacierized with some of the longest valley glaciers outside the polar regions. Abundant evidence exists for the former advance of glaciers at various times during the Quaternary.

This paper reports new information on the present and past distribution of glaciers in Kashmir. This is compared with previous work from elsewhere in the northwest Himalaya. The results are evaluated in terms of the major controls on glaciation, namely climate, topography and tectonic uplift.

## GLACIER–CLIMATE RELATIONSHIPS

### Methods

Glacier fluctuations result from changes in mass balance. If there is an increase in net accumulation, the glacier will tend to thicken and advance. Conversely, an increase in net ablation will cause a glacier to thin and recede.

Much of the work on the distribution of glaciers has used firn lines and equilibrium lines. The firn line is defined as the line which separates ice from snow at the end of the ablation season (Flint, 1971). On temperate glaciers, this corresponds with the equilibrium line, which is defined as the line or zone on a glacier where annual accumulation equals annual ablation and where, consequently, annual mass balance equals zero. In many studies of glaciation patterns, the position of the firn line is mapped as a surrogate for the equilibrium line, on the assumption that the two coincide.

Various methods have been used to locate the equilibrium line in the field, and on maps and aerial photographs. They include direct observations, mass–balance measurements, methods based on area measurements with an assumed accumulation area ratio (AAR; e.g. Meier and Post, 1962), and height methods using toe and headwall altitudes and an assumed toe–headwall altitude ratio (THAR; e.g. Osmaston, 1975).

An indirect indication of the equilibrium line altitude (ELA) is given by the glaciation threshold (GT; Bruckner, 1886; Østrem, 1961; Porter, 1977). The GT is defined as the arithmetic mean of the altitudes of the lowest glacier-clad summit and the highest ice-free summit within an area, excluding peaks of unsuitable shape to support glaciers (Østrem, 1961). Since the GT is based on summit heights, it will always be higher than the ELA.

Changes in the height of the equilibrium line and the GT within an area are frequently used to study climatic change in glaciated regions. Several authors have used cirque-floor altitudes to approximate equilibrium lines of former glaciers (e.g. Meierding, 1982). In spite of the fact that cirque altitudes and ELAs are coincident only for small glaciers, trends in cirque altitudes can provide useful information about past gradients of glacierization (e.g. Peterson and Robinson, 1969; Flint, 1971). In areas where glacial deposits have been well mapped, it may be possible to calculate values for the past GT using similar methods as for the calculation of the modern GT (e.g. Andersen, 1968; Wahraftig and Birman, 1965; Porter, 1977).

Although ELAs cannot be measured directly for former glaciers, methods are available for their estimation. If the geometry of former glaciers can be determined by detailed mapping of moraines and if a former AAR can be assumed, the past ELA of the glacier can be approximated using area methods. If the former geometries cannot be determined accurately, but toe and headwall altitudes are available, the past ELA may be estimated using an assumed THAR. If lateral moraines are preserved within a valley, their position can be used to locate the equilibrium line, since lateral moraines occur only in the ablation zone of the glacier. Furthermore, lateral moraines delimit the former extent of the ice up the valley side and so provide important information about the past geometry of the glacier.

## Glacier–climate relationships

### Precipitation

This is an important control on glacierization, although the relationship is far from simple. In general terms, greater precipitation gives rise to larger glaciers. However, much depends on the form of the precipitation and the season in which it falls. Thus, in areas where precipitation is high but where much of it falls as rain in the ablation season, little contribution is made to the mass of the glacier. Conversely, in arid environments, a small amount of snowfall during the accumulation season may make a relatively large contribution to glacier mass. In the Himalaya, particularly the more arid parts, low precipitation may be recorded at the valley-floor level, whereas at higher altitudes, particularly above the snow line, snowfall may be much higher (e.g. Goudie *et al.*, 1984b; Collins, 1988).

### Temperature

Because solar radiation at the glacier surface controls ablation, there is a strong correlation between mean annual temperature and the extent of glacierization on a regional scale. Despite this, glaciers can and do exist in locations that may appear unfavourable, owing to the influence of other factors. It is often thought that ablation-season temperature exerts a strong control on glacierization. However, Porter (1977) found a low correlation between the height of the GT and the ablation-season temperature, for the Cascade range in the USA.

The fact that both precipitation and temperature control mass balance hampers attempts to interpret past changes in glacierization. Clearly, if meaningful palaeoclimatic conclusions are to be drawn from patterns of past glacierization, some independent indication of palaeotemperature and/or palaeoprecipitation is required.

## Altitude and aspect

Glaciers will not develop unless upland areas of sufficient altitude exist. On a gross scale, this critical altitude will decrease with increasing latitude although variations exist owing to the influence of other variables on a local scale. The effects of solar radiation on glacierization are pronounced, particularly in mid-latitude areas. Thus, in the northern hemisphere, north-facing slopes tend to have larger glaciers. Moreover, northeast-facing slopes, which are in the lee of prevailing winds in the northern hemisphere mid-latitudes, are the most favourable sites for the accumulation of ice.

## Tectonic uplift

Patterns of past glaciation can be modified substantially by subsequent tectonic uplift, which will decrease the measured ELA depression below the actual value and will also raise the altitude of formerly glacierized cirques. If uplift is un-identified and not corrected for, palaeoclimatic inferences drawn from the palaeoglaciation data may be incorrect. Tectonic uplift also has direct effects on glaciation as well, by placing a greater part of the massif within the accumulation area of the glacier. This would lead to a progressive increase in the size of glaciers through time. Adjacent ranges may, depending on atmospheric circulation patterns, be placed in a rainshadow, thereby reducing the magnitude of glaciers through time.

## THE STUDY AREAS

Although records of glaciers have been made for many parts of the northwest Himalaya, the areas considered in this study are restricted to those where information on both past and present ELAs is available. Recent data from Kashmir (Holmes, 1988) are compared with existing information on glaciers in Ladakh and Zanskar (Burbank and Fort, 1985; Osmaston, in press), the Nanga Parbat massif (Finsterwalder, 1937; Kick, 1980; Owen, 1988a), Shroder *et al.*, 1989), the Karakoram range (Kick, 1964; Visser and Visser-Hooft, 1935–8; Derbyshire *et al.*, 1984; Owen 1988a), and Swat Kohistan (Porter, 1970) (Figure 3.1).

### Kashmir

Kashmir (Figure 3.1) is an intermontane basin which developed during the late Cenozoic within a series of thrust faults. It has a mean basin-floor altitude of ~ 1600 m. The Great Himalayan range, rising to a maximum altitude of ~ 5500 m, bounds the basin to the northeast. The Pir Panjal range, rising to only 4500 m, lies to the southwest.

The Pir Panjal range prevents much of the summer rainfall associated with the

*Table 3.1* Summary precipitation data for the study areas.

| Location | Annual precipitation (mm) | Duration of record (years) |
|---|---|---|
| Srinagar | 658.9 | 50 |
| Sonamarg | 1815.5 | 49 |
| Gulmarg | 1308.6 | 2 |
| Dras | 673.0 | 41 |
| Kargil | 264.5 | 41 |
| Leh | 92.6 | 50 |
| Skardu | 159.7 | 47 |
| Gurez | 1314.8 | 31 |
| Astor | 402.7 | 7 |
| Bunji | 158.3 | 7 |
| Gilgit | 134.0 | 47 |
| Kalam | 596.9 | 1 |

Data from Indian Meteorological Department Records with the exception of Kalam, data for which come from Porter (1970). The location of the meteorological stations is shown on Figure 3.1.

southwest monsoon from reaching Kashmir. The mean annual precipitation at Srinagar, in Kashmir basin, is 659 mm, of which 24 per cent falls during summer (Table 3.1). This compares with Jammu, to the south of the Pir Panjal, where 67 per cent of the mean annual precipitation (1116 mm) falls during the summer monsoon (July–September). Abundant meteorological data, mainly from basin-floor stations, show that annual precipitation in Kashmir ranges from 584 to 1229 mm, although there is no clear spatial pattern. Temperature is measured at few sites. At Srinagar, mean monthly maximum ranges from 9.7 °C (January) to 35.5 °C (July); mean monthly minimum ranges from −6.7 °C (January) to 14.5 °C (July). The mountain flanks are cooler and wetter than the basin floor, although few data are available to quantify the differences.

The present glacier cover of Kashmir is sparse; as little as 100 km² (Holmes, 1988). On the Pir Panjal flank, glaciers are restricted to the southwestern extremity in the headwater region of the Rembiara and Vishav rivers, where they descend to altitudes of 3900–3700 m. On the Himalayan flank, glaciers are more widespread and occur extensively in the Sind and Liddar catchments. The snouts of these glaciers typically lie at altitudes of 4400–3800 m.

Burbank (1982) has shown that the mountain ranges surrounding Kashmir have undergone substantial uplift during the late Cenozoic. The focus of uplift changed from the Himalayan margin to the Pir Panjal margin at ~ 1.7 myr BP. Since ~ 400,000 yr BP, the Pir Panjal range has been uplifted 3.5–10 mm yr⁻¹ (Burbank, 1982). Uplift rates of the Himalayan flank are less well constrained but are thought to have been low during the latter part of the Quaternary (Table 3.2).

*Table 3.2* Selected rates of tectonic uplift for the study areas.

| Location | Method | Rate of uplift $(mm\ yr^{-1})$ | Period $(myr\ BP)$ | Reference |
|---|---|---|---|---|
| Ladakh | Palaeontology | ~ 1.0 | 3.0–0 | Lakhanpal *et al.* (1983) |
| Pir Panjal | Palaeomagnetism | 3.4–10.0 | 0.4–0 | Burbank (1982) |
| Dir and northernmost Swat | Fission track | 0.24 | 15.0–0 | Zeitler (1985) |
| Northern Swat | Fission track | 0.39 | 9.4–0 | Zeitler (1985) |
| Southern Swat – Hazara | Fission track | 0.20 | 18.1–0 | Zeitler (1985) |
| Nanga Parbat – Haramosh massif | Fission track | 4.50 | 0.7–0 | Zeitler (1985) |
| Kulu–Mandi Belt | Mineral cooling temperatures | 0.7–0.8 | 25.0–0 | Mehta (1980) |

## Ladakh and Zanskar

The area of Ladakh lies to the east of Kashmir (Figure 3.1). It consists of several sub-parallel mountain ranges which rise to more than 5000 m. The mountains include the Ladakh and Zanskar ranges. Few climatic data are available, but the region experiences a high altitude, arid climate because it lies in the rainshadow of the Great Himalaya. This aridity is illustrated by the lack of vegetation. Because of the high altitude, there is considerable solar radiation. Maximum summer temperatures are high, although a large diurnal temperature range occurs. Winter temperatures are commonly low. Available meteorological data suggest an eastward decrease in mean annual precipitation from Sonamarg, which is in the Great Himalayan range but within Kashmir basin (1815 mm), through Dras (673 mm) and Kargil (264 mm) to Leh (93 mm) (Table 3.1). As with other parts of the Himalaya, these figures relate to valley floors; precipitation at higher altitudes is undoubtedly greater.

The distribution of glaciers in Ladakh is poorly known. Burbank and Fort (1985) described glaciers on the southern side of the Ladakh range and the northern side of the Zanskar range. These mountains, oriented northwest–southeast, rise to 5400–5700 m and flank the Indus river. Here, numerous small cirque glaciers were found, particularly about 5100 m and on northwest–northeast facing slopes.

Studies on uplift of Ladakh are limited. There is some evidence to suggest modest rates of uplift during the mid- to late Tertiary (Lakhanpal *et al.*, 1983; Table 3.2), although Burbank and Fort (1985) provided geomorphological evidence to suggest that late Quaternary uplift in their study area was limited.

## Nanga Parbat

This is a large, heavily glacierized massif to the north of Kashmir, within the upper Indus drainage basin (Figure 3.1). In the vicinity of the massif itself, the peaks are > 8000 m and the relief is extremely high. Meteorological observations have not been made in this region. However, Kick (1980) suggests that below 2500 m precipitation is low, particularly on the more arid northern side, where mean annual precipitation may be < 120 mm. At altitudes above 4500 m, precipitation is much higher and this promotes active glacierization. Kick (1980) mapped 61 glaciers on the Nanga Parbat massif, based largely on the map produced by Finsterwalder on an expedition in 1934. This map shows that, on the southern side of Nanga Parbat, glaciers descend to 3600–3500 m whereas, on the northern side, they descend to ~ 3200 m. Detailed mapping of the Raikot glacier on the northern slopes of the massif, which descended to 3212 m in 1985 (Gardner, 1986), is consistent with Kick's observations.

Recent fission-track evidence has shown that the Nanga Parbat massif has been uplifted substantially during the Quaternary. Rates of uplift have increased during the last 7 myr, from ~ 0.5 mm yr$^{-1}$ to as much as 5 mm yr$^{-1}$, according to some estimates (e.g. Zeitler, 1985), during the mid- and late Quaternary (Table 3.2).

## Karakoram range

This region, to the north of the Nanga Parbat massif, is one of considerable altitude and relief (Figure 3.1). The Lesser Karakoram, the southernmost part of the range, has some peaks rising to over 7000 m and many rise over 6000 m. To the north, in the Greater Karakoram, many of the peaks exceed 7000 m. Further north still, the Lupghar group and the Ghujerab mountains have many peaks over 6000 m.

Few climatic data are available for these areas and long-term temperature records exist only for Gilgit, which is to the south of the Rakaposhi range in the Lesser Karakoram. Total annual precipitation is low, ~ 130 mm (Table 3.1). This is generally associated with westerly depressions during March–May, although pulsatory extensions of the southwest monsoon occasionally lead to summer rainfall. Precipitation apparently decreases from southwest to northwest (Goudie et al., 1984b), although this is complicated by a considerable increase with altitude, especially above the snowline. Temperatures vary considerably: during summer, maximum temperatures exceed 35 °C whereas, in winter, freezing occurs down to 1500 m altitude (Goudie et al., 1984b).

The Karakoram range contains some of the world's longest glaciers outside the polar regions. More than 100 glaciers measure longer than 10 km and the Batura glacier, in the Greater Karakoram, is 59 km long. The glaciers extend to low altitudes, commonly below 3000 m, indicating very active glacierization.

The occurrence of deeply incised valleys and uplifted sediments in the

Karakoram range suggests that rapid uplift has occurred during the Quaternary. This is confirmed for part of the Hunza by Zeitler's (1985) work, which indicates accelerating uplift during the late Cenozoic for the Nanga Parbat–Haramosh massif and the Hunza, up to 5 mm yr$^{-1}$ during the late Quaternary (Table 3.2).

## Swat Kohistan

This area, drained by the Swat river, lies to the west of the Karakoram range (Figure 3.1). In the headwater region, peaks exceed 6000 m; relief and altitude decrease southward. Few systematic climatic data are available for Swat Kohistan, although observations by Porter (1970) for the period July 1967–June 1968 indicate a mean annual temperature of about 10.5 °C, with a mean monthly minimum in January (1.4 °C) and maximum in July (22 °C). The annual temperature range was −10 to +32 °C. Annual precipitation was thought to be in the range 760–890 mm although, for the period of observation, 597 mm were recorded, falling mainly between March and May (Table 3.1).

The glacier distribution of Swat Kohistan is not well known, although total coverage is thought to be no more than 130 km$^2$ (Porter, 1970). The glaciers are mainly of the small cirque type and, although some extend beyond the cirques, few are longer than 2 km. Glaciers originating from north-facing cirques terminate as low as 3350 m, whereas other glaciers descend to 3800–4500 m.

Late Quaternary uplift of Swat Kohistan has been limited. Zeitler's (1985) fission-track evidence suggests rates substantially less than 1 mm yr$^{-1}$ during the late Cenozoic (Table 3.2).

## PATTERNS OF GLACIATION IN KASHMIR

### Modern glaciers

Modern glaciers were investigated using information from Survey of India 1:63,360 scale maps produced before independence in 1947. These maps depict relief, drainage and glacier cover with acceptable accuracy. Although all altitudes and distances on the maps are in imperial units, these have been converted to metric units for this study.

GTs were determined for each 5′ east × 5′ north quadrangle using the method of Østrem (1961). The value for the GT was plotted at the quadrangle centre and isoglacihypses (lines of equal GT) were drawn. A height method was used to determine the ELA for each glacier, since the area of each glacier as depicted on the maps is quite generalized. Toe and headwall altitudes are shown clearly for many glaciers. Previous work on the glaciers of Kashmir has not provided sufficient detail for a value of the THAR to be calculated, so a value of 0.4 (Meierding, 1982) was employed. Mean values of ELAs for north-facing (N ± 30 °) glaciers in each 5′ × 5′ quadrangle were calculated and iso-glacihypses of equal ELA drawn.

Calculations of the GT yielded 40 data points. There is a clear distinction in the height of the GT between the Pir Panjal range and the Great Himalayan range. Broadly, the GT rises from 4100–4500 m in the Pir Panjal range to 4600–4700 m in the Great Himalayan range. The GT was also determined for several quadrangles further to the east, beyond the Kashmir watershed, where it is generally higher than 4700 m (Figure 3.2).

This general pattern is best explained by the existence of a precipitation gradient. There are two major sources of precipitation received in Kashmir: one from the south and southeast, associated with summer monsoon disturbances; the other from the west, associated with winter westerly depressions. In either case, the glacierized part of the Pir Panjal range would receive more precipitation than the Great Himalayan range. The aridity of Ladakh, for example, is well shown by the abrupt change in vegetation seen when crossing the watershed from Kashmir.

The pattern of isoglacihypses also shows topographic control. The glacierized part of Kashmir runs broadly in a southwest–northeast direction and this follows the axis of high mountains. Thus, although other parts of Kashmir may have the climatic potential to support glaciers, the mountains are too low. Therefore, the apparent rise in GTs from southwest to northeast reflects control by both precipitation and topography.

The gradient of the GTs is indicated by the spacing of the isoglacihypses. The gradient of the GT is consistently high on the Pir Panjal flank, ranging from 20 to 68 m km$^{-1}$. On the Himalayan flank, the gradient varies from 5 to 68 m km$^{-1}$. Whereas the lower values of the GT gradient are well within the range of figures quoted for other areas, the higher gradients exceed published values by a factor of two (e.g. Porter, 1977). It is possible that the values for Kashmir are incorrect, owing to errors in the topographic information on the maps. However, the highest gradients are, in most cases, constrained by several data points and are a feature of both mountain flanks. Therefore, some reason other than data error is required to explain the large GT gradient.

The gradients quoted by Porter (1977) refer to mid- to high latitude, maritime mountain ranges. The highest gradient, $\sim 25$ m km$^{-1}$, occurs in mid-latitude ranges, in areas of high precipitation. In these maritime regions, precipitation is high on the coast and declines inland, leading to a sharp GT gradient. Locally, precipitation often interacts with topography. For example, low GT gradients are frequently found in association with cols in maritime ranges which are oriented broadly perpendicular to the direction of airflow. This is because the cols allow greater inland penetration of moist, maritime air masses (e.g. Porter, 1977). The effect of this is to produce bulges in the isoglacihypses.

On the Pir Panjal flank, the isoglacihypses follow the topography quite closely, although they show two quite pronounced bulges which are associated with two cols in the headwater regions of the Rembiara and Vishav valleys (Figure 3.2). These cols presumably allow greater penetration of moist air over the Pir Panjal crest. However, the steepness of the GT gradient indicates a marked decrease in precipitation north and northeastward, suggesting that the rainshadow effect of the Pir Panjal range must be very strong.

The pattern of isoglacihypses on the Himalayan flank is more difficult to explain. The presence of pronounced bulges suggests that precipitation and topography may be controlling the GT, especially since the bulges coincide partially with the two major valleys that drain the Himalayan flank. The steepest GT gradient occurs close to the watershed with Ladakh. This is to be expected, since the watershed marks a very narrow zone of pronounced precipitation decline.

In summary, the rainshadow effect of the Pir Panjal range and the location of the high altitude mountains appear to have produced a southwest–northeast rise in the GT, with local variations, including very high GT gradients in some areas. Although this may in part be a product of poor-quality data, topographic variations appear to produce local, steep, precipitation gradients which, in turn, affect the GT.

Isoglacihypses of mean ELAs yielded 20 data points. ELAs lie 100–200 m below GTs, a reflection of the fact that the latter are based on summit heights. The steady rise in the ELA from the Pir Panjal flank to the Himalayan flank is similar to that for the GT, although the gradient is more gentle (up to

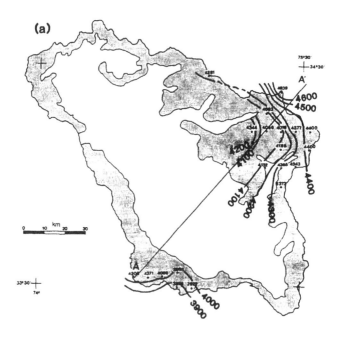

*Figure* 3.2 Isoglacihypses of (a) the equilibrium-line altitude; (b) the glaciation threshold; and (c) cirque altitudes in Kashmir. Shaded area is land over 2500 m altitude. All altitudes shown are in metres.

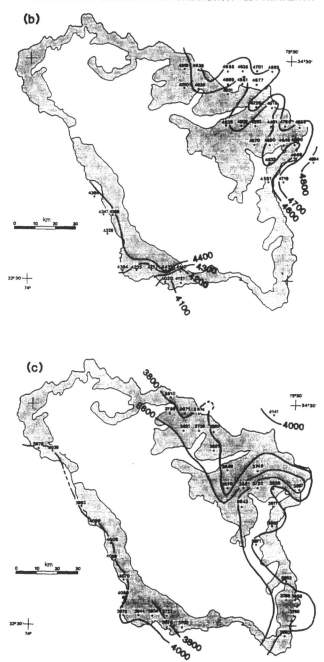

~ 45 m km$^{-1}$) (Figure 3.2; Table 3.3). The pattern of isoglacihypses is less complex than that of the GT. This is a result of the fewer data points and the smoothing produced by the averaging procedure used to calculate the mean ELA.

## Former glaciers

The pattern of former glaciation in Kashmir is indicated by two sources of evidence: the altitudes of empty, north-facing cirques and the depression of the ELA associated with maximum ice advance. Mean altitudes of cirques facing N ± 30° were calculated for each 5' × 5' quadrangle and isoglacihypses drawn in the same manner as for GTs and ELAs. ELA depressions were calculated for maximum ice advances in the Sind and Liddar valleys on the Great Himalayan flank and the Ningle valley on the Pir Panjal flank.

Determination of mean cirque altitudes yielded 38 data points. Both the isoglacihypses (Figure 3.2) and the cross-section of cirque heights (Figure 3.3) show that cirque altitudes are higher on the Pir Panjal flank than on the Great Himalaya, the difference in mean cirque altitude being about 150 m. The spread of values is a reflection of the fact that geological structure and palaeoglaciation affect cirque altitude. Furthermore, some of the features on the maps may have been wrongly identified as cirques. However, the overlap of cirque heights between the two flanks is only ~ 100 m altitude. Thus, it is likely that the greater altitude of the Pir Panjal cirques is a reliable observation and not the result of errors in the data.

Depressions of the ELA were calculated from glacial geological mapping of the former maximum extent of glaciers (Table 3.4). Maximum ELA depressions were 700 m below present values in the Ningle valley, 750 m in the Liddar valley, and 800 m in the Sind valley. The value for the Ningle valley, however, must be taken with some reservation because there are currently no glaciers in the catchment. Thus, a value for the modern ELA had to be taken from the Rembiara and Vishav catchments in the southwest part of Kashmir.

The pattern of cirque altitudes is the reverse of that for modern glaciation as shown by the GT and ELAs. Initially, this suggests a reversal of precipitation gradients, since lower cirques would be expected with larger glaciers which would, in turn, occur in areas of higher precipitation. This hypothesis is supported by the lower ELA depression on the Pir Panjal flank than that on the Himalayan flank. However, it is unlikely that the areas to the east of Kashmir would have been a source of precipitation during a glacial maximum. These areas, which include the Tibetan plateau and Tarim basin, are currently quite arid. Evidence from ocean cores taken from the northern Indian Ocean (e.g. Cullen, 1981; Duplessy, 1982) strongly suggests that glacial conditions over south Asia were arid, with a weakened southwest monsoon and a stronger, colder and drier northeast monsoon. Although there is evidence to suggest that that westerly storm tracks in winter over Central Asia may have been stronger during the last glacial maximum, and may have been associated with wetter

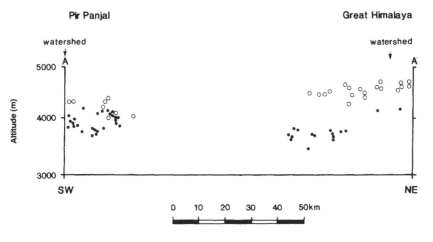

*Figure 3.3* Cross-section of Kashmir basin, showing the height of the glaciation threshold (open circles) and cirque altitudes (closed circles). Location of the section is shown on Figure 3.2 (a).

conditions (e.g. COHMAP, 1988), precipitation gradients over Kashmir would not have been reversed.

The absence of any plausible climatic explanation of the change in glaciation gradients makes it necessary to look for some other cause. One such explanation might be differential tectonic uplift within the basin. If uplift has followed cirque formation, the cirques would be higher today than when they were formed. This hypothesis is supported by uplift rates for the Pir Panjal range (Table 3.2). Burbank (1982) has shown that the Pir Panjal range has been uplifted at rates of 3.5–10 mm yr⁻¹ over the last 400,000 yr. During the same period, uplift of the Great Himalayan flank was limited. This would explain why the Pir Panjal

*Table 3.3* The height of the equilibrium line in the northwest Himalaya.

| Location | ELA (m) | N | Reference |
|---|---|---|---|
| Kashmir: Pir Panjal range | 4052 ± 177 | 24 | Holmes (1988) |
| Kashmir: Great Himalaya | 4287 ± 215 | 51 | Holmes (1988) |
| South side Ladakh range | 5200–5400 | | Burbank and Fort (1985) |
| North side Zanskar range | 5200–5400 | | Burbank and Fort (1985) |
| North side Nanga Parbat | 4763 ± 223 | 45 | Kick (1980) |
| South side Nanga Parbat | 4678 ± 279 | 26 | Kick (1980) |
| North side Rakaposhi range | 5000 ± 0 | 6 | Visser (1928) |
| North side Greater Karakoram | 5050 ± 35 | 5 | Visser (1928) |
| South side Rakaposhi range | 5107 ± 169 | 14 | Visser (1928) |
| Northeast Greater Karakoram | 5575 ± 89 | 8 | Visser (1928) |
| Swat Kohistan | 4115 ± 152 | | Porter (1970) |

*Table 3.4* Present and past equilibrium-line altitudes in the northwest Himalaya.

| Location | Present ELA (m) | ELA at maximum advance (m) | $\Delta$ELA (m) | Reference |
|---|---|---|---|---|
| Kashmir | | | | |
|   Sind valley | 4300 | 3500 | 800 | Holmes (1988) |
|   Liddar valley | 4050 | 3300 | 750 | Holmes (1988) |
|   Ningle valley | 3900 | 3200 | 700 | Holmes (1988) |
| Ladakh range | 5300 | 4300 | 1000 | Burbank and Fort (1985) |
| Zanskar range | 5300 | 4700 | 600 | Burbank and Fort (1985) |
| Hunza valley[a] | 5000 | 3900 | 1100 | From info. in Owen (1988a) |
| Nanga Parbat[b] | 4650 | 3700 | 950 | From info. in Owen (1988a) |
| Swat Kohistan | 4200 | 3100 | 1100 | Porter (1970) |

[a] Based on Batura glacier; Borit Jheel advance of Derbyshire *et al.* (1984).
[b] Based on Raikot glacier (northern side of Nanga Parbat); Borit Jheel advance of Derbyshire *et al.* (1984).

cirques are higher than those in the Great Himalayan range. Furthermore, it would explain the lower depression of the ELA in the Pir Panjal flank compared to that in the Great Himalaya.

## PATTERNS OF GLACIATION ELSEWHERE IN THE NORTHWEST HIMALAYA

### Ladakh and Zanskar

Burbank and Fort (1985) observed numerous small (0.5–2 km²) cirque glaciers on the southern side of the Ladakh range and the northern side of the Zanskar range in the vicinity of Leh. They estimated that steady-state ELAs for glaciers in northeast–northwest facing cirques were 5200–5400 m in both ranges (Table 3.3).

ELAs were reconstructed for the last glacial maximum advance, known informally as the Leh stage (Fort, 1983), using a height method and a THAR of 0.4. The results show an ELA depression of ~ 1000 m below present values in the Ladakh range but only 600 m below present values in the Zanskar range (Table 3.4).

Burbank and Fort (1985) maintained that similarities between the two adjacent ranges ruled out climatic, altitudinal or tectonic explanations of the differences in ELA depressions. They concluded that differences in the bedrock configuration of the two ranges were responsible, with the molasse strata of the Zanskar range forming a topographic constraint to the downward movement of glaciers during the last glaciation. This explanation, however, is not universally accepted (e.g. Osmaston, pers. comm., 1988).

## Nanga Parbat massif

The modern glaciers of this massif have been studied in some considerable detail by Kick (1980, 1986). The results of this work indicate that the mean ELA for north-facing glaciers rises from ~ 4550 m on the south side of the massif to > 4660 m on the northern side (Table 3.3). If glaciers of all orientations are included in the analysis, a similar pattern emerges, although mean ELAs are ~ 100 m higher. These data suggest that a south–north precipitation gradient exists in the Nanga Parbat region, which is reflected in the ELAs of the current glaciers.

Detailed studies of former glacier advances are limited. However, a crude estimate of the lowering of the equilibrium line can be made for the northern side of the massif by using the glacial geological mapping of Owen (1988a) and Shroder *et al.* (1989) in the middle Indus valley. Tills deposited by glaciers originating from Hunza and the Nanga Parbat massif have been mapped down valley to Shatial at 1500 m altitude. From this, together with the estimate of the modern ELA, a tentative maximum lowering of the ELA of 950 m can be made This figure is likely to have been modified considerably by subsequent uplift of the massif and so represents a minimum value (Table 3.4).

## Karakoram range

Data from Visser (1928) in Mercer (1975) indicate a rise in the modern ELA to a maximum of > 5000 m to the east of the Hunza valley and to the north of the Skardu basin. The overall height of the equilibrium line in the Karakoram range reflects the aridity of the area as a whole, particularly when compared to the Kashmir basin and, to a lesser extent, the Nanga Parbat massif. The data also suggest a decrease in precipitation in a northeasterly direction, as reflected by the rising equilibrium line (Table 3.3).

Information on former equilibrium lines in this area is lacking and any estimates are complicated by rapid Quaternary uplift. However, Owen (1988a) and Shroder *et al.* (1989) have mapped tills from a glacier that extended down the Hunza valley as far as Shatial, in the middle Indus valley. These tills were correlated with the Yunz advance of Derbyshire *et al.* (1984), which predates 139,000 yr BP. The advance of a glacier to Shatial involved a lowering of the equilibrium line to ~ 4020 m, which is ~ 980 m below present levels. For the Borit Jheel advance, which was the last major phase of trunk-valley glaciation in the Hunza valley (Derbyshire *et al.*, 1984), glaciers extended to the Indus–Astor confluence, to an altitude of ~ 1300 m. This represents a former ELA of ~ 3400 m, and a lowering of the equilibrium line of ~ 1100 m (Table 3.4). The Borit Jheel advance has been dated to 40,000–50,000 yr BP.

Although a high degree of accuracy cannot be claimed for these figures, they do indicate very active glacierization in the Hunza valley during former ice advances. Furthermore, since the observed depression of the equilibrium line will have been reduced by subsequent tectonic uplift, the figures quoted must be regarded as minimum values. The rapidity of late Quaternary uplift of this region is indicated qualitatively by the fact that deposits of the penultimate (Yunz) advance occur several hundred metres above the valley floor at Shatial (Owen, 1988a) and by the fact that the depression of the ELA is less for the Yunz glaciation than for the later Borit Jheel advance.

## Swat Kohistan

Mean, steady-state ELAs for the northern part of the Swat valley range from ~ 3960 m to 4200 m and show a slight northward rise (Porter, 1970), which may reflect a north–south precipitation gradient (Table 3.3). With the exception of Kashmir, modern ELAs in the Swat region are low compared with most other parts of the northwest Himalaya for which data are available. Porter (1970) determined ELAs of former glaciers by using an area method (with an AAR of $0.6 \pm 0.1$). An ELA depression of ~ 1000 m was calculated for the oldest (Laikot) advance, with marginally lesser depressions for the two younger (Gabral and Kalam) advances (Table 3.4). No geochronometric dates, however, are available for any of these advances.

Porter (1970) argued that the similarity in the ELA depression of each of the three advances suggested that late Quaternary tectonic uplift of northern Swat had been limited. This conclusion was subsequently supported by fission-track evidence from the Swat area (Zeitler *et al.*, 1982b; Table 3.2).

## DISCUSSION

Problems with data relate to differences in methods used, uncertainties in the chronology of glacial advances and difficulties in quantifying the amount of tectonic uplift. A variety of different methods has been used to estimate the present and past ELAs referred to in this paper. This raises the problem of compatibility. A further problem lies in the fact that some of the older studies do not state which method was used. The effects of different methods and types of evidence are difficult to evaluate. The results of many of the studies, however, are not likely to be highly accurate.

The lack of detailed chronologies for the glacial advances in the northwest Himalaya makes it difficult to establish the contemporaneity of changes in ELAs. Clearly, meaningful conclusions can be drawn for the whole of the region only if ELA changes have adequate chronological control. Thus, any conclusions drawn should bear this point firmly in mind.

Although the uplift history of the northwest Himalaya is quite well known, rates of uplift are often insufficiently constrained to allow quantitative estimates of changes in ELAs. Furthermore, spatial variations in the rates of uplift are poorly known in many areas. Thus, the effects of tectonic uplift on patterns of glaciation can be determined only qualitatively.

The location of modern glaciers in the northwest Himalaya are controlled by topography, with the high massifs of the Karakoram range and Nanga Parbat being heavily glacierized. Regional variations in precipitation appear to control modern ELAs. The areas studied fall into three groups based on the height of the modern equilibrium line and mean annual precipitation (Tables 3.1, 3.3).

Group 1 comprises the Pir Panjal and Great Himalayan ranges of Kashmir together with Swat Kohistan, with ELAs of 4000–4300 m. In these areas, the mean annual precipitation varies from 600 mm to > 1000 mm. Group 2 comprises Ladakh and Zanskar, the Nanga Parbat massif and much of the Karakoram range. In these areas, ELAs range from 4600 m to 5100 m. Mean annual precipitation is substantially lower than in group 1 areas, between ~ 90 mm and 130 mm. Group 3 comprises the north side of the Karakoram range to the northeast of Skardu basin. In this area, mean ELAs exceed 5500 m and mean annual precipitation is substantially less (although by an undetermined amount) than at Gilgit, which is further to the west. Thus, despite the fact that actual precipitation in glacier accumulation areas is likely to be greater than the amounts quoted, which are from valley-floor meteorological observations, a good correspondence appears between mean annual precipitation and the height of the equilibrium line in the northwest Himalaya.

The results from this study accord quite well with the patterns of snowline altitudes shown on maps produced by von Wissmann (1959). However, the actual altitudes themselves seem too high in some areas. For example, in Swat Kohistan, von Wissmann's maps indicate that the snowline rises from 4600 m to 4800 m to the north and, in Ladakh, the snowline is mapped at 5600 m (Table 3.3). Elsewhere, there is greater accordance with von Wissmann's findings and those reported in this study.

Maximum depressions of the equilibrium line for the areas that have been studied in the northwest Himalaya range from 600 m to 1100 m below present values (Table 3.4). The smallest maximum depression is for the Zanskar range, close to Leh. However, results from the nearby Ladakh range (Burbank and Fort, 1985) suggest that, without the bedrock constraints on glaciation previously discussed, the depression of the equilibrium line may have been as high as 1000 m.

Depressions of the equilibrium line of 700–800 m occurred in the Pir Panjal and Great Himalayan ranges of Kashmir. In neither of these areas were glacial advances constrained by topography. However, the Pir Panjal range has experienced uplift during the late Quaternary, which will have reduced the observed ELA depression compared to the actual value.

In the Hunza valley, the northern side of the Nanga Parbat massif, the southern side of the Ladakh range and in Swat Kohistan, the maximum depression of the equilibrium line was 900–1100 m. In both the Hunza valley and the Nanga Parbat massif, these figures are likely to have been reduced from the actual values, as a result of tectonic uplift during the late Quaternary. Relative tectonic stability in Swat Kohistan and the Ladakh range during this period, however, means that the ELA depressions quoted for these areas are likely to be close to actual amounts.

The results show, overall, that glacierization was extremely active in parts of the northwest Himalaya during the last glaciation, except where the advance of glaciers was topographically constrained. Limited dating control of glacial advances, which is available for the Hunza valley (Derbyshire *et al.*, 1984), for the northern side of Nanga Parbat (Owen, 1988a) and for Kashmir (Holmes, 1988) shows that the last advance was out of phase with the last maximum global ice volume. Rather limited glacierization in Kashmir is more difficult to explain. For the Pir Panjal range, the depression of the equilibrium line was clearly modified by tectonic uplift during the late Quaternary. However, in the Great Himalayan flank, which was tectonically inactive during the late Quaternary, some other explanation is required. Possibly, the Pir Panjal acted as a major barrier to incoming moisture, which may have come from westerly disturbances. However, this is impossible to substantiate in the absence of further independent palaeoclimatic data.

## CONCLUSIONS

Patterns of modern glacierization indicate the strong influence of precipitation. Hence, modern ELA gradients reflect precipitation gradients.

Maximum depressions of the equilibrium line varied in magnitude. The limited glaciation of the Zanskar range has been attributed to topographic constraints on the advance of glaciers. In the Pir Panjal range of Kashmir, late Quaternary uplift has clearly reduced the observed ELA depression compared to actual amounts. Limited ELA depression in the Great Himalayan range of Kashmir is more difficult to explain: it may be the result of changes in the location and intensity of precipitation sources. Glaciation of Swat Kohistan involved substantial depressions of the ELA. However, in the absence of tectonic uplift in this area, these figures are likely to approach actual values. Particularly large ELA depressions occurred in the Hunza valley and on the northern flank of the Nanga Parbat massif. These figures are also minimum values, because both of these areas have undergone rapid and substantial late Quaternary uplift. The reasons for such active glacierization remain speculative: however, it may be the result of the extremely high gradients in both of these regions, which are also cited in explanation of the highly active glacierization in these areas today.

Limited chronological data suggest that the maximum depressions of the ELA in some parts of the northwest Himalaya were out of phase with global ice-volume changes. They may have occurred during the middle part of the late Quaternary, rather than toward the end. Long-term precipitation changes associated with fluctuations in the South Asian monsoon have been cited as a control on the timing of glacial advances in the northwest Himalaya during the late Quaternary. However, further work on the chronology of glaciation in this region is required for a clearer relationship to emerge.

# 4

# REVISED CHRONOSTRATIGRAPHY OF THE LATE CENOZOIC BUNTHANG SEQUENCE OF SKARDU INTERMONTANE BASIN, KARAKORAM HIMALAYA, PAKISTAN

*Vincent S. Cronin and Gary D. Johnson*

## ABSTRACT

The Bunthang sequence constitutes the sedimentary record of the late Cenozoic infilling of the Skardu basin, an ~ 250 km² intermontane basin within the Karakoram Himalaya mountains of Pakistan. Previous palaeomagnetic studies of the 1.24 km thick Bunthang sequence identified only reversely magnetized strata and were unable to collect samples from a 330 m thick interval in the middle Bunthang sequence. Although chronometric resolution of these data was not good, the age of the Bunthang sequence was bracketed between 3.2 myr and 0.73 myr. Four additional palaeomagnetic sample sites were established in the middle Bunthang and basal upper Bunthang sequence to supplement the 72 reliable sites previously established in the upper and lower Bunthang sequence. One of the sites in the middle Bunthang sequence is normally magnetized. The remaining 75 sites are reversely magnetized.

The Bunthang till is considered to be approximately the same age as the Jalipur till, and no older than the Olduvai subchron (1.88–1.66 myr), based upon qualitative evaluation of field relationships. Tentative correlation of the Bunthang reversal with the Jaramillo subchron (0.91–0.98) yields a minimum average rate of deposition of 3.5 mm yr⁻¹ over ~ 350,000 yr to deposit the entire Bunthang sequence. Statistical analysis independently estimated a deposition rate of 3.4 mm yr⁻¹ over ~ 368,000 yr. The best estimate for the period of Bunthang sequence accumulation is 1.25–1.1 myr to 0.73 myr. The Bunthang strata are among the oldest late Cenozoic sedimentary rocks in the internal zone of the Himalaya and are correlative with glaciation, rapid uplift and consequent slope instability downstream that may have temporarily raised the base level of the Indus river in the Skardu basin.

# INTRODUCTION

The young sedimentary strata observed along much of the Indus valley in the Himalaya mountains of northern India and Pakistan are typically related to fluvial processes, glaciation or mass wasting of local slope areas and include fluvial sands, moraines, till, landslide debris and sediment impounded during ephemeral blockages of the Indus by glaciers or landslides. Palaeomagnetic analysis of fine-grained sedimentary deposits in the Indus valley has shown that these strata are, in general, normally magnetized, indicating that they were deposited after the last major reversal of Earth's magnetic field occurred at 0.73 myr. Thermoluminescence dating of glacial lakebeds and carbon-14 dating of wood fragments in moraines show that these deposits from the Indus, Gilgit and Hunza river valleys are generally less than a few hundred thousand years old (Shroder *et al.*, 1989). Older sedimentary strata are rare in this environment of active erosion, transportation, temporary deposition and repeated remobilization.

The Indus valley broadens at its confluence with the Shigar river valley to form the Skardu valley (Figure 4.1). The Bunthang outcrop on the northeast side of the Skardu valley is a sequence of glacial, fluvial, lacustrine and coarse alluvial strata that provides the most complete sedimentary record of the early glaciation and subsequent basinal sedimentation of the Skardu valley. Initial studies of the physical and magnetic polarity stratigraphy (MPS) of the 1.24 km thick Bunthang sequence established that deposition of the upper part of the sequence occurred before 0.73 myr, based upon the occurrence of reversely magnetized mudstone in all 72 reliable sample sites that were established in both the upper and lower parts of the sequence (Cronin, 1982; Johnson, 1986; Cronin *et al.*, 1989). The upper limit for the age of the Bunthang sequence is based upon the occurrence of glacial till at its base. The Bunthang till has not been dated; however, it is undoubtedly younger than ~ 3.2 myr, when the Cenozoic continental glaciation of Europe, Asia and North America began (Berggren and van Couvering, 1974; Fillon and Williams, 1983). No volcanic ash or other independently datable material was recognized in the Bunthang sequence that might assist in establishing its geochronology. Prior to the work described herein, the best estimate for the age of the Bunthang sequence was sometime within the Matuyama reversed chron, between 2.47 and 0.73 myr (Cronin *et al.*, 1989).

The initial palaeomagnetic surveys provided a dense array of samples from the upper and lower parts of the Bunthang sequence (Cronin, 1982; Johnson, 1986); however, an ~ 330 m thick interval in the middle Bunthang sequence was not sampled owing to logistical difficulties and the generally coarse grain size within this interval, which is unsuitable for palaeomagnetic analysis. Many of the outcrops of the middle Bunthang sequence are near-vertical walls of conglomeratic sandstone and boulder conglomerate. This broad gap in sampling, along with the reversely magnetized character of all sites from the upper and lower Bunthang sequence, made it difficult to correlate between the Bunthang sequence

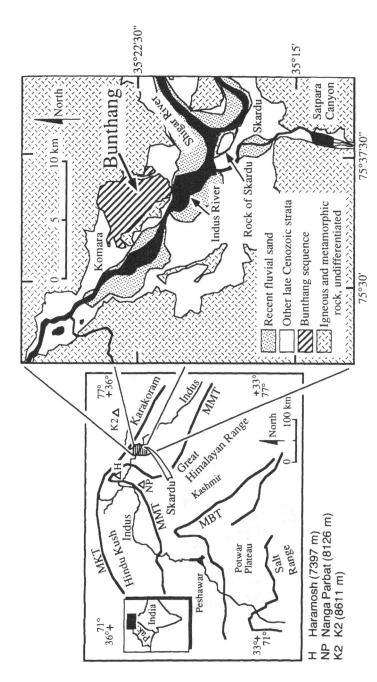

*Figure 4.1* Location maps showing the distribution of several major intermontane basins (halftone) within the structural framework of the northwest Himalaya, northern Pakistan and India.

The map area near Skardu is expanded at right to show the location of the Bunthang outcrop within Skardu valley.

MPS and the geomagnetic time scale (Mankinen and Dalrymple, 1979; Kent and Gradstein, 1986).

Fieldwork conducted in the fall of 1986 established four additional palaeomagnetic sites: two within the middle Bunthang sequence, and two at the base of the upper sequence that had previously been described by Johnson (1986). The new samples were subsequently analysed using thermal demagnetization techniques. The purpose of this paper is to describe the results of the palaeomagnetic analysis of samples from the four additional sites within the Bunthang sequence, to synthesize the new data with the results of earlier work (Cronin et al., 1989), and to revise the temporal correlations that had previously been suggested.

## Setting of Skardu basin

The town of Skardu is the capital of Baltistan, which is administered by Pakistan. The elevation of the Indus and Shigar rivers at Skardu is ~ 2130 m; however, the peaks that surround Skardu valley range from 4500 to 5800 m in elevation. Upstream from Skardu are several of the largest alpine glaciers on Earth, including the Baltoro, Biafo Gyang and Chogo Lungma glaciers. Surrounding these glaciers are some of Earth's highest mountain peaks: K2 (8611 m), Gasherbrum (8068 m) and related peaks, Masherbrum (7821 m), and many others. High relief and active glaciation lead to vigorous erosion and the transport of enormous volumes of sediment through Skardu valley from the area upstream. Less than 50 km downstream from Skardu, the great massif containing Nanga Parbat (8126 m), Haramosh (7397 m) and Rakaposhi (7754 m) lies transverse to the Indus river. The Nanga Parbat–Haramosh massif (NPHM) has uplifted at a rapid rate during the past 5–10 myr (Zeitler, 1983, 1985; Zeitler et al., 1982a, 1986, 1989), which may have caused occasional ponding at the Indus river at Skardu (Cronin et al., 1989).

Skardu basin lies within the Kohistan–Ladakh terrane, which is interpreted to have originated as a magmatic arc that formed above a Tethyan subduction zone and was later accreted on to Eurasian continental lithosphere. Final closure of the Tethys Ocean along the Indus–Tsangpo Suture occurred in the late Cretaceous-early Tertiary (e.g., Tahirkheli et al., 1979; Bard et al., 1980; Andrews-Speed and Brookfield, 1982; Petterson and Windley, 1985; Searle et al., 1987, 1989). Current knowledge of the structural setting of the Skardu basin is summarized by Cronin (1989). The sparse seismicity around Skardu basin, compared with the surrounding regions, indicates that Skardu is located on the relatively passive structural element within the Himalayan thrust prism (Cronin, 1989). The metamorphic bedrock of Skardu basin is predominantly Katzara schist (Zannetin, 1964; Casnedi and Ebblin, 1977; Tahirkheli, 1982), which is locally intruded by tonalites–granodiorites of the Ladakh batholith. Radiometric ages for the Ladakh batholith range from ~ 105 myr to ~ 37 myr (Searle et al., 1989), although younger leucogranites along the western edge of the NPHM have been reported (Zeitler, pers. comm., 1990).

The underthrust Indian continental lithosphere is inferred to be present at depths of $> 20$ km below the Kohistan–Ladakh terrane at Skardu (e.g., Powell and Conaghan, 1973; Seeber and Armbruster, 1981; Seeber et al., 1981; Ni and Barazangi, 1984; Coward and Butler, 1985). Northwest of Skardu, an allochthonous slice of Indian continental lithosphere is exposed at the surface within the NPHM (e.g., Wadia, 1931; Zeitler et al., 1989). This unusual exposure is related, at least in part, to northwest-directed reverse faulting along the Raikot fault (Madin et al., 1989; Cronin, 1989). Uplift of the NPHM continues today, based upon recent deformation along the Raikot fault (Lawrence and Ghauri, 1983; Madin, 1986; Madin et al., 1989; Shroder et al., 1989).

The Indus river flows along one of several prominent northwest-directed drainage lineaments that are observable on aerial imagery of the northwest Himalaya (Casnedi, 1976; Ebblin, 1976; Cronin, 1989; Cronin et al., 1990). These drainage lineaments are interpreted to be structural or lithologic discontinuities. Some lineaments have been correlated with observed faults or joint sets; however, field investigation of many of these lineaments has not yet been undertaken. Although the Indus river course varies in a step-like manner through the NPHM, it maintains its generally northwest trend. Where the Indus river turns southwest at its confluence with the Gilgit river, the extension of the northwest drainage lineament is occupied by the Gilgit river. The apparent continuity of this drainage lineament across the NPHM suggests that the course of the Indus river is antecedent to the uplift of the NPHM (Cronin, 1989).

Current understanding of late Cenozoic history of the northwest Himalaya relative to the Bunthang sequence at Skardu can be summarized in five stages as follows. First, the Indus river localized along northwest-trending structural discontinuities within the Himalayan thrust prism prior to the uplift of the NPHM. Second, uplift of the NPHM began after 5–10 myr. Third, the Indus river valley at Skardu was glaciated after 3.2 myr. Glaciers probably flowed into the Skardu area from local cirques, from the Shigar valley, from the Indus valley upstream of Skardu and from the Deosai plateau northward down Satpara canyon. Early glaciation resulted in the broadening of the Skardu valley, and in the deposition of till. Glaciation of the Skardu area has continued into the Holocene. Fourth, base level downstream from Skardu rose. This resulted in the deposition of the $\geq 1.24$ km thick Bunthang sedimentary sequence in the Skardu basin prior to 0.73 myr. Factors controlling base level at Skardu during the deposition of the Bunthang sequence may have included damming of the Indus valley downstream of Skardu by glaciers or landslides, accelerated uplift of the NPHM, subsidence in the Skardu basin relative to areas downstream, or some combination of these mechanisms (Cronin, 1982; Cronin et al., 1989). Fifth, base level downstream from Skardu fell relative to the Skardu basin. Reduction in relative base level is most likely to have been caused by accelerated erosion in the Indus valley downstream from Skardu, decreased rate of uplift along the NPHM, or a combination of both. This led to the erosion of the Bunthang sequence throughout most of Skardu basin. At present, the Indus river

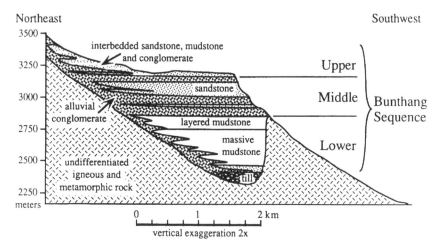

*Figure 4.2* Schematic cross-section of the Bunthang outcrop on the northeast side of Skardu valley. Vertical exaggeration is 2×. This most completely preserved outcrop of the Bunthang sequence is buttressed on the southwest by an outlier of Katzara schist, which appears to have preserved these strata from glacial and fluvial erosion.

appears to be incising its course through the NPHM at least as fast as the NPHM is rising. The Indus river gradient through the NPHM from Rondu to Jaglot is a relatively steep 10.8 m km$^{-1}$ owing to the delicate interplay of uplift and erosion (Seeber and Gornitz, 1983; Cronin *et al.*, 1989).

## PHYSICAL STRATIGRAPHY

The Bunthang outcrop is an ~ 1.3 km thick succession of young sedimentary deposits that is unusually well preserved on the northeast wall of Skardu basin (Figures 4.1, 4.2). The Bunthang till, which is locally exposed at the base of the Bunthang outcrop, is overlain by the succession of fluvial sandstone, lacustrine mudstone, alluvial fanglomerate and till that comprises the Bunthang sequence (Cronin, 1982). The basinward edge of the Bunthang outcrop is buttressed by a large outlier of Katzara schist, which has protected the sedimentary strata from erosion by the Indus river and by glaciers that may have occupied the Skardu valley subsequent to Bunthang deposition. Glaciers that originated in cirques to the north and northeast of the Bunthang outcrop flowed on either side of Bunthang. The Oro canyon till was deposited by a late Pleistocene glacier that extended from Oro canyon northeast of Komara and truncated the Bunthang sequence on the north side of the Bunthang outcrop. Elsewhere in Skardu and Shigar valleys, glaciation and fluvial erosion have removed all but isolated vestiges of Bunthang-equivalent strata. Hence, the Bunthang sequence provides a unique opportunity to study the basin-fill history of Skardu basin and to understand the evolution of intermontane basins in the interior zone of the Himalaya.

96

The Bunthang sequence was deposited unconformably on top of the Bunthang till, which is in contact with the Katzara schist where the base of the till is exposed (Figure 4.2). The Bunthang till is a coarse boulder conglomerate that is > 250 m thick where it is exposed near Komara and on the Rock of Skardu. The Bunthang till has not been dated but has been tentatively correlated with an early advance of the middle glacial interval of Shroder *et al.* (1989) and the Yunz glaciation of Zhang and Shi (1980). This correlation suggests that the Bunthang till is younger than the Jalipur till, which is exposed near Chilas, ~ 200 km downstream from Skardu. The Jalipur till is thought to be much less than 2 myr old, based on fission-track ages that cluster between 1.0 and 1.9 myr for detrital zircons in it (Shroder *et al.*, 1989; Zeitler, 1983). These ages are indistinguishable from zircon ages in the nearby source terrane of the NPHM. Therefore, all that is known with certainty about the age of the Jalipur till is that it is older than adjacent glacial deposits in the Indus valley and that it is quite young (Zeitler, 1983, and pers. comm., 1990; Shroder *et al.*, 1989).

The Bunthang sequence was first mapped by Dainelli (1922), who considered the strata to be composed of morainal material from the second of his four Himalayan glacial stages, along with lacustrine mudstone from the third glacial stage. Cronin (1982) named the Bunthang sequence, based on the Balti name for the largest outcrop of the sequence in the Skardu basin, and provided a general description of the sequence. Johnson (1986) described the upper Bunthang sequence in detail. The physical stratigraphy of the Bunthang sequence has been summarized by Cronin *et al.* (1989). The principal elements of the Bunthang sequence are as follows, in stratigraphic order (Figure 4.2).

### Lower Bunthang sequence

Basal sandstone overlain by mudstone, with a total thickness of ~ 450 m.

1   Approximately 50 m of conglomeratic sandstone, overlying the Bunthang till, is interpreted to have been deposited in a braided stream environment. Seven reliable palaeomagnetic sites were established within muddy interbeds in this sandstone.

2   Approximately 300 m of massive mudstone is interpreted to have been deposited in a glacial lake. The fine-grained particles were produced by mechanical weathering and are similar to modern glacial flour. Anomalous clasts within the mudstone are interpreted as glacial dropstones. Sixteen reliable palaeomagnetic sites from undeformed material were established in the massive mudstone.

3   Approximately 100 m of rhythmically bedded mudstone and sandstone is interpreted as a series of glaciolacustrine turbidites. Individual turbidites are between ~ 1.1 m and 5 m thick, and typically contain a scoured base overlain by graded sand that fines upward into a laminated mudstone. Glacial dropstones were observed. At a few stratigraphic horizons, coarse

boulder conglomerate to ~ 2 m in thickness is present within the turbidite sequence. These are interpreted to have originated as debris-flow deposits. Nine reliable palaeomagnetic sites were established in the massive mudstone. This distinctive layered mudstone marks the top of the lower Bunthang sequence.

## Middle Bunthang sequence

Dominantly conglomerate with interbedded sandstone to a total thickness of ~ 330 m (Figure 4.3).

1   The lowest of the middle Bunthang units is ~ 80–95 m of dark conglomeratic sandstone, interbedded with thin ( < 5 m), lighter coloured sandstone beds and thick boulder conglomerates with angular clasts to several m in diameter. The boulder conglomerates may be debris-flow deposits. The base of this set of beds is locally channelled into the underlying mudstone. These strata thin basinward and thicken toward the mountains, and are interpreted to have been part of an alluvial fan that prograded on to the older lacustrine muds in the interior of Skardu basin.

2   Overlying the fanglomerate is ~ 20 m of light sandstone that has not been sampled, owing to its position between two cliff-forming conglomerates. This sandstone appears to thicken basinward and thin toward the mountain front, and is tentatively interpreted to be a fluvial sandstone.

3   Approximately 50–65 m of dark conglomeratic sandstone, similar to the lowest unit of the middle Bunthang sequence, is interpreted to have been part of a second, stratigraphically higher alluvial fan. This unit is capped by a pronounced boulder conglomerate that is interpreted to be a debris-flow deposit.

4   The second fanglomerate is overlain by ~ 90–120 m of conglomeratic or granular sandstone, locally interbedded with mudstone or coarse boulder conglomerate. The boulder conglomerates are interpreted to be debris-flow deposits. Two palaeomagnetic sample sites were established in thin muddy interbeds within these strata. Site A is located at the base of an ~ 30 m high cliff that is composed dominantly of granular to pebbly, fine- to medium-grained sandstone. The cliff is capped by a prominent boulder conglomerate. Site B is located on a slope above the cliff and ~ 2 m above a thin conglomerate that was traced to the opposite side of the canyon. The continuity of the overlying conglomerate indicates that site B had not been disturbed by creep or other slope instability.

5   The top of the middle Bunthang sequence is marked by ~ 45 m of dark conglomeratic sandstone, interpreted to have been part of a third alluvial fan. A pronounced boulder conglomerate lies near the top of this unit, which is locally overlain by a coarse channel sandstone.

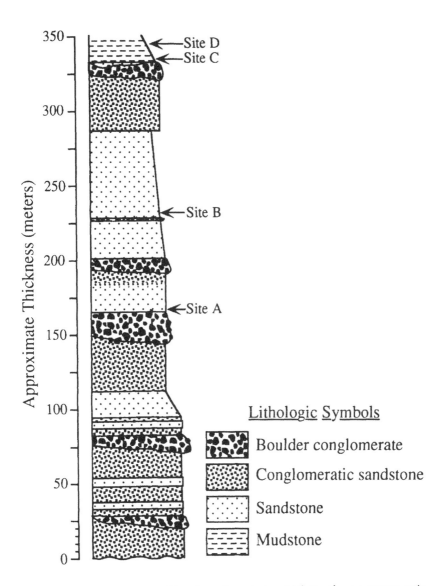

*Figure 4.3* Schematic stratigraphic column through the middle Bunthang sequence, as it is exposed on the northwest side of Bunthang in Ghothamal canyon, above Komara. Palaeomagnetic sample sites A and B are in the middle Bunthang sequence, while sites C and D are at the base of the upper Bunthang sequence.

## Upper Bunthang sequence

Fine- to coarse-grained, locally conglomeratic sandstone at the base, overlain by mudstone, with a total thickness of ~ 460 m. Forty-two reliable palaeomagnetic sample sites have been established in the upper Bunthang sequence.

1   Approximately 20 per cent of the 310 m of sandstone in the upper Banthang sequence is a medium- to coarse-grained, locally conglomeratic lithic arkose, interpreted to have accumulated by lateral accretion in a high energy braided stream.

2   Approximately 80 per cent of the sandstone in the upper Bunthang sequence is a fine- to medium-grained sandstone that grades vertically into a laminated mudstone. These units are interpreted to have accumulated by vertical accretion in a lower energy fluvial environment. Palaeocurrent directions parallel the flow of the modern Indus river.

3   The uppermost ~ 150 m of the Bunthang sequence is dominated by massive mudstone, interpreted to be glaciolacustrine.

## MAGNETIC POLARITY STRATIGRAPHY

### Previous work

MPS has been successfully used as a tool in stratigraphic correlation and age determination within the Himalayan molasse of northern Pakistan and India (e.g., Barndt et al., 1978; G.D. Johnson et al., 1979; Opdyke et al., 1979; N.M. Johnson et al., 1982; Burbank and Johnson, 1982, 1983; Burbank and Tahirkheli, 1985; Cronin et al., 1989). Studies that employed MPS in other Himalayan basins have generally been able to utilize fossil evidence or absolute ages from bedded volcanic ash as independent chronostratigraphic markers to correlate uniquely the observed magnetic stratigraphy with the magnetic polarity time scale. Similar datable markers have not been identified within the Bunthang sequence.

Cronin et al. (1989) described the results of palaeomagnetic sampling of the upper and lower Bunthang sequence (Cronin, 1982; Johnson, 1986). In these earlier studies, sample sites were established every 10 m (measured vertically through the section with a Jacob's staff and Abney level) where suitable material was present. Three or more oriented samples of unweathered mudstone were collected at each site and analysed in accordance with procedures described by Johnson et al. (1975). Seventy-two sites yielded reliable palaeomagnetic data (class I or class II data as defined by Tarling, 1971) after thermal demagnetization. All 72 sites from these initial studies were reversely magnetized. These results established that the Bunthang sequence is older than the most recent reversal of Earth's magnetic field at 0.73 myr (Mankinen and Dalrymple, 1979; Kent and Gradstein, 1986). Unfortunately, the lack of a documented reversal within the sequence did not permit a more precise correlation with the magnetic polarity time scale.

## Analysis of new data

Two types of magnetic behaviour were observed during the stepwise thermal demagnetization of samples from the four new sites established in the Bunthang sequence. Zijderveld projections of the progressive thermal demagnetization of representative samples from the four sites are shown in Figure 4.4 (Zijderveld, 1967). Samples from sites A and D reveal a simple, one-component remanence, in which heating to 400 °C removed 60–75 per cent of the natural remanent magnetic (NRM) moment of the samples. This magnetic behaviour is similar to that described for samples from the lower Bunthang sequence, in which the magnetic remanence is carried by magnetite with perhaps some maghemite (Johnson, 1986; Cronin et al., 1989). Representative samples from sites B and C were stripped of a lower-temperature magnetic phase (probably goethite) during the initial thermal demagnetization steps, causing an increase in magnetic moment from the NRM value. For the sample from site B, the intensity of the magnetic moment dropped to below the NRM value only after partial demagnetization to 620 °C, at which step the sample's magnetic intensity was only 22 per cent of the NRM. The site C sample lost 50 per cent of its NRM moment only after partial demagnetization at > 670 °C. The magnetic remanence is interpreted to be carried by hematite for the site B and C samples. Sites B and C are similar to the magnetic character of samples from the lowermost units of the upper Bunthang sequence that were analysed by Johnson (1986).

In order to remove any post-depositional magnetic overprinting, each of the three samples per site was partially demagnetized at two temperatures (550–60 °C and 620–50 °C) in order to remove approximately 50 per cent of the NRM moment from each sample. All samples from the sites B, C and D displayed reversed polarity (southerly declination, upward inclination) after demagnetization (Table 4.1). All samples from site A displayed normal polarity (northerly declination, downward inclination) after demagnetization. Sites A, B, C and D are all class I sites, as defined by an R value that is > 2.62 for 3 samples (Tarling, 1971). Class II sites show good agreement between at least two of the three samples within the site, so the site polarity is not in doubt, but have an R value that is < 2.62.

The site mean magnetic orientation after thermal demagnetization for each of the 76 class I and class II sites within the Bunthang sequence is plotted on an equal-area stereonet in Figure 4.5. Lower Bunthang sites show a greater dispersion in the orientation of their site mean magnetic vectors, compared with upper Bunthang sites; however, the magnetic polarity of all upper and lower Bunthang sites is unambiguously reversed. Only site A from the middle Bunthang sequence displays normal polarity.

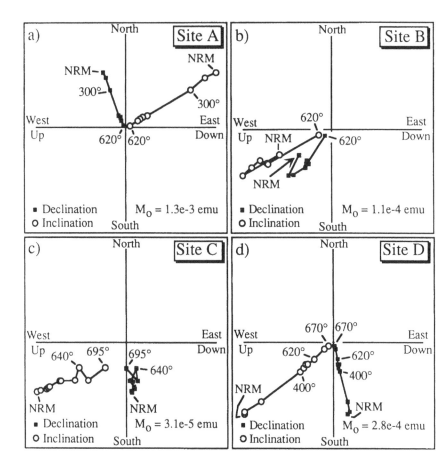

*Figure 4.4* Zijderveld projections of the thermal demagnetization of representative samples from each of the four newly sampled sites. NRM was measured for all samples, which were then partially demagnetized at temperatures of 200 °, 300 °, 400 °, 500 °, 560 °, 580 ° and 620 °C. In addition, samples from sites C and D were partially demagnetized at 640 ° and 670 °C, and the site C sample was also treated at 695 °C.

(a) Samples from site A remain normally magnetized (positive [downward] inclination; norther declination) as magnetic moment decreases with increasing temperature.
(b)–(d) Samples from sites B, C and D remain reversely magnetized during thermal demagnetization.

Plotted values are normalized to the magnetic moment for the corresponding untreated sample ($M_O$).

*Table 4.1* Site mean palaeomagnetic statistics for the four new sample sites in the Bunthang sequence. Sites A and B are in the middle Bunthang sequence.

| Site | Mean declination | Mean inclination | No. of samples | R | K | α95 | Class | VGP latitude | Polarity |
|------|------|------|------|------|------|------|------|------|------|
| A | 315 | 50 | 3 | 2.979 | 97.1 | 12.57 | I | 52.3 | normal |
| B | 237 | −49 | 3 | 2.920 | 25.2 | 25.07 | I | −42.0 | reversed |
| C | 171 | −43 | 3 | 2.953 | 48.1 | 19.01 | I | −77.6 | reversed |
| D | 165 | −45 | 3 | 2.987 | 165.4 | 9.61 | I | −75.3 | reversed |

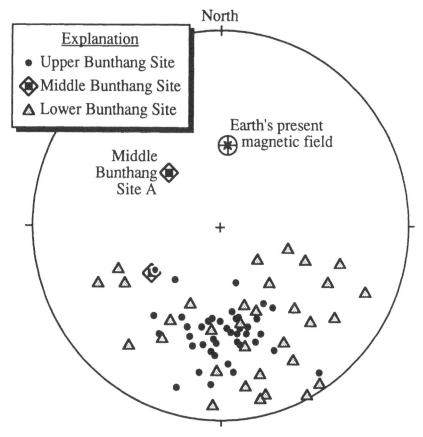

*Figure 4.5* Equal-area stereonet displaying site mean values for all 76 class I and II sites in the Bunthang sequence. Middle Bunthang site A is normally magnetized; all other sites are reversely magnetized and plot in the southern half of the stereonet.

### Interpretation of new data and synthesis

The single-site reversal that has been documented at site A within the middle Bunthang sequence improves our ability to correlate with the magnetic polarity time scale; however, that correlation still lacks the level of resolution that is desirable for definitive interpretation (Figure 4.6). In the absence of other chronometric data, four lines of evidence can help to limit the likely correlations. First, the occurrence of the Bunthang till at the base of the Bunthang sequence limits possible correlations to within the last 3.2 myr, during the late Cenozoic glacial period. Second, we assume that the reversal documented at site A is a major reversal, and not a spurious event related to a temporary or local instability in the magnetic field. This assumption allows possible correlations between the

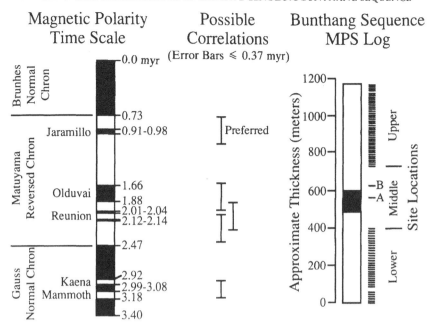

*Figure 4.6* Comparison of the MPS of the Bunthang sequence with the magnetic polarity time scale (Kent and Gradstein, 1986; Mankinen and Dalrymple, 1979). The five most likely correlations are indicated by error bars ≤ 0.37 myr in length, corresponding to the statistically predicted duration of Bunthang sequence deposition (Johnson and McGee, 1983).

reversal at site A and any normal subchron within either the Matuyama or Gauss chrons that is younger than 3.2 myr.

Third, a statistical model of geomagnetic reversals (Johnson and McGee, 1983) is used to estimate the duration of deposition for the Bunthang sequence, given the number of reliable sample sites ($N = 76$), the observed number of magnetic reversals ($R = 1$), and the magnetic reversal frequency characteristic of the late Neogene ($\tau = 120,000$ yr). The model is based upon the statistical properties of variations in the polarity of the Earth's magnetic field, and the stochastic nature of the palaeomagnetic sampling process. The probability ($p$) that two adjacent sites show opposite polarity is given by

$$p = \frac{R}{(N-1)}.$$

The duration of deposition ($\Delta t$) for a sedimentary sequence is estimated to be

$$\Delta t = \frac{-\ln(1 - 2p)}{2} \; \tau N \pm 2\sigma$$

where

$$2\sigma = 2\sqrt{[p(1-p)(N-1)]}\ \frac{\Delta t}{R}$$

(Johnson and McGee, 1983). For the Bunthang sequence, the duration of deposition is estimated to be 123,251 ± 244,853 yr at the 95 per cent confidence interval. The Bunthang sequence is predicted to have been deposited in less than ~ 0.37 myr. This estimate means that it is unlikely that the 450,000 year-long normal subchron at the end of the Gauss chron is recorded at site A. The minimum average rate of deposition for the 1240 m Bunthang sequence is predicted to be ~ 3.4 mm yr$^{-1}$.

At this stage, there are five possible correlations between the normal event recorded at site A and the magnetic polarity time scale (Figure 4.6). The Bunthang sequence could contain the Jaramillo, Olduvai, later Reunion, earlier Reunion, or both Reunion events within the Matuyama reversed chron, or the normal subchron between the Kaena and Mammoth events within the Gauss normal chron. The fourth line of evidence that helps to limit the likely correlations is the tentative correspondence between the Bunthang till and the Jalipur till. The Jalipur till is the oldest glacial deposit yet recognized in the Hunza, Gilgit, or Indus river valleys west of the NPHM (Shroder *et al.*, 1989). The very young ages of detrital zircons in the Jalipur till indicate that the till is much less than 2 myr old (Zeitler, 1983). The Jalipur and Bunthang tills are assumed to be no older than the Olduvai event (1.88–1.66 myr; Kent and Gradstein, 1986). The lack of stratal continuity from Skardu to Jalipur and the current lack of absolute ages makes it impossible to do more than suggest that the Bunthang and Jalipur tills may be approximately the same age. The Bunthang and Jalipur tills are considered roughly coeval because they are the oldest tills in their respective areas along the Indus valley, and both are associated with multiple younger tills. If the Jalipur till is < 1.88 myr old, and the Bunthang till is approximately the same age as the Jalipur till, then the only normal subchron that is likely to correlate with middle Bunthang site A is the Jaramillo event, between 0.91 and 0.98 myr.

The correlation between site A and the Jaramillo event yields a minimum average deposition rate for the ~ 625 m of middle and upper Bunthang sequence above site A of ~ 3.5 mm yr$^{-1}$, measured at the present state of compaction. There is ~ 615 m of middle and lower Bunthang sequence below site A, for a total of ~ 1240 m in the Bunthang sequence. At a minimum average rate of deposition of 3.5 mm yr$^{-1}$, it would take less than ~ 0.35 myr to deposit the entire Bunthang sequence. This compares very favourably with the statistical prediction of the duration of Bunthang sequence deposition: less than ~ 0.37 myr at a rate of ~ 3.4 mm yr$^{-1}$. The stratigraphic distance between the lowest reversed site above site A (site B) and the highest reversed site below site A is ~ 230 m. It would take ~ 66,000 yr to deposit 230 m of strata at a rate of 3.5 mm yr$^{-1}$. The duration of the Jaramillo normal subchron is ~ 70,000 yr, so the gap in palaeomagnetic sampling within the middle Bunthang sequence is approximately large enough to span the entire Jaramillo event. It is likely that the finer sediments of the upper and lower Bunthang sequence were deposited at a more rapid rate, and the coarser strata of the middle Bunthang sequence were deposited at a less rapid rate than the average for the entire sequence. Most of the middle Bunthang sequence below site B may have been deposited during the Jaramillo event and be normally magnetized.

## CONCLUSIONS

A single normally magnetized site has been identified within the middle Bunthang sequence, along with 75 reversely magnetized sites in the upper and lower Bunthang sequence. The normally magnetized site is tentatively correlated with the Jaramillo event of 0.91–0.98 myr, based primarily on (a) constraints for the age of the Bunthang till at the base of the Bunthang sequence; and (b) the assumption that the normally magnetized site corresponds to a recognized normal subchron and not to an ephemeral variation in Earth's magnetic field. Statistical modelling of the palaeomagnetic results for the Bunthang sequence predicts that the entire sequence was deposited in $< 0.37$ myr, at an average rate of $\sim 3.4$ mm yr$^{-1}$, measured at current state of compaction. The minimum average rate of deposition for the 625 m of reversely magnetized strata above site A is $\sim 3.5$ mm yr$^{-1}$, based on the correlation of the normally magnetized site with the Jaramillo event. This would place the age of the base of the Bunthang sequence at $\sim 1.25$–1.1 myr, which is consistent with the estimated age of the Bunthang and Jalipur tills. The top of the Bunthang sequence was probably deposited between 0.91 and 0.73 myr. If the reversal at site A is an ephemeral event other than the Jaramillo subchron and if the Bunthang till is no older than the Olduvai event (1.88–1.66 myr), then the age of the Bunthang sequence must be between 1.66 and 0.73 myr.

The revised estimates for the age of the Bunthang sequence suggest that the basin phase at Skardu was coincident with rapid uplift of the NPHM, as well as with consequent mass wasting owing to slope instability and glaciation. These mechanisms, acting singly or in concert, could have elevated the base level of the Indus river downstream from Skardu, thereby causing basinal sedimentation at Skardu.

107

# 5

# QUATERNARY AND HOLOCENE INTERMONTANE BASIN SEDIMENTATION IN THE KARAKORAM MOUNTAINS

*Lewis A. Owen and Edward Derbyshire*

## ABSTRACT

The deposition of sediments within the valleys and intermontane basins of the Karakoram mountains is complex, dominated by the glacial and mass movement processes. Valley fills and terraces reach many hundreds of metres in thickness comprising till, glaciofluvial, glaciolacustrine, fluvial, debris-flow, flowslide, lacustrine and aeolian sediments. Holocene and Pleistocene sediments deposited in the Karakoram valleys and basins are described in terms of their sedimentological characteristics, their landforms and the processes that produced them. Six main sedimentary environments are recognized.

## INTRODUCTION

Vast thicknesses of Quaternary and Holocene sediments occur within the valleys of the Karakoram mountains. These comprise debris-flow, mudflow, till, glaciofluvial, fluvial, lacustrine and aeolian sediments and form thick valley fills which frequently exceed 700 m thick. These valley fills provide evidence for palaeoenvironment change during the Quaternary, which includes at least three extensive valley glaciations. In addition, they provide an insight into the types of processes which have led to the formation of the present Karakoram landscape. Elucidation of the evolution of the valley fills and their depositional environment within the Karakoram mountains and similar high mountain belts requires an understanding of the nature of the interaction between tectonics, uplift, climate, Quaternary glaciations, catastrophic land-forming events and allocyclic sedimentary processes. This paper discusses the formation of these valley fills and the associated landforms in terms of the relative roles of tectonics, climate and allocyclic changes.

*Table 5.1* Exhumation, denudation and uplift rates for selected areas across the trans-Himalayan mountains

| | Method of Determination | Region | Source |
|---|---|---|---|
| *Exhumation rate (mm yr$^{-1}$)* | | | |
| 0.1–4.5 | Fission track | Nanga Parbat Himalaya | Zeitler (1985) |
| 0.7–0.8 | Mineral cooling temperature | High Himalaya, India | Mehta (1980) |
| 1.4–2.1 | Kyanite grade metamorphism | Suru valley, western Zanskar | Searle (1991) |
| 1.2–1.6 | Mineral cooling ages | Baltoro area, High Himalaya | Searle (1991) |
| *Denudation rates (mm yr$^{-1}$)* | | | |
| 2.0 | Growth of Bengal Fan | Ganges/Brahmaputra watershed | Curray and Moore (1971) |
| 1.8 | Sediment yields | Hunza watershed | R.I. Ferguson (1984) |
| 2.56–5.14 | Sediment yields | R. Tamur watershed | Seshadri (1960)[a], Ahuja and Rao (1958), Williams (1977)[a] |
| 1.43–2.50 | | R. Sun Kosi watershed | Williams (1977)[a], Pal and Bagchi (1974)[a] |
| > 0.65 | Sediment yields | Annapurna range | Fort (1987) |
| *Uplift (mm yr$^{-1}$)* | | | |
| 5.0–12.0 | Flexual modelling of the lithosphere | Northern Pakistan | Seeber and Gornitz (1983) |
| 0.1–1.5 | Levelling of river terraces | Nepal | Iwata (1987) |

[a] In Ramsay (1985).

## KARAKORAM MOUNTAIN PROCESSES AND ENVIRONMENT

Recent work has shown that the Karakoram mountains are one of the most rapidly rising mountain belts in the world, averaging ~ 2 mm yr$^{-1}$ across the whole region and resulting in peaks 7000–8000 m above sea level (Zeitler, 1985; Gornitz and Seeber, 1981; Seeber and Gornitz, 1983; Lyon-Caen and Molnar, 1983; R.I. Ferguson, 1984; Owen, 1989a; Table 5.1). Denudation is intense, involving cryogenic and chemical weathering (Hewitt, 1968b; Whalley *et al.*, 1984; Goudie *et al.*, 1984a; Goudie, 1984); glacial erosion (Goudie *et al.*, 1984a; Li Jijun *et al.*, 1979); fluvial incision (R.I. Ferguson, 1984; R.I. Ferguson *et al.*, 1984) and mass movement processes (Brunsden and Jones, 1984; Brunsden *et al.*, 1984).

In addition, the mountains are incised by some of the world's greatest rivers (Indus, Gilgit, Hunza, Shyok) with sediment loads among the highest known, e.g. the Hunza river has a sediment yield of 4800 t km$^{-2}$ yr$^{-1}$ (Ferguson, 1984). Moreover, the modern glaciers, covering at least 20 per cent of the land area, are of high activity type (i.e. they have high ablation and accumulation rates: Derbyshire, 1981; Derbyshire and Owen, 1990) and include some of the largest outside the polar regions.

The combination of high denudation and uplift rates has produced the greatest relative relief on Earth (with the exception of the oceanic landscapes), with valley floors averaging 1500 m in altitude and peaks lying 7000–8000 m above sea level. Hewitt (1989b) showed that as a result of such a large range in height there is a relatively well defined altitudinal organization of geomorphic processes and depositional environments within the Karakoram mountains. He identified four altitudinal zones:

Zone I    Perennial ice climate, generally > 5500 m altitude.
Zone II   High alpine tundra (humid), ~ 4000–5500 m altitude.
Zone III  Subalpine/montane (seasonal drought), ~ 3000–4000 m altitude.
Zone IV   Submontane/cool steppic (semi-arid), generally < 3000 m altitude.

The climates of the region are transitional between central Asian and monsoonal south Asian types, varying considerably with altitude, aspect and local relief. The Karakoram valley floors are essentially deserts with a mean annual precipitation of < 150 mm, most of this occurring over short periods in summer as heavy storms. Extreme diurnal maximum temperatures in summer exceed 38 °C while winter temperatures fall below 0 °C even in the valleys (Goudie et al., 1984a). Dust storms are common, occurring about once a week, enhanced in the summer by katabatic effects. Vegetation is altitudinally controlled (Paffen et al., 1956). It is scarce along the valley floors and on the lower valley-side slopes, which are of desert steppe type, and is replaced at higher levels by temperate coniferous trees and then by alpine meadow vegetation. Human-made irrigation systems, dependent to a large degree on glacier meltwaters, are extensive along the valley floors and form the basis of an agriculture in which wheat, corn, barley, potatoes and deciduous orchards are important.

Catastrophic events are frequent in the Karakoram mountains, notably the large debris flows and the flooding produced by natural dams associated with landslides and rapid glacial advances (Mason, 1929, 1935; Brunsden and Jones, 1984). Such events are capable of rapid and violent erosion and widespread resedimentation of unconsolidated deposits which may dramatically reshape the landscape.

## NEOTECTONICS

Although the Karakoram mountains are one of the most tectonically active areas in the world, there is little evidence of neotectonic deformation of the valley-fill

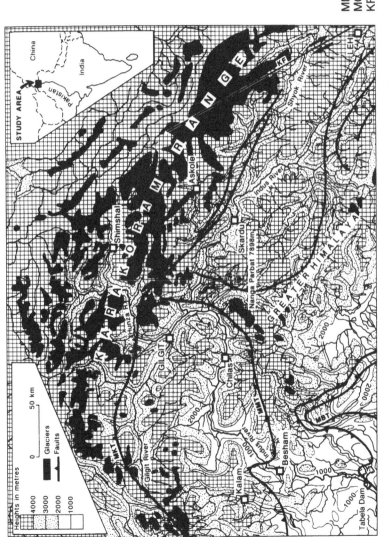

*Figure 5.1* Topographic and drainage map of the Karakoram mountains, western Himalaya and adjacent foreland basins showing the major geological structures. Note the rectangular drainage pattern and the antecedent drainage of the Gilgit, Shigar and Indus rivers.

MMT  Main Mantle Thrust
MCT  Main Central Thrust
KF    Karakoran Fault

sediments. Neotectonic processes probably have little influence on valley sedimentation, though they may have been important in initiating large-scale mass movements (Ambraseys *et al.*, 1981; Butler *et al.*, 1989). Tectonics, however, are important in controlling the formation and configuration of the mountains and valleys which clearly control the distribution of substantial thicknesses of valley-fill sediments (Owen, 1989a; Searle, 1991; Figure 5.1). Drainage patterns north of the Great Himalaya are rectilinear with a strong NNW–SSE and SW–NE orientation. Some of these rivers, which include the Shigar, Hunza and Gilgit, are antecedent to the mountain grain (Figure 5.1). Owen (1989a) suggests that these represent the development of drainage systems early in the structural evolution of the mountains. Also of note is the deflection of the Indus river around the Nanga Parbat massif and its east–west trend which flows concordant to the Great Himalaya for several hundred km. Approximately 80 km east of Nanga Parbat the Indus river flows south, antecedent to the grain of the Great Himalaya. This drainage implies differential uplift of the Nanga Parbat massif and Great Himalaya with respect to the Karakoram mountains, which inhibit the direct southern flow of the Indus making it flow parallel to the trend of the Great Himalaya and deflecting it around the Nanga Parbat massif. Several active fault zones have been recognized which deform Quaternary sediments in the Karakoram mountains and the Great Himalaya of northern

*Plate 5.1* A major neotectonic thrust in the middle Indus valley near Lichar. High grade Nanga Parbat gneiss has been thrusted at least 200 m over unlithofied glaciofluvial sands and gravels.

Pakistan (Owen, 1989a; Searle, 1991). The Raikot fault zone is the most intensely studied (Madin, 1986; Butler and Prior, 1988; Owen, 1989a). The deformation is complex and involves folding, thrusting (Plate 5.1), shearing and tilting of valley-fill sediments (Owen, 1989a).

Various studies (Table 5.1) have provided evidence on the magnitude of uplift of the Himalayas and Karakoram mountains. Of particular note is the work of Zeitler (1985) using fission-track and $^{40}Ar/^{39}Ar$ ages. Interpretation of cooling ages which range from < 0.5 myr to > 80 myr BP suggests that during the late Tertiary, long-term exhumation rates at least doubled, from < 0.2 mm $yr^{-1}$ to in some cases well over 0.5 mm $yr^{-1}$. He also showed that exhumation rates have strong systematic regional variations reflecting greater exhumation in the eastern and northern areas, especially in the Nanga Parbat–Haramosh massif and the Hunza regions which may have had exhumation rates in the order of several mm $yr^{-1}$. However, Zeitler does not resolve the problem of differentiating uplift and denudation, so that true uplift rates cannot be calculated until denudation rates are known.

Throughout the Karakoram mountains a series of high surfaces or rounded ice-smoothed slopes can be recognized and traced over many tens of km. The highest surface is > 5200 m and probably represents a late Tertiary palaeorelief. Derbyshire *et al.* (1984) pointed out that this must be younger than 8.6 myr BP because the granodiorites into which the bench has been eroded at Karimabad have a radiometric age of 8.5 myr BP. A lower surface at a height of 4100–4200 m was recognized by Paffen *et al.* (1956) in the middle Hunza valley, which he considered to represent 'pre-Pleistocene relief'. Derbyshire *et al.* (1984) recognized a similar surface in the upper Hunza valley at a similar height but renamed it the 'Patundas surface' and assigned its formation to the Shanoz glaciation. Two lower surfaces can be recognized at heights of ~ 3000 and 2700 m and were probably formed during two major glaciations which Derbyshire *et al.* (1984) assigned to the Yunz and Borit Jheel glaciations respectively. Similar surfaces have been recognized throughout the Karakoram mountains and Tibet. Although most of these surfaces have clearly been eroded by ice and have scattered erratics and tills on them, the large altitudinal range in heights and the continuity of the surfaces suggest that tectonic uplift may have been important in their formation. Chinese workers recognize similar surfaces in Tibet (Li Jijun *et al.*, 1979; Sun and Wu, 1986) and attribute them to a combination of changes in the Quaternary climate and punctuated tectonic uplift, which Sun and Wu (1986) call tectonoclimatic events. Owen (1989a) indicated that a broad correlation may be made between Tibet and the Karakoram on the basis of these planation surfaces. The correlation, however, is very tentative because of the poor dating constraint throughout the region, and the possibilities of different styles of glaciation across this broad region. These large surfaces indicate that long-wave, long-term regional warping is important in the formation of the Karakoram mountains and that discrete movements along fault zones have only a localized effect.

## VALLEY-FILL SEDIMENTS AND SEDIMENTARY PROCESSES

Six main sedimentary environments/systems are important in formations of landforms and valley fills within the Karakoram mountains. These include:

(a) Glacial system;
(b) Glaciofluvial and fluvial system;
(c) Mass movement system;
(d) Lacustrine system;
(e) Debris flow–alluvial fan system;
(f) Aeolian system.

The glacial system dominates the environment and greatly influences the fluvial, aeolian and mass movement system (Figure 5.2). An understanding of the paraglacial environment and paraglacial processes, therefore, is of considerable importance in understanding the landscape evolution.

## THE GLACIAL DEPOSITIONAL ENVIRONMENT

The Karakoram glacial depositional environment is complex, interacting with the paraglacial, proglacial and periglacial environments (Mason, 1929; Hewitt, 1961, 1964, 1967, 1968b, 1969; Batura Glacier Investigation Group, 1976, 1979; Goudie *et al.*, 1984a; Perrott and Goudie, 1984; Li Jijun *et al.*, 1979; Owen *et al.*, in press). A large variety of landforms are produced, including complex 'ablation valleys', lateral moraines, ice-contact fans, subglacial landforms and outwash plains (Owen, 1988a; Owen and Derbyshire, 1988; Derbyshire and Owen, 1990). Owen and Derbyshire (1988) and Derbyshire and Owen (1990) classify the glaciers into two main types on the basis of the sediment landform association in the lower reaches and on the proglacial plains. These are:

1   Ghulkin type, consisting of ice-contact fan sedimentation dominated by glaciofluvial debris-flow and debris-slide processes (Figure 5.3). This is characterized by an ice-contact cone which may reach several hundreds of metres in height and consists of a chaotic assemblage of tills deposited by meltout processes and resedimented by debris flow and slide processes (Plate 5.2). The cone is highly dissected by steep torrent meltwater channels. Meltwater fans often develop at the mouth of the incised meltwater channels. Many ancient examples can be seen along the southern margin of the Skardu basin, and ice-contact sediments at Dainyor form one of the most impressive dissected valley fills in the Karakoram mountains, which exceeds 700 m in thickness (Owen and Derbyshire, 1988; Derbyshire and Owen, 1990).

2   Pasu type, dominated by hummocky moraines and glaciofluvial outwash plains typical of the valley glaciers described by Boulton and Eyles (1979; Plate 5.3 and Figure 5.4). Small dissected end moraines, lateral moraines and subglacial tills are common.

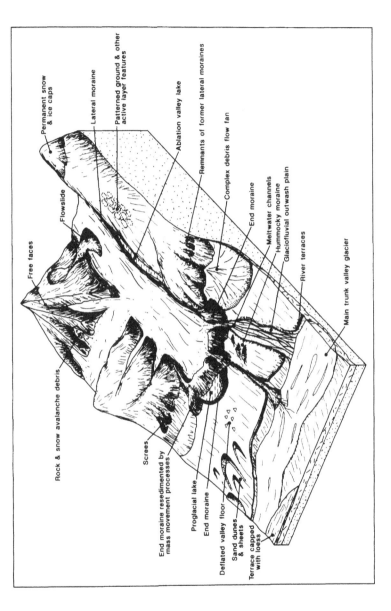

Permanent snow
& ice caps

Lateral moraine

Patterned ground & other
active layer features

Ablation valley lake

Remnants of former lateral moraines

Complex debris flow fan

End moraine

Meltwater channels

Hummocky moraine

Glaciofluvial outwash plain

River terraces

Main trunk valley glacier

Flowslide

Free faces

Rock & snow avalanche debris

Screes

End moraine resedimented by
mass movement processes

Proglacial lake

End moraine

Deflated valley floor

Sand dunes
& sheets

Terrace capped
with loess

*Figure 5.2* Schematic diagram showing the environmental and geomorphological settings with the main landform types in the Karakoram mountains. Note the close proximity of the different environments and the interaction between the different sedimentary systems.

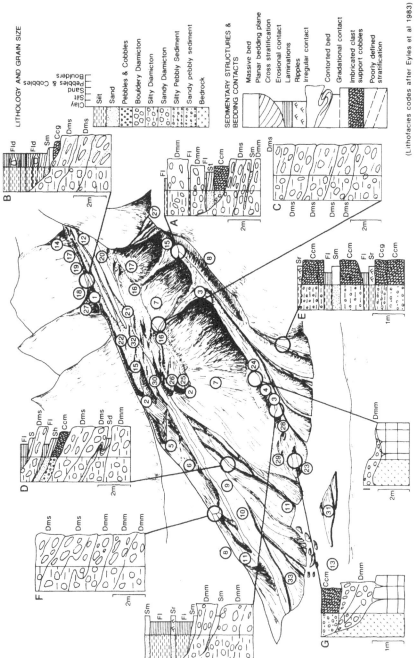

(Lithofacies codes after Eyles et al 1983)

**LITHOLOGY AND GRAIN SIZE**

Clay
Silt
Sand
Pebbles & Cobbles
Boulders

Silt
Sand
Pebbles & Cobbles
Bouldery Diamicton
Silty Diamicton
Sandy Diamicton
Silty Pebbly Sediment
Sandy pebbly sediment
Bedrock

**SEDIMENTARY STRUCTURES & BEDDING CONTACTS**

Massive bed
Planar bedding plane
Cross stratification
Erosional contact
Laminations
Ripples
Irregular contact
Contorted bed
Gradational contact
Imbricated clast support cobbles
Poorly defined stratification

*Figure 5.3* Main landforms and lithofacies associated with a Ghulkin type glacier.

1 Truncated scree;

2 Lateroterminal dump moraine;

3 Outwash channel drained laterally;

4 Glaciofluvial outwash fan;

5 Slide moraine;

6 Slide-debris flow cones;

7 Slide-modified lateral moraine;

8 Abandoned lateral outwash fan;

9 Meltwater channel;

10 Meltwater fan;

11 Abandoned meltwater fan;

12 Bare ice areas;

13 Trunk-valley river;

14 Debris flow;

15 Flow slide;

16 Gullied lateral moraine;

17 Lateral moraine;

18 Ablation valley lake;

19 Ablation valley;

20 Supraglacial lake;

21 Supraglacial stream;

22 Ice-contact terrace;

23 Lateral lodgement till;

24 *Roche moutonné*;

25 Fluted moraine;

26 Diffluence col;

27 High level till remnant;

28 Diffluence col lake;

29 Fines washed from supraglacial debris;

30 Ice-cored moraines;

31 River alluvium;

32 Supraglacial debris;

33 Dead ice.

*Plate 5.2* Oblique lateral view of the Ghulkin glacier in the upper Hunza valley. The Karakoram Highway and the village provides the scale. Compare this plate to Figure 5.3 for detailed explanations.

These glacial processes result in a gradational series of sediments, from subglacial tills of lodgement type, through subglacial and supraglacial meltout tills, to tills resedimented by debris-flow and debris-slide processes in both the Pasu and Ghulkin type glaciers. Sedimentary characteristics are broadly similar, all being very poorly to poorly sorted, positively skewed, depleted of fines ( < 10 per cent clay), and having angular clasts. These properties result from a glacial system dominated by supraglacial deposits derived from rockfalls and avalanches.

Study of the glacial sediments indicates that the area has undergone three main glaciations during the Pleistocene and at least five minor advances in the Karakoram and Nanga Parbat areas (Derbyshire *et al.*, 1984; Owen, 1989a).

Some of the valley fills, which comprise tills and glaciofluvial sands and gravels, have been deformed by glaciotectonic processes (Owen, 1988b; Owen and Derbyshire, 1988). The most impressive examples are present in the Skardu basin, where ice advanced northward from the Deosai plateau into the Skardu basin and overrode and pushed up terraces composed of floodplain sediments (Owen, 1988b). Owen (1989a) showed that it was important to distinguish between neotectonic deformation and deformation produced by glacial processes in order to elucidate the role of glacial processes and also to formulate appropriate glacial histories and neotectonic models for the Karakoram mountains.

118

*Plate 5.3* View of the snout of the Pasu glacier and its proglacial lake from hummocky moraines in the upper Hunza valley. Note the high lateral moraines in the distance and the thin veneer of subglacial tills on the bedrock in the middle ground. Compare this plate to Figure 5.4 for further explanations.

## Glaciofluvial and fluvial environment

Glaciofluvial processes are greatly influenced by the strong seasonal and diurnal variation in ablation rates controlled by large temperature contrasts. Glaciofluvial channels adjacent to ice fronts are characterized by steep slopes and frequent channel abandonment and new channel formation as ice tunnels open and close within the high activity glaciers. The main rivers throughout the Karakoram mountains are directly fed by these glaciofluvial streams and melting snows in spring, thus they respond in a similar fashion to seasonal and diurnal variations of the glaciofluvial streams. In the Indus catchment, for example, sediment output and river discharge may have increased 500–1000 and 20–50 times respectively during the daytime in the summer because of the ablation of glacial ice (R.I. Ferguson, 1984; R.I. Ferguson *et al.*, 1984). This, therefore, controls the mode of deposition of the sediments and the sediment type. True fluvial sediments deposited by the main rivers such as the Indus and Gilgit show a better degree of sorting than the glaciofluvial sediments and are characterized by sheets of sands and imbricated cobbles. Changes in the position of meander belts in some of the basins, e.g. Skardu, help to produce impressive terraces and/or chute cutoffs, adding to the variety of terrace types. The main rivers change

119

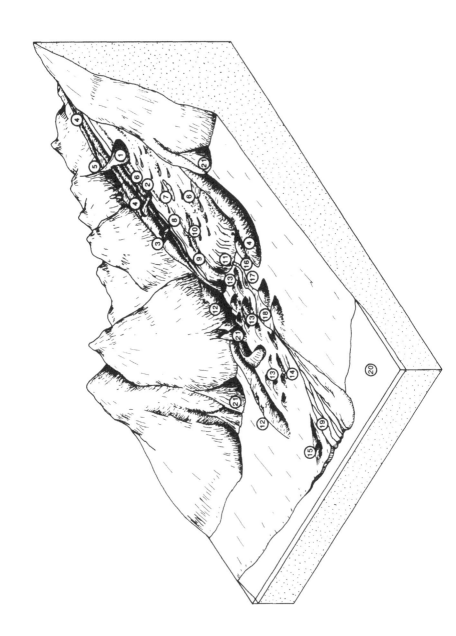

*Figure 5.4* Main landforms associated with a Pasu type glacier.

1  Rock avalanche/flowslide deposit;

2  Landsliding from a lateral moraine;

3  Ablation valley;

4  Lateral moraine;

5  Steep rocky valley sides;

6  Supraglacial marginal pond;

7  Supraglacial pond;

8  Supraglacial meltwater stream;

9  Surface of glaciers covered in supraglacial debris;

10  Crevasses and shears exposing fresh glacial ice;

11  Subglacial meltwater tunnel;

12  Ancient lateral moraines;

13  Hummocky moraines;

14  Fluted moraines;

15  *Roche moutonné*;

16  Glacial eroded bedrocks with small deposits of subglacial tills;

17  Meltwater stream;

18  Proglacial lake;

19  Sandur;

20  Main valley river;

21  Scree slopes and talus cones.

character along their length and include meandering to braided to incised gorge sections. This is probably a complex function of uplift, bedrock lithology and adjacent geomorphological features such as glaciers, landslides and others. Many ancient examples of these sediments can be recognized through the region but they comprise a minor component of the valley fills.

## MASS MOVEMENT PROCESSES AND ASSOCIATED ENVIRONMENTS

Mass movement is a major process in the Karakoram mountains (Goudie *et al.*, 1984; Brunsden and Jones, 1984; Brunsden *et al.*, 1984; Owen, 1991), resulting in deposits of diamictons, many of which are resedimented till or previous mass movement deposits (Owen *et al.*, in press). Mass movement deposits range in scale from isolated small failures to extensive areas of complex movement. Owen (1988a, 1991) recognized nine main categories, which include: rockfall, rock-slides and debris slides, debris flows, mudslides, flowslides, rotational slips, slumps and creep (Figure 5.5).

*Plate 5.4* Large debris cones in the Bagrot valley; note houses for sale. The cones were produced by high magnitude slope failures which blocked the valley allowing lakes to build up. Silts were deposited in these lakes which can now be seen as the planar bedded deposits in the central middle ground. These sediments were subsequently eroded, which now forms an impressive gorge which separates the village in the foreground from the large cones.

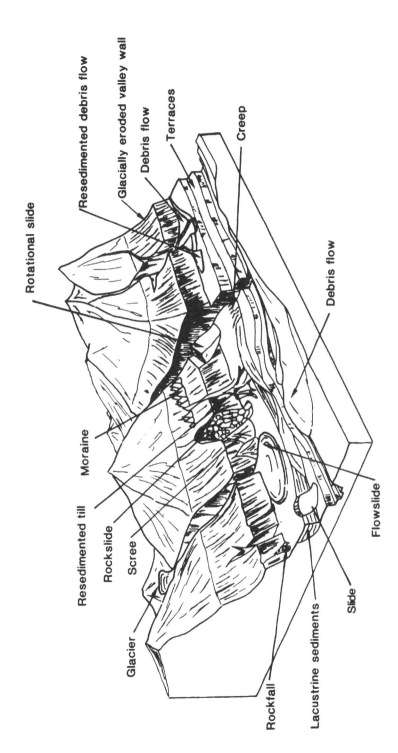

Rotational slide

Resedimented debris flow

Glacially eroded valley wall

Debris flow

Terraces

Creep

Debris flow

Resedimented till

Moraine

Rockslide

Scree

Glacier

Flowslide

Rockfall

Lacustrine sediments

Slide

*Figure 5.5* The main types of mass movement processes and landforms in the Karakoram mountains.

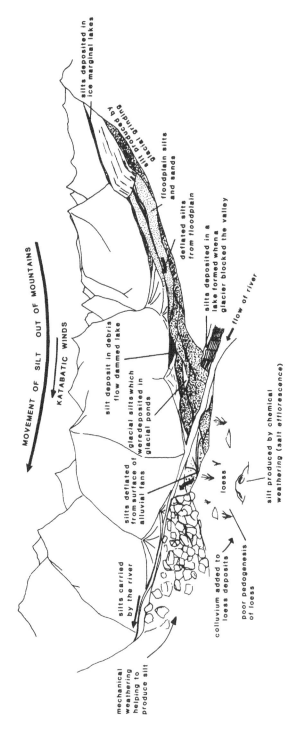

*Figure 5.6* Schematic diagram showing the main sources, movement and deposition of silts in the Karakoram mountains. Silts are formed by a variety of processes including cryogenic and chemical weathering, glacial grinding, and aeolian and fluvial attritions (thick quantities of several different types of lakes). These are later eroded by fluvial, glacial and aeolian processes, and the silts are carried out of the mountains by the major rivers.

Large-scale failures in the gorge sections of trunk valleys are common. These are characterized by steep (15°–30°) cone-shaped fans (Plate 5.4). Many of these may have blocked the valley, helping to form a lake which backed up the valley allowing lacustrine sediments to be deposited. Upon breaching of the debris dam and draining of the lake, the lacustrine silts are incised and the debris cone is incised at its toe, forming impressive terraces and dissected sediment fills many tens of m thick. Owen (1989a, 1991) described good examples of these throughout the Karakoram mountains.

## LACUSTRINE ENVIRONMENT

Deposits of lacustrine sediments are frequent throughout the Karakoram mountains. These sediments may have been deposited in several different types of lakes (Figure 5.6). Owen (1988a, 1989b) recognized four main types of such lakes, including:

(a)  Lakes formed behind tributary valley glacial ice that dammed the main valley. Contemporary examples can be seen in the Shaksgam valley (Deiso, 1980; Burgisser *et al.*, 1982).

(b)  Lakes formed behind moraines or within moraine complexes. Instances occur in association with many of the contemporary Karakoram glaciers.

(c)  Lakes formed within glacially eroded bedrock depressions. A good example is Borit Jheel in the upper Hunza valley.

(d)  Lakes resulting from blockage by debris flows, rockfalls and landslides. Many examples of these exist and range in size from several hundred m (e.g. Khunjerab river, summer 1987) to many tens of km long.

The lacustrine sediments are usually planar laminated or rippled fine-grained clastic sediments, dominantly of silt size grade, containing quartz, feldspars, micas, chlorites and illite, but few other clay minerals.

Good examples of Quaternary glaciolacustrine silts can be seen stretching 10 km northwest of Gilgit town. These form two main terraces representing sedimentation in a glacial lake that formed during the Borit Jheel glaciation (Derbyshire and Owen, 1990) when ice advanced down the Hunza valley and blocked the Gilgit valley. Derbyshire *et al.* (1984) and Goudie *et al.* (1984a) describe locations in the Hunza valley where Pleistocene and Holocene glacial advances blocked the main valley allowing lakes to form and lacustrine silts to be deposited. Burgisser *et al.* (1982) described similar sediments in the middle Indus and Gilgit valleys.

Sediments deposited within lakes dammed by mass movements were described by Owen (1988a, 1989b, 1991). One of the most impressive historical landslide dams formed during the winter of 1857/8 at Serat Pungurh in the Hunza valley. It produced a lake which backed up the Hunza for about 10 km (Plate 5.5; Becher, 1859; Mason, 1929). The dam was breached six months later and resulted in a catastrophic flood which Mason (1929) referred to as the 'Second

*Plate 5.5* Terraces comprised of lacustrine silts at Serat in the middle Hunza valley. These silts were deposited in a lake which formed when a landslide blocked the Indus in 1857/8. [*Editor*: Although these beds at first glance appear to be caused by the landside, their considerable thickness, convoluted bedding and complex interfingered gravel lenses upstream would have been precluded by the short time during which the lake was known to have been impounded and suggest instead a longer-lived glacial dam origin.]

Great Indus Flood'. During this six-month period, as much as 10 m of silt accumulated and now forms the lowest of three sets of terraces (Owen, 1988a). These high deposition rates characterize the Karakoram lacustrine sedimentary environments. The massive Lichar failure of 1841 (Shroder, this volume p. 23) did not produce such silt deposits even in the high discharge Indus river. Floodwaters produced by such catastrophic breaching of the dam may destroy earlier terraces and lead to redeposition of huge quantities of sediment. Mason (1929) provides graphic descriptions of the effect of such processes in the Indus catchments and recent examples were described by Vuichard and Zimmerman (1987) in Nepal. These examples help to highlight the role of high magnitude–low frequency events in the evolution of the Karakoram landscape.

## DEBRIS FLOW–ALLUVIAL FAN ENVIRONMENT

Drew (1873) was first to describe 'alluvial fans' in the valleys of the upper Indus river which he attributed to fluvial sedimentation (Plate 5.6). Detailed studies by Owen (1988a, 1991) and Derbyshire and Owen (1990) showed that they comprise dominantly of debris-flow deposits derived from till, often with inliers of lacustrine, till, fluvial, glaciofluvial and aeolian sediments. These authors

*Plate 5.6* View of 'alluvial' fans in the Gilgit valley near Jutial (10 km southeast of Gilgit).

showed that many of these features are relict forms created in a relatively short period after deglaciation. They recognize a major phase of fan development in the Gilgit area ~ 60,000 yr BP and the upper Hunza valley area ~ 47,000 yr BP, soon after deglaciation of those regions. This was a result of large-scale resedimentation of till from long steep slopes by debris-flow processes, owing to their metastable state upon ice retreat. Following rapid deposition of the fans, fluvial incision produced fan-head trenching while fan-toe truncation helped to create the terrace forms. The surfaces were modified by small-scale debris-flow, fluvial and aeolian processes.

## AEOLIAN ENVIRONMENT

Dust storms are commonly induced by temperature differences between the cold glacial ice and warm valley floors. These katabic winds deflate the terrace and fan surfaces of their fine sand and silt components and redeposit them as small dunes or as part of large dune fields, such as those near Skardu. However, loess is rare and deposits are < 1 m thick. Where present it is quickly reworked and eroded (Owen *et al.*, in press).

# VALLEY FILL AND LANDSCAPE EVOLUTION OF THE KARAKORAM MOUNTAINS

The Quaternary evolution of the Karakoram mountains is complex. Three main factors control the evolution of the Karakoram landscape and valley-fill sedimentation (Owen, 1988a, 1989b): (a) tectonics; (b) earth surface processes; and (c) climatic change.

The sediments and facies associations that comprise the valley-fill successions are controlled by glacial, glaciofluvial, fluvial, mass movement, lacustrine and aeolian processes. The glacial system dominates, strongly influencing the other sedimentary systems. These sedimentary systems must be considered in their paraglacial context. Clearly, the evolution of the Karakoram mountains is a consequence of the tectonic collision between the Indian and Eurasian continental plates. The initial positive relief and uplift occurred during the Eocene denudation (probably marine and fluvial erosion) helped to produce early erosion surfaces, which may be represented today by the coincidence of peaks at an altitude of ~ 6000 m.

With progressive uplift of the Karakoram region a primeval draining system developed, draining the mountains in a SSE and WSE direction (Owen, 1989a). With increasing uplift the rivers became incised, forming an antecedent drainage. With increased altitude firn formed, leading to the formation of glaciers. The timing of the first glacier and the onset of the first glaciation in the Karakoram mountains is not known. The glacial system then began to dominate the environment, which would have lead to a different suite of processes controlling denudation. The dynamic tectonism would have helped to intensify the role of denudation, conforming to the 'principle of antagonism' (Scheidegger, 1979) and hence accelerating the supply of detritus helping to produce thick valley fills. This may have also resulted in an early Pleistocene or a pre-Pleistocene plantation surface ( < 8 myr BP) in the Hunza valley and Baltistan, which is now at a height of 4100–4200 m. This has scattered erratics on it and possibly provides evidence for the first glaciation in this region.

Evidence for Quaternary and recent faulting was localized and deformation and uplift probably occurred as regional warping. A series of surfaces throughout the region implies that the uplift was sporadic. The relative importance of tectonics and climate (i.e. glaciation) in the formation of these surfaces, however, is difficult to determine. Chinese workers (Sun and Wu, 1986) attribute similar surfaces in Tibet to tectonoclimatic events, when they are unable to resolve the formative process. There is no apparent structural control on large embayments and extensive valley areas where the thickest valley fills are to be found. Their formation is attributed to a consequence of localized intense glacial erosion, where several glaciers confluenced, and fluvial erosion. The Skardu and Gilgit areas provide good examples of this.

Owen (1988a, 1989a) suggested that differential uplift of the Great Himalaya and the Nanga Parbat–Haramosh massif along the southern edge of the study region modified the drainage systems, creating the east–west drainage of the Indus and its constriction through the Nanga Parbat–Haramosh massif which, according to Zeitler (1985), has the maximum uplift rates.

Progressive uplift restricted the southern Asian monsoon from penetrating into the Karakoram mountains, resulting in increased aridity. It would also have had profound effects on global circulation and climatic changes through most of High Asia. However, the timing and influence of this climatic change on the glaciations within the Karakoram mountains is not known.

Climatic changes throughout the Quaternary resulted in at least three major glaciations and five minor advances (Derbyshire et al., 1984). During the three major glaciations extensive glaciers flowed down the valleys of the Karakoram. The glacial system produced thick deposits of till, both on the valley sides and on the valley floors. Upon deglaciation much of the valley-side till was resedimented by debris-flow processes en masse to form extensive fans. Meltwater from the retreating ice also deposited glaciofluvial sediments. Further glaciations or advances of a fluctuating snout may have eroded previously deposited sediments or deformed them by glaciotectonic processes. Climatic changes may have been responsible for changes in the type and quantity of sediment. For example, shortly after deglaciation, paraglacial sedimentation would have been very important whilst, during glaciations, cryogenic and glacial processes would have probably been more dominant.

Throughout the evolution of the mountains high magnitude–low frequency events may have had a profound influence on valley sedimentation. Debris flows and landslide failures temporarily blocked some of the valleys and rivers and produced lakes which would become infilled with sediment, further contributing a component to the valley fills. Upon breaching of the dam and emptying of the lake, the floodwaters may have eroded and resedimented large quantities of material.

Allocyclic processes including fluvial, glaciofluvial, glacial, mass movement, aeolian, lacustrine, colluvial and weathering continually modified the landscape, providing a continuous supply of sediment to the valley fills. Presently, the Karakoram is dominated by the glacial systems, considerably influencing the other sedimentary systems. Fluvial sedimentation contributes little to the valley-fill sedimentation; rather it is important as a transporting system essentially carrying sediment through the mountains out to the foothills and into the foreland basins along limited corridors such as the Indus valley. Collins (1988) showed that the sediment loads in the main trunk rivers are considerably higher than sediment loads produced by the glacial systems and came to a similar conclusion that the rivers are actively eroding previously deposited valley-fill sediments and bedrock within the gorge sections. This suggests that the rivers are degrading the landscape to a considerable degree, no doubt mainly as a result of rapid uplift. This active incision and reworking of valley fills complicates the unresolved dilemma of uplift versus fluvial incision when considering the exhumation rates produced by Zeitler (1985).

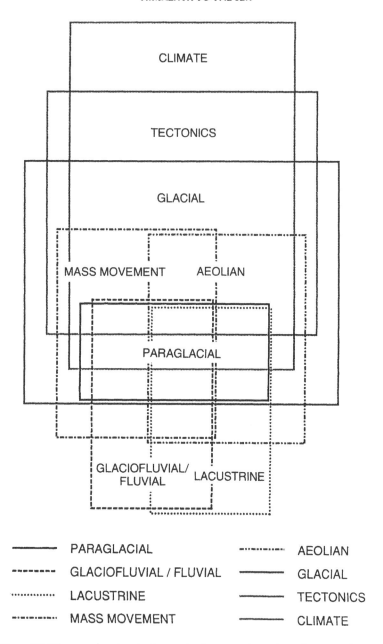

*Figure 5.7* Diagram showing the interaction between the main environmental systems that contributed to the evolution of the Karakoram landscape. *Each box represents a distinct system. Overlapping boxes indicate the interaction between the different systems. Note the large degree of overlap incorporated into the paraglacial system, highlighting its important role in the understanding of Karakoram environmental systems.*

## CONCLUSIONS

There are considerable problems in reconstructing the Quaternary histories and resolving the relative roles of tectonics, earth surface processes and climatic change in landscape evolution of the Karakoram mountains from the valley-fill successions and the geomorphology. The difficulty stems from poor dating constraints on the timing of uplift and glaciation; few quantifiable measurements of the rate of uplift and denudation; few process studies to provide information on rates and modes of sedimentation; and little information on sediment budgets.

This paper has provided information on the style of sedimentation in the valley fills present in the Karakoram mountains. In addition, it provides a tectonic and climatic framework in which the valley fills may be considered. Emphasis is placed on the importance of the glacial system and the role of paraglacial processes in this environment. Figure 5.7 summarizes the interaction between the environmental systems that are important in the landscape evolution of this region. The overlapping boxes represent the interaction between the various systems. The degree of overlap, however, is not to scale and the interaction of the various processes needs to be quantified in future work.

# 6

# QUATERNARY GLACIATION OF THE KARAKORAM AND NANGA PARBAT HIMALAYA

*John F. Shroder, jr., Lewis Owen and Edward Derbyshire*

## ABSTRACT

Controversy and agreement on extent and timing of Pleistocene glaciation in the western Himalaya are highlighted or resolved in this work through recognition of three main glacial stages as well a number of later stades. The early glaciation or Shanoz stage includes glaciated high-level erosion surfaces, the Bunthang till, and perhaps other tills as well. The Jalipur till or tillite is both glacially and tectonically indurated and deformed; its age remains uncertain. The middle glaciation or Yunz stage includes a number of tills, perhaps had two major phases or stades, and extended the farthest down the Indus of any glaciation. The last glaciation is subdivided into several stades, the most important of which is the Borit Jheel. Further work is necessary to resolve remaining controversy and uncertainty.

## INTRODUCTION

Modern study of the Quaternary glacial history of the Himalayan mountains is still in its infancy. Apart from the obvious difficulties of access, the interpretation of many of the sedimentary sequences frequently initiates contentious debate and much remains to be learned of the extent and timing of the Pleistocene glaciations. Much speculative correlation has been done to make limited sense out of discontinuous and poorly exposed sections. Furthermore, some speculation has been in violation of the principle of parsimony in which Occam's Razor dictates that we keep our explanations simple if we lack reasonable data. Questions of precedence and nationalistic competition also have impinged adversely upon this work, wherein valid prior *observations* going back nearly a century and even those of more recent vintage have not been taken sufficiently into account because of both the generally admitted need to reinterpret prior efforts and too hurried publication. In addition, little is known about glacial dynamics in the Himalaya, thus making it difficult to interpret ancient sediments and to interrelate them or to reconstruct past glacial systems. This is particularly so in the Karakoram mountains and the Nanga Parbat Himalaya.

Fieldwork for this study was initiated by Shroder (JS) in 1978, by Derbyshire (ED) in 1980, and by Owen (LO) in 1985; all of us have revisited key localities at least three times since. The 1988 Leicester Symposium on the Neogene of the Karakoram and Himalaya resolved some of our divergent views and led to this collaboration. Field efforts by JS in 1991 and 1992 and future reassessment of our findings are expected to advance our work. This paper examines the extent and timing of glaciations in these mountains, and resolves or highlights some of the controversies resulting from past work (Owen, 1988a; Shroder et al., 1989).

## PREVIOUS WORK

The earliest descriptions of the Quaternary geology of the western Himalaya were by Moorcroft and Trebeck (1841), Vigne (1842), Drew (1875) and Oestreich (1906). Many other early explorers provided only some descriptions of sediments and geomorphological features but did not explain significance. The first detailed studies were made in a series of expeditions between 1911 and 1914, during De Filippi's Italian Expedition to the Himalaya, Karakoram, and Chinese Turkistan; efforts were concentrated on the upper Indus river valley area west of Skardu (Dainelli, 1922, 1934, 1935; Figure 6.1). Dainelli proposed a four-fold glacial sequence, influenced by work in the Alps, and attributed the first stage to the Mindel and the fourth to a post-Wurm (possible Buhl) advance. This was later modified by de Terra and Paterson (1939), who equated the fourth glaciation to the Wurm of Europe as well. Dainelli had also speculated on the effects of glaciation in the middle Indus and Hunza valleys, but had no field evidence for the extensive glaciations that he postulated. As Hewitt (1989a) has recently elaborated, however, Dainelli's work stands alone as regards consistency of development, and detail, depth and attempted synthesis; no one has yet demonstrated an overall grasp of Karakoram glaciation sufficient to challenge his major theses.

Norin (1925) believed, on the basis of 'bottom moraines' within terrace sequences, that ice of the last Pleistocene glacial maximum (Shigar glacial subepoch) filled the whole of the Indus drainage system and descended to altitudes of ~1675–1830 m on the Kashmir slope. He also believed that most of the valleys in Baltistan were originally filled with several hundreds of m of drift from this glaciation. This view of an extensive glaciation was favoured by many, especially Cotter (1929) and Coulson (1938) who attributed large boulders on the Potwar plateau to 'Punjab erratics' left from ice that extended down the Indus valley during the maximum advance (Figure 6.1). None of these authors had travelled along the lower Indus gorge, however, and it is now believed that ice did not extend out of the Himalaya and that the Punjab erratics are boulders carried down the Indus valley by catastrophic flooding, probably in iceberg rafts (Burbank, 1983; Shroder, 1987; Butler et al., 1988; Shroder et al., 1989a,c).

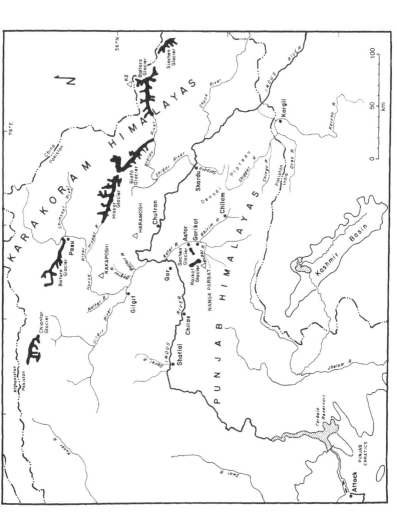

*Figure 6.1* Index map of the study area. Only a few of the largest modern glaciers are included.

The first description of deposits in the middle Indus valley near Nanga Parbat was provided by Misch (1935), in part to help to provide information on the extent of glaciation in the Karakoram and Great Himalaya. He assigned a group of folded and faulted sandstones and diamicts to the 'Jalipur sequence' and interpreted the diamicts as tillites of late Tertiary–early Pleistocene age. Further work in the Hunza area by Paffen *et al.* (1956) and Schneider (1959) suggested a three-fold glacial sequence with upper and lower terrace remnants plus a moraine sequence close to the present glacial fronts. They emphasized a relic 'pre-Pleistocene relief' above 4000 m with a shoulder below 3000 m which was similar to that described by Grinlinton (1928) east of the Liddar valley in Kashmir. Similar features were also recognized on the north and northwest side of Haramosh and on the south sides of the Gilgit–Kohistan range. These surfaces later were reconfirmed in the Hunza valley (Derbyshire *et al.*, 1984; Figure 6.1). These workers recognized a surface at 4100–4200 m altitude in the Pasu area which they referred to as the 'Patundas surface', remnants of which occur all along Hunza valley. They also recognized a higher surface at 5200 m, e.g. a rock bench on Mirschiken (5445 m) opposite Aliabad in lower Hunza. Derbyshire *et al.* (1984) indicated that this must be younger than 8.6 myr, in view of the Rb/Sr age of the Karakoram granodiorite there (Desio *et al.*, 1964), and attributed a Pliocene age to the surface and a Pleistocene age to the Patundas surface. They also suggested that the Patundas surface could be broadly correlated with the Potwar plateau at 500 m, which is of early Pleistocene age. A higher surface north of the Main Boundary Fault must be Pliocene in age as it truncates the Muree Series, now dissected into a broad valley, and is correlated with the highest planation surface in Hunza.

Li *et al.* (1979) also recognized similar surfaces on the Tibetan plateau; a late Eocene planation surface above 6000 m, the plateau surface itself, and a lower surface with fossil evidence which suggests that it developed in a warm humid environment now preserved at an altitude of 4500–5000 m. Below the lower planation surface occurs a third surface of early Pleistocene age that is made up of broad valleys and lake basins unused by modern valley systems. Ambraseys *et al.* (1981) also recognized abraded surfaces on valley shoulders in the lower Indus gorge and speculatively attributed these to three distinct glacial advances. Sound evidence in the form of tills or erratics appears lacking there, however, and the features are probably only the product of fluvial erosion resulting from the rapid but sporadic uplift of the range. It seems likely that the recognition of such surfaces throughout large areas of High Asia has much to do with the particular combination of uplift and glaciation. Chinese workers have related the surfaces to tectonoclimatic cycles (Sun and Wu, 1986) and have recognized three main cycles during the Neogene.

Porter's (1970) work in the Swat Himalaya produced a glacial chronology that provided the first study using modern stratigraphic and sedimentologic methodology. This three-fold sequence seemed to support an apparently similar three-fold sequence in Hunza (Schneider, 1959; Wiche, 1959a,b; Batura Glacier

Investigation Group, 1979, 1980; Derbyshire *et al.*, 1984; Shi and Zhang, 1984). Similarly, Osmaston (in press) has recognized three glaciations in the Zanskar area of the Ladakh Himalaya. Because of climate differences between adjacent regions, however, these sequences might not be contemporaneous; the Swat Himalaya receiving moisture from the monsoon, for example, whereas the others are in the variable rainshadows of the Great Himalaya to the south. Furthermore, both Porter (pers. comm. to JS, 1986) and Rendell (this volume) now believe the Swat glaciations to be quite young, perhaps only middle or even late Pleistocene and Holocene.

Care must be taken when considering the number of glaciations within these tectonically and geomorphologically active areas. First, glacial advances tend to destroy evidence of previous, less extensive glaciations. Second, Gibbons *et al.* (1984) used a probability model to show that the likely number of surviving moraines will vary considerably given a total number of 32 possible glaciations according to the deep sea record (Shackleton and Opdyke, 1973): a 32 per cent chance for three and a 90 per cent chance for between one and four. Further complicating the story is the influence of tectonism. With increased uplift, glaciations on the Himalayan flanks, e.g. in Swat or on the Nanga Parbat massif, would be expected to become more extensive as a consequence of the increased area above the perennial snowline. The relationship is not so simple, however, in the case of the high Karakoram where moisture penetration into the source areas of the glaciers would be inhibited. These problems are rarely mentioned in the early works and have still to attract the attention that they deserve.

Research using modern methods and concepts on Quaternary geology in the Nanga Parbat and Karakoram Himalaya first began in the 1970s and early 1980s with the opening of parts of the Karakoram Highway (known informally as the 'KKH') (Batura Glacier Investigation Group, 1979, 1980; Miller, 1984; Shroder, 1985, 1989a, 1989c; Shroder *et al.*, 1986, 1989). Members of the Chinese Academy of Sciences in 1974 and 1975 were first concerned with the upper Hunza valley, and especially with the threat posed to the KKH by the Batura glacier. They recognized several stages of moraine deposition from the last Pleistocene glaciation (termed the Dali glaciation in western China: Shi and Wang, 1981) to the present, with older tills present at altitudes above 3500 m on the mountain sides north and south of the Batura glacier. Zhang and Shi (1980) first cited evidence for at least three Pleistocene glacial stages in the Batura region and termed them Shanoz (oldest), Yunz and Hunza.

Derbyshire *et al.* (1984) followed this scheme of the Chinese workers and recognized eight further glacial phases of varying magnitude. Standard methods of relative age dating were used, coupled with absolute dates on the sediments of the area, using $^{14}$C and thermoluminescence (TL) dating. The TL dating was among the first done for samples from the Himalaya but few details of method were published owing to space restrictions. Questions of reliability of these and other TL dates from the Himalaya (Shroder *et al.*, 1989) indicate that greater standards of TL date methodology should be maintained and published in future studies. High background radiation producing partial trap saturation and incomplete solar bleaching in deep and therefore darker mountain valleys with turbid glacial waters, in addition to unknown but important later water-saturation histories, mean that caution is advised for all TL dates from the Himalaya until these problems are resolved. At present we accept with appropriate reservation all prior published TL dates from the Himalaya, pending further work.

In any case, Derbyshire et al. (1984) recognized a series of isolated and deeply weathered erratics on summit surfaces, such as the previously mentioned Patundas surface, above 4150 m altitude and attributed these to the Shanoz glaciation, which was considered to be early Pleistocene in age. The second and third glaciations, Yunz and Borit Jheel, were recognized from weathered till remnants on benches in the main Hunza valley. The Yunz glaciation was considered to be older than 139,000 (TL) yr BP and was correlated with the penultimate Pleistocene glaciation, while the Borit Jheel glaciation was dated at ~ 50,000 (TL) yr BP and was thus a possible correlative with the Hunza glaciation of the Chinese work as well as a phase or stade in the last Pleistocene glaciation.

Derbyshire et al. (1984) considered the Yunz and Borit Jheel to represent extensive valley glaciations in which glaciers extended down the main Hunza valley into the Gilgit and Indus valleys. Desio and Orombelli (1971, 1983) had already noted five scattered glacial tills and moraines in the middle Indus and Gilgit valleys and considered that they generally decreased in age up valley following trunk-valley glaciation as far down valley as Sazin, although they also recognized several younger cross-valley glacial moraines, especially around Nanga Parbat. On the other hand, Paffen et al. (1956) had argued that late Pleistocene snowline depressions in the Hunza valley were in the order of 1000 m but that, owing to the very steep slopes, areas were not large enough to maintain trunk glaciers into the Hunza valley itself. In contrast, Wiche (1959a) had also estimated a snowline depression of 1200–1300 m for the area northwest of Chilas in the mountains south of Gilgit. These consist of more extensive areas in the altitude range of 3000–5000 m and so were considered to have the potential to sustain extensive glaciation, including maintenance of ice in the trunk valleys.

Derbyshire et al. (1984) also recognized a fourth glacial stage (Ghulkin I) represented by 'expanded foot' as well as the former existence of minor valley glaciers younger than 47,000 (TL) yr BP. Following the prior Chinese work, four other minor advances were also recognized, with the last representing minor oscillations during the nineteenth and twentieth centuries.

Shroder et al. (1989) and Shroder (1989a,c) followed the prior publication of Derbyshire et al. (1984) and extended that chronology out of Hunza and down the Gilgit and middle Indus rivers. The basic three-fold glacial sequence was recognized, based on objective, relative age dating criteria, including those previously established by Derbyshire et al. (1984). Rather than using the prior geographic terminology from Hunza, however, they used general temporal terminology in early, middle and last glaciations in order to be able to incorporate other work on a more regional basis. Shroder (1989a,c) and Shroder *et al.* (1989) also determined that the early glaciation was represented by the indurated lower Jalipur tillites and the heterogeneous upper Jalipur valley fill I sedimentary rocks that are folded, overturned or overridden by movement along the Raikot fault at the base of Nanga Parbat. The middle glaciation was thought to be represented by two tills, the early middle (M1) and the late middle (M2) glaciations. The two tills were intercalated within variable sediments of valley fill II and valley fill III, including thick lacustrine units dipping as much as 43° along the fault. Both were equated to the Yunz glaciation. The last glaciation was seen to consist of 3–4 or more separate advances (L1, L2, L3, L4) that left morainal topography on the tills. The L2, or early last glaciation, was equated to the Borit Jheel of Derbyshire *et al.* (1984) and the Dak Chauki moraine at Gilgit of Desio and Orombelli (1971). Many of these last glaciations were also seen to have been of the 'expanded foot' or cross-valley moraine type that dammed the Indus river in several places below the Gilgit confluence.

A major problem facing Quaternary geologists working in the Karakoram and Himalaya has been the timing of the onset of glaciation. Recent work on oceanic sediments indicates that global cooling occurred during the Pliocene (Shackleton and Opdyke, 1973) but, of course, this does not necessarily imply terrestrial glaciation. Evidence for pre-Pleistocene glaciations has been found, however, e.g. ~ 5 myr BP in southern Patagonia (Mercer *et al.*, 1975), and even older in other locations. In the older literature on South Asia, Woldstedt (1965) had regarded Morris' (1938) main boulder bed in the Soan terraces of Pakistan as indicating the first glaciation of South Asia, while de Terra and Paterson (1939) had placed the first glaciation in the Tatrot zone, with maximum extent of Pleistocene ice occurring in the second glaciation, its till merging with the boulder conglomerate in the piedmont. It is now clear from palaeomagnetic studies, however, that the Tatrot zone is probably much older than the first glaciation in the Himalaya, the deposit probably representing vigorous uplift of the mountains rather than severe climate change.

Palaeomagnetic dating of lacustrine sediments above tills of the Qiangtong Group near the Kun Lun pass in the north part of the Qinghai–Tibet plateau has yielded an age of 2.7–1.4 myr BP and the Qugo Group west of Tanggula pass in the central part of the plateau has been dated at between 4.32 and 2.2 myr BP (Qiang and Ma, 1979), all suggesting the onset of major glaciations there after ~ 2.4 myr BP. The palaeomagnetic studies of Heller and Lui (1984) show that the initiation of loess accumulation ion the Loess plateau of Central China started ~ 2.4 myr ago. Given the apparent coincidence between the onset of hemispheric glaciation and the beginning of loess deposition, a maximum age of ~ 2.4 myr for the first glaciation in High Asia is suggested, thus probably also restricting the age of the glacial sediments in the Himalaya to < 2.4 myr. Pollen studies from a variety of sites in Kashmir also indicate a progressive cooling of climate after 2.9 myr (Agrawal, 1984) and from this the first glacial event was placed at ~ 0.7 myr.

The oldest known glacial deposits within the Karakoram mountains are the high-level scattered erratics of the Shanoz glaciation (Derbyshire et al., 1984) which were shown to be younger than 8.6 myr BP. Cronin (1982), Johnson (1986) and Cronin et al. (1989) have shown that the Bunthang sequence in the Skardu basin includes a basal till overlain by > 1000 m of fluvial and lacustrine sediments that are magnetically reversed throughout. This suggests that the basin filling took place after a glacial advance and prior to the Burnhes normal chron, i.e. between 3.3 and 0.78 myr (newly revised palaeomagnetic dating of 1991); thus the onset of glaciation was prior to ~ 780,000 yr BP. Owen and Derbyshire (1988) noted that part of the Bunthang sequence has been glaciotectonically deformed on its edges and therefore questioned the prior palaeomagnetic procedures, but because Cronin et al. (1989) and Cronin and Johnson (this volume) worked only with nearly horizontal and undeformed sediments (pers. comm. to JS, 1990) their work stands as a valid and useful contribution to the question of old glaciations in the Himalaya. A problem with the Bunthang sequence as an old unit continues to be that the Indus river has not downcut much below it. This can be explained, however, by rapid uplift of the Nanga Parbat–Haramosh massif across the course of the Indus a few km downstream from the Bunthang outcrops.

Shroder et al. (1989) believed that the Jalipur tills or tillites represent another set of oldest glacial deposits in the Indus valley near Nanga Parbat, which Olson (1982) had indicated must be less than 2 myr BP in age, based on Zeitler's fission-track dates (Zeitler, 1985). Shroder et al. (1989) tentatively equated the Jalipur to the Bunthang, but this is now known to be incorrect as Burbank (pers. comm. to JS, 1992) has found the Jalipur to be normally magnetized.

In summary, a three-fold glacial sequence seems to be indicated throughout the areas studied, to which are added a series of minor advances up to the present. A well defined series of sediments and landforms occurs throughout the main valleys of the Karakoram and Nanga Parbat Himalaya. Their chief characteristics are described herein to help to provide a state-of-the-art review as well as to form a framework for the study of the regional glacial history. The basic chronology of Derbyshire et al. (1984), with modifications from the doctoral study of Owen (1988a), the work of Shroder et al. (1989), and the new JS fieldwork of 1991 and 1992, is adopted through the rest of this chapter.

## THE SEQUENCE OF SEDIMENTS AND LANDFORMS

### Pre-Pleistocene planation surfaces–early glaciation–Shanoz stage

The previously mentioned highest surface > 5200 m in Hunza was thought to be a former valley floor (Derbyshire et al., 1984). This pre-Pleistocene surface can be traced into Baltistan at a similar height (Owen, 1988b). The differential and rapid uplift of the Nanga Parbat–Haramosh massif (NPHM), however, makes the surface difficult to recognize in this area, where strong erosion (Shroder, 1989a,c) and extensive ice cover would have destroyed most evidence. The Silver Saddle at ~ 7000 m near the top of the Nanga Parbat could be a remnant, however. The old surface is difficult to recognize but also occurs as some accordant summit plains in the areas of Kohistan south of Gilgit and west of NPHM where altitudes are rarely in excess of 5000 m.

The weathered tills and residual clusters of boulders on the high-level Patundas surface represent the oldest till remnants of the Shanoz phase (Zhang and Shi, 1980; Derbyshire et al., 1984). Owen (1988a) traced similar surfaces down the Hunza valley and into the Gilgit valley. In the lower Gilgit valley, they are present as glacial benches at a height of ~ 2600 m upon which occur reworked glacial diamictons. Similar surfaces with scattered erratics and weathered bouldery diamictons on them occur in the Haramosh area at a height of ~ 3200 m (Madin, 1986; Owen, 1988a). The difference in height between the Hunza and Gilgit surfaces is probably a function of irregular former land surfaces, variable gradients of the former glaciers, and differential uplift.

Similar surfaces, again with till remnants, can be recognized along the Shigar valley northwest of Skardu, at a height of 3200–3500 m (Dainelli, 1922; Owen, 1988a). These can be correlated with the surfaces of several buttes in the Skardu basin, representing a major valley glaciation which extended from the Shigar valley. Owen (1988a) attributed these surfaces to the Shanoz glaciation.

The Deosai plateau, south between Nanga Parbat mountain and the Skardu valley, is a rolling upland of ~ 400 km$^2$ which averages over 4000 m in altitude (Figure 6.1). The plateau is surrounded by higher peaks ranging close to 5000 m from which extensive glaciers descended in Pleistocene. Dainelli (1922) found evidence for an ice cover on Deosai during his designated first glaciation, a strong glaciation with an ice cap during his second glaciation, and only a thin ice cover during his third glaciation. Fieldwork by JS in 1991 confirms this assessment of at least two major older glaciations, with a third younger one at the higher elevations close to the surrounding mountain sources. Most of the older ice seems to have spread out widely on the Deosai before sending outlet glaciers into the Satpara drainage on the north and the Shigar drainage to the southeast, which itself discharges into the upper Indus above Skardu. Some of this older ice may also have moved northwest through the Sheosar lake basin and over the low ( ~ 40 m) rim at Chachor pass, which is tributary to the Chilum, Khirim and Astor valleys.

Shroder (1989a,c) and Shroder et al. (1989) suggested that the Jalipur sequence was deposited during the Shanoz glaciation. On the basis of glacial facies associations, glacially striated clasts, dropstones and sand grain surface textures (Higgins, 1986), Shroder et al. (1989) showed that the basal diamict of the Jalipur units was in fact a glacial till or tillite. Similarly, Owen (1988a), with additional evidence from microfabric studies, macrofabrics and macrostructural evidence, showed it to be a lodgement till. His macrostructural work, however, was confounded by deformation produced by apparently sedimentologically penecontemporaneous seismicity and glaciotectonism, together with major movement along the Raikot fault at the base of Nanga Parbat. In fact both the Jalipur till or tillites and the overlying thick ( > 100 m) Jalipur alluvial and colluvial facies of valley fill I (Shroder *et al.*, 1989) originally were defined and recognized in the field by their induration to rock and their extensive deformation. This induration of the overlying valley fill I nonglacial facies either shows the antiquity of the Jalipur (since it cannot have been produced by glacial lodgement pressures, inasmuch as at least some of it clearly postdates the Jalipur ice) or was produced by considerable age and/or tectonic pressure. Furthermore, some of the overlying Jalipur valley fill I nonglacial sediments are directly overlain by thick sequences of non-indurated younger tills and other later valley-fill sediments, further indicating the relative antiquity of the Jalipur units.

## Deformation and induration of the Jalipur sequence

The timing and nature of deformation and induration of the Jalipur sequence are the most difficult and contentious issues of our past research. Misch (1935), Olson (1982), Lawrence and Ghauri (1983), Lawrence and Shroder (1984), Shroder *et al.* (1989) and Owen (1988a) have done considerable work on various

*Plate 6.1* Jalipur alluvial facies sandstone dipping unbroken NW 60° for > 150 m into the Indus river below. Such large monoclinal amplitudes preclude the likelihood of glaciotectonism here. (Photograph: Shroder 1991)

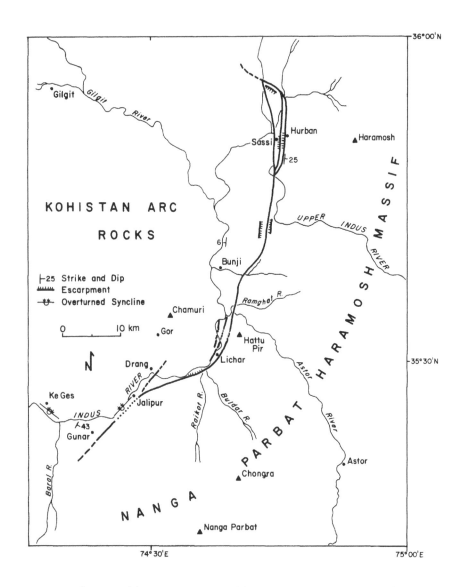

*Figure* 6.2  Index map of the type section area of the Jalipur formation outcrops (after a map by R.D. Lawrence). *Multiple traces of the Raikot fault are shown by heavy line; young scarps are indicated by hachures. The Ke Ges and Patro synclines are located by overturned fold symbols. Tilted Pleistocene lake beds and fluvial bedding planes are shown by strike and dip symbols. Compare with Figure 1.7.*

aspects of the sequence, yet many of our conclusions are remarkably different. In each case some essential aspects were not discussed and need to be addressed in future work in order to achieve a better consensus.

It seems probable that Jalipur sediments were preserved best where they were deformed, indurated, deeply buried or overthrust and hence at least partly protected from later erosion. It is also possible, however, that other Jalipur equivalents may not yet be recognized where they are not lithologically as indurated or as tectonically affected. In any case, the known lowermost Jalipur tillite is caught in several fault shear zones which are part of the Raikot fault system (Lawrence and Ghauri, 1983). For example, at Drang, Lichar and below Hattu Pir, gneissic bedrock has been thrust at a high angle to the north or west over the tillite. Between the old Jalipur resthouse and Lichar, the Jalipur sandstones are steeply dipping to near vertical up to 150 m above the Indus (Plate 6.1). At both Ke Ges and Jalipur the sequence has been overturned in tight synclines (Misch, 1935).

The Raikot fault trace is commonly made up of many strands at the surface; those recognized to date by Lawrence (pers. comm. to JS, 1988) are shown in Figure 6.2. The main bedrock fault separates high metamorphic grade gneisses derived from Indian crustal rocks (Wadia, 1931; Misch 1949) from norites, metanorites, gabbros, metavolcanics and granites of the Kohistan island arc (Jan et al., 1984; Coward et al., 1986). Most deformation of the Jalipur sequence appears to be associated with motion on this fault. Lawrence (pers. comm. to JS, 1988) considers that the Raikot fault has mainly right-lateral strike-slip displacement. The total vertical offset on the fault is > 15 km (Madin et al., 1989), which is far larger than any motion recorded by the Jalipur, indicating that the fault was active prior to Jalipur deposition as well as afterwards. Fission-track, K-Ar, and Ar-Ar studies in the Nanga Parbat area (Zeitler et al., 1982a; Zeitler, 1985; Treloar et al., 1989) indicate that uplift of the massif has continued, presumably along the Raikot fault, and at an accelerating rate from 2 myr ago; i.e. throughout the Pleistocene and Holocene, to a current figure of $\sim 5.5$ mm yr$^{-1}$. In spite of such major offset, Owen (1989a) felt that only minor surface folding of the Jalipur was in evidence where he mapped.

Discrimination between tectonic and glacial deformation and induration in the Jalipur is important in working out the Quaternary history of the region. Syndepositional tectonic features, however, can easily be confused with syndepositional ice-contact push and collapse structures, especially in poorly exposed sections. Owen (1989a) was able to discriminate successfully between some of these different features in the Jalipur sequence, but much remains to be done. Furthermore, evidence for earthquake-developed sedimentologic features occurs as well in the upper Jalipur valley fill. For example, such structures occur at the west end of the Ke Ges outcrop, at Ame Ges and near Jalipur village where there are ball-and-pillow structures, sometimes referred to as 'quake sheets'. These are layers of intraformational conglomerate and convoluted bedding confined between planar, undeformed silt and sand layers (Olson, 1982).

*Plate 6.2* The Ke Ges syncline with steeply dipping Jalipur till or tillite forming the steep cliff across the Indus river. The curvilinear outcrops to the right are Jalipur sandstone which overlies the basal tillite and forms the core of the overturned syncline.
(Photograph: Shroder, 1984)

Widespread sandstone dikes and local breccia-bearing litharenites interpreted as slope-failure debris in the upper Jalipur may also be related to syndepositional seismic activity. In contrast, swirled bedding with ice-rafted dropstones in the Jalipur, best seen near Patro (Figure 8 of Shroder *et al.*, 1989), are evidence for ice-contact deposition in a kame terrace, ablation valley or similar environment.

In the southwestern part of the area, Owen (1989a) noted only an open anticline–syncline fold pair above what he thought might be a ramping thrust, whereas Lawrence and Shroder had noted something quite different (Lawrence, mapping with and pers. comm. to JS, 1988). Instead, the Jalipur appears folded into two overturned synclines. Axial traces of these synclines are about at 90° to each other. At Ke Ges the synclinal axis is subhorizontal and trends approximately northwest (Plate 6.2). The Indus river flows parallel to the fold axis beside the structure (Figure 2 of Misch, 1935). The lower Jalipur diamictite at the base of the fold on the southwestern limb is slightly overturned. This fold is the only major structure in the Jalipur so far known that has no clear genetic relationship to the Raikot fault. Owen (1988a) considered this deformation to be ice-contact related, but the involvement of the thick overlying nonglacial alluvial facies and the high amplitude of the fold with its axial trace parallel to the valley trend indicates that further assessment is warranted.

144

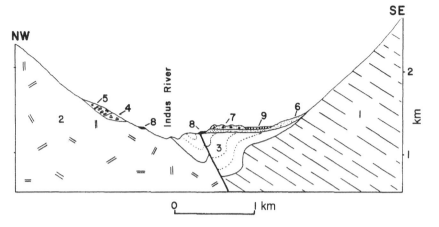

*Figure 6.3* Overturned syncline in Jalipur sediments at Patro near Jalipur village along trace of the Raikot fault (after an unpublished drawing by R.D. Lawrence). Units are:

1 Nanga Parbat paragneisses;
2 Chilas complex of Kohistan arc sequence;
3 Jalipur tillite and valley-fill sediments;
4 Middle glaciation (probable Yunz-equivalent) till;
5 Middle glaciation (probable Yunz-equivalent) valley-fill sediments;
6 Late glacial proglacial outwash;
7 Late glacial till and moraine of probable Patro side-valley glacier;
8 Lake beds of probable late glacial age;
9 Holocene fan gravel.

At Patro, southwest of Jalipur, a second overturned fold occurs. The section illustrated (Figure 6.3) is a composite, constructed from limited field data by Lawrence (mapping with and pers. comm. to JS, 1988); the actual situation may be more complex than is shown. The Indus river channel there is cut into bedrock of the Chilas complex on the northwest side of the valley. Immediately across a bedrock ridge is the Jalipur-age Indus channel filled with lower Jalipur glacial deposits consisting of till, the swirled, dropstone-bearing lacustrine units previously mentioned, and other units. The lowest till sits on slickensided or striated norite bedrock that indicates some sliding of the lowest Jalipur unit against the bedrock during flexural slip folding of the Jalipur. Slickensides plunge at N 10°–30° W straight down the bedrock slope perpendicular to both the fold and the valley axis, an improbable condition if this high amplitude fold were the result of glaciotectonism and the slickensides were instead glacial striae. To the east in a discontinuous outcrop, upper Jalipur sandstones are overturned to the southeast. Lawrence (pers. comm. to JS, 1988) mapped a strand of the Raikot fault just east of this outcrop in an area covered by younger sediments. Still farther to the east, upper Jalipur sandstones dip 20° NW parallel to the exhumed Jalipur valley bottom that makes the current south wall of the Indus valley.

This valley wall is underlain by Nanga Parbat gneisses that dip SW 20°–30 . The foliation planes are strongly slickensided in many places, indicating that part of the reverse motion of the Raikot fault system took place by slip along the foliation. The main trace of the Raikot fault system must be beneath the Jalipur syncline. The dip slip motion shown in Figure 6.3 is opposite to that of the bedrock. Lawrence (pers. comm. to JS, 1988) considers that this must reflect the strike-slip component of motion also known to exist on the fault.

In spite of the above evidence for preservation of the Jalipur units by downfolding and downfaulting, Owen (1988a) rightly emphasized that if the Jalipur tillites were indeed deposited during the early glaciation of Shanoz times, because of the rapid regional uplift, the units should also be expected to occur high up on the valley sides, in at least a few places. Instead, all of the Jalipur sequence is recognized only in the valley floor or near present river level, some 1500–2000 m below glaciated high-level erosion surfaces upstream only a few tens of km. Either the Indus river downcutting below the Jalipur has been retarded by uplift across its course downstream, or the Jalipur is very young, as Owen (1988a) thought. In this case, however, the induration and extensive tectonic deformation of the Jalipur would seem to preclude a young age. In addition, the apparent offset of late glacial moraines from the Darel valley to the south side of the Indus near Shatial, discussed below, may be evidence of downstream tectonism that has slowed Indus downcutting.

Shroder et al. (1989) also argued that the lack of Nanga Parbat gneiss within the Jalipur tillite reflected its early age, being deposited at a time when the Nanga Parbat gneiss had not yet risen sufficiently high to act as a source of gneiss clasts. Owen (1988a) offered an alternative explanation to the fact that tills at Bunar (Figure 3, section C of Shroder et al., 1989) have markedly different lithologies and structural attitudes. He proposed that the action of glacial lodgement processes along the valley floor provided a Jalipur till dominated only by Kohistan island are rocks. On the other hand, although chaotic and turbulent surge behaviour of ice and entrained debris is common in Himalayan glaciers (Hewitt, 1969; Shroder, in press a), another important characteristic of Himalayan glaciers is the widespread dominance of the supraglacial and englacial debris paths in which debris is transported passively rather than in the basal (traction) zone. Because of this possibility, Nanga Parbat gneiss might not have been carried into the lodgement tills, with such erratics being confined only to the supraglacial path. If this were in fact the case, the Jalipur tillite could be contemporaneous with the Bunar till of the early middle glaciation of Shroder et al. (1989), and therefore not equivalent to the early Shanoz glaciation. Instead, the Jalipur tillite would be younger, and perhaps equivalent to the Yunz or Borit Jheel glaciations. This is not viewed as likely by one of us (JS), given the considerable non-glaciotectonic deformation of the Jalipur and the lack of comparable deformation in the early middle sequence, but does remain possible.

Little other age constraint exists for the Jalipur units, except that, as Olson (1982) has pointed out, they must be < 2 myr old because of the presence of

Nanga Parbat clasts in the Jalipur valley-fill sediments with young cooling ages (Zeitler, 1985). In addition, as noted above, they are normally magnetized, which makes them less than $\sim 0.78$ myr old. Similarly, there is little more constraint on the lower till at Bunthang in the Skardu basin (Cronin and Johnson, 1989) which, because directly overlying sediments are reversely magnetized, must be older than $\sim 0.78$ myr.

Moreover, Dainelli (1922) first characterized the previously mentioned benches at $\sim 3200$–$3500$ m altitude along the Shigar valley as the 'terrace level representing the old valley level' of his first glacial expansion. Owen (1988b) thought that, together with surfaces of a similar height on the southern side of the Skardu basin (Chunda ridge, Karpochi and Blukro rocks, and Strongdokma ridge), all of these terraces represented an early extensive glaciation of the Shigar valley and Skardu basin, with a presumed equivalence to the Shanoz glaciation in Hunza. He further felt that a later glaciation had eroded away much of this glaciated surface, deepening the Shigar and Skardu valleys and producing an extensive surface at a lower elevation. Either this glaciation or yet another later advance, which also placed tills on top of Karpochi and the other rocks and ridges, might be correlative with the lower Bunthang till of Cronin and Johnson (1989).

In any case, the earliest glaciation of the entire study area probably represents a large regional valley ice mass that extended down the main Hunza and Gilgit river valleys. Similarly, extensive valley ice also moved southeast down the Shigar valley and merged with ice moving north from the Deosai plateau to flow together down the Indus to the east of Skardu (Norin, 1925). Whether or not this ice merged with other ice streams coming in from the Stak and other valleys to flow toward Haramosh is as yet unknown. High terrace surfaces on Haramosh, however, are consistent with ice movement down the Indus from Skardu as well as from the flanks of Haramosh itself. In the middle Indus valley below the Gilgit confluence, the high surfaces can be traced only as far as Bunji. If the Jalipur sequence is indeed younger than these surfaces, then there is no evidence known yet for Shanoz glaciation past Bunji.

To add to the questions about early glaciations, recent discovery by JS of a Jalipur-like till with lithologies from Nanga Parbat at Gorikot in the Astor valley introduces other possibilities as well. This well-indurated till or tillite is exposed as isolated outcrops of a well cemented and resistant diamict. Its induration and cavernous weathering cause it to resemble the basal lodgement till of the Jalipur. On the other hand, it is not sheared and occurs well above (150–200 m) the valley bottom, both indications that its induration is possibly because of antiquity and not lodgement pressures. The age, source area connections to Deosai, Rupal or other ice, and the overall age and importance of the Gorikot till are yet to be determined. Nevertheless, our working hypothesis at present is that the till is as old as or older than the Yunz or middle glaciation, and that ice from here probably descended to the Indus valley.

147

## Middle glaciation–Yunz stage

Compact tills resorting on benches at altitudes of 3200–3650 m on the diffluence cols between the Batura, Pasu and Ghulkin glaciers (Derbyshire *et al.*, 1984) represent the Yunz stage of glaciation in the Karakoram. These tills are also present elsewhere in the Hunza valley; e.g. near Serat at 3000 m. Derbyshire *et al.* (1984) showed that these deposits were older than 139,000 TL yr BP and they considered this stage to be a phase of major ice extension filling and overtopping high diffluence cols and entering the trunk valley. Owen and Derbyshire (1988) showed that ice extended into the Gilgit valley. This ice produced glacially eroded surfaces with a stepped morphology typical of alpine glaciated slopes. A well defined break in slope of ~ 300 m above the present river level can be traced for 5–6 km east of Gilgit near Jutial. The diamictons on these features are probably redeposited tills. Northeast of Dainyor, a palaeovalley is separated from the main valley by a 300 m high rock bar perhaps produced by Mani ice entering the Gilgit valley and being diverted by the main Gilgit ice stream. At the base of this valley, a well consolidated lodgement till is present and overlain by lacustrine silts dated at > 100,000 TL yr BP by Shroder *et al.* (1989). The palaeomagnetism of these same lacustrine silts measured by Owen (1988a) has shown declinations very close to the present field, so they would appear younger than 780,000 yr BP.

These deposits at Dianyor are all overlain by a massive recessional or end moraine (Dak Chauki), everywhere with the same post-maximal weathering characteristics, which extends from the top of the section, > 600 m above the valley floor, down to and across the valley. This moraine was first noted by Desio and Orombelli (1971). Because Derbyshire *et al.* (1984) originally had terminated their Borit Jheel glaciation near Nanga Parbat, where no clear end moraine seemed to exist, Shroder *et al.* (1989) thought that this Dak Chauki moraine was the Borit Jheel terminus. Subsequently the Borit Jheel terminus was thought by Owen (1988a) to be further down the Indus valley near Chilas, and this Dak Chauki moraine to be but a recessional moraine in the Borit Jheel sequence.

East of Batkor Gah near the head of the valley, a smooth asymmetrical surface with some till remnants can be traced southwards into the main Gilgit valley at ~ 2500 m, and within the valley large deposits of till can be seen which are probably of the same age. Erratics on the till surface have a post-maximal rock varnish (Derbyshire *et al.*, 1984), indicating a large time span for the development and weathering of the rock varnish. Little evidence of till from the main valley ice occurs within the Gilgit valley itself. Owen and Derbyshire (1988) thought that most of the till within the Gilgit valley was redeposited by debris-flow processes soon after deglaciation, which would account for the general absence of diamictons with characteristic till fabrics.

At Farhad Bridge on the north side of the lower Gilgit valley near its confluence with the Indus, just above the old road from Gilgit, are two tills separated by 120 m of bedded glaciofluvial sediments (Owen, 1988a; Shroder *et al.*, 1989). Shroder *et al.* (1989) attributed the lower till to their early middle glaciation and the upper till to their late middle glaciation, both correlative to the Yunz glaciation of Zhang and Shi (1980) in Hunza, and perhaps also to Dainelli's (1922) second and third glacial expansions in Skardu. On the other hand, Owen (1988a) felt that, in view of the low altitudes and exposed position of the Farhad Bridge section compared with other deposits attributed to the Yunz glaciation, the deposits were probably younger, perhaps Borit Jheel in age. Moreover, stacked tills in this location might not necessarily indicate successive glaciations but could represent sub-and supraglacial till emplacement (Boulton and Eyles, 1979) or perhaps even different phases of nearly contemporaneous sedimentation of tills and fluvial deposits into a south-facing marginal ablation valley (Hewitt, this volume).

Multiple tills from a middle glaciation are also present near Bunar and Drang. At Bunar, ~ 280 m of till with clasts from Hunza and Nanga Parbat, with post-maximal rock varnish on the surface boulders, is overlain by 60 m of tilted lake beds which, in turn, are overlain by 145 m of younger till and untilted glaciofluvial sediments (Shroder *et al.*, 1989). The lower till is considered to be Yunz in age but if the Borit Jheel glaciation came this far, as thought by Owen (1988a), then the upper till may be of that age. The alternative explanation originally expressed by Shroder *et al.* (1989) is that both tills are phases of the middle glacial Yunz ice. This Bunar section, at the mouth of the Diamir Canyon access to northwest Nanga Parbat, is especially important because it is one of the few unfaulted exposures with Jalipur tillite and sandstones in direct contact with the overlying less lithified tills and ablation valley sediments of the known middle glaciation. Perhaps if the Jalipur is indeed middle glaciation (Yunz) it is equivalent to the tills at Shatial, as Shroder (1985) originally thought.

Between Chilas and Shatial, two sets of well consolidated tills are present. The older set, exhibiting post-maximal rock varnish, lie 100–150 m above the present river level. The clasts are mainly of angular norite derived from local bedrock on the valley floor; the tills occur in stoss-and-lee forms (Shroder *et al.*, 1989), indicating a down-valley ice movement direction, and the till has lodgement characteristics (Owen, 1988a). The second set of tills occurs near river level and is less weathered and finer grained. These tills are dominantly of lodgement type, indicating an ice movement direction across or oblique to the valley line. This set is considered to be younger than Yunz in age and is discussed further on. Associated with the older till sets are the outcrops of till along the north side of the valley. These relate to main trunk-valley ice depositing lodgement tills together with lateral moraine at its margins. At a height similar to the later moraine tops ( ~ 1500 m), glacially eroded benches can be recognized. They can be traced into neighbouring valleys, suggesting that ice from these tributaries once entered the Indus valley as well. The low altitude of the lodgement tills ( ~ 100 m above river level), representing the valley floor of middle glacial or Yunz time, is perhaps explained by the lower uplift rates ( $<1$ mm yr$^{-1}$) in this are compared to the nearby NPHM (Zeitler, 1985).

Owen (1988a) and Shroder *et al.* (1989) recognized tills near Shatial several hundred m above the valley floors, reaching altitude differences of up to 1500 m, probably representing the maximum extent of Yunz ice in the Indus valley (Desio and Orombelli, 1971; Derbyshire *et al.*, 1984; Shroder *et al.*, 1989). A problem with these deposits is that on the north side of the Indus the old tills have no morainal landforms and are overlain by thick ablation valley outwash and alluvial deposits, whereas on the south the deposits are a series of moraines with much younger surfaces. For example, the lower moraine has only ~ 0.5 m of loess on it and the upper has ~ 1 m of loess. In addition, a young and slightly offset cross-valley moraine dam, as opposed to a moraine down the Indus valley, helps to explain the plentiful young (38,100 TL yr BP) and thick lake deposits on both sides of the Indus directly upstream that extended nearly to Chilas (Shroder *et al.*, 1989). Without a glacial dam from the Darel valley opposite, there appears to be no other possible blockage to produce the lake beds. Furthermore, the younger moraines are oriented perpendicular to the Indus valley axis and therefore appear to have been emplaced from the Darel valley. Long-continued tribal unrest in the Darel valley, however, has up till now precluded fieldwork to investigate this hypothesis further.

Directly west of Shatial, the main Indus valley becomes narrow and engorged within a few km. No evidence of till occurs below here and it is reasonable to suggest that the early stage of the main middle glacial or Yunz valley ice did not extend much further down valley than Shatial, supporting the judgements of Desio and Orombelli (1971), Derbyshire *et al.* (1984), Owen (1988a) and Shroder *et al.* (1989).

The late stage of middle glacial or Yunz ice probably terminated at Gor directly northwest of Nanga Parbat. Massive weathered tills occur here beneath younger moraines. If this is the terminus of that stage, and if the Jalipur is actually early Yunz in age, then many of the tills above the Jalipur units downstream from Gor would probably be late glacial in age and presumably from Nanga Parbat.

Upstream in the upper Indus gorge past the Gilgit river confluence, the evidence for middle glacial or Yunz stage glaciation is more sketchy, probably because of the high erosion associated with active tectonism across the Raikot and other faults there. U-shaped erosion forms in the Indus valley between Haramosh and the Gilgit confluence, and glacially eroded benches on the east side of the Indus valley near Jaglot, suggest that ice probably extended to these areas from the Haramosh massif by way of the Indus gorge. Whether or not ice extended from the Skardu valley to Haramosh at this time is as yet unknown. Strong U-shapes and thick tills occur, however, from Skardu to the NPHM syntaxis fault near Chutran, close to Haramosh. Probably the rapid uplift of NPHM and associated erosion has destroyed evidence of glaciation across it.

In the Shigar valley northwest of Skardu, glacially eroded surfaces bearing till with post-maximal varnish occur at a height of ~ 2800 m (Owen, 1988b). These can be correlated with a series of buttes in the Skardu basin that have similar till-draped surfaces. These were produced by ice from the Braldu and Chogo Lungma glaciers that deepened the Shigar valley and played a part in the excavation of the main Skardu basin.

## Last glaciation–Borit Jheel stage

Much evidence for extensive glaciation occurs in tills deposited below 3000 m altitude in the Hunza valley (Derbyshire *et al.*, 1984). The tills are predominantly of lodgement and subglacial meltout origin and were assigned by those writers to the Borit Jheel stage and correlated with the Hunza stage of Zhang and Shi (1980). The tills occur in the diffluence coll of the Batura–Pasu glacier, on the slopes of the Pasu–Ghulkin diffluence col, around Borit Jheel just above 2500 m altitude, and at the base of the Batura 'drift plateau'. The till-like occurrence beneath ~ 20 m of debris flow south of Gulmit village has since been re-interpreted as ancient flow debris (Owen, 1988a). Glaciolacustrine silts overlie lodgement tills of the Minipin glacier on a bench at 2500–2600 m altitude: although this till has been dissected, it can still be traced as far as the till plateau in the upper Hunza. Glaciolacustrine silts overlying these tills in the Batura area give dates of ~ 50,000 ± 2500 TL yr BP, and silts beneath till of the Ghulkin I stage have been dated at 65,000 ± 3300 TL yr BP (Institute of Geology, Academia Sinica: in Derbyshire *et al.*, 1984). This glaciation was believed to represent the last full glacial maximum, with ice occupying diffluence cols, and tributary glaciers coalescing into extensive Hunza valley ice streams, with ice extending perhaps as far as the Astor–Indus confluence, although there is little evidence for such a terminus there. Owen (1988a) and Derbyshire and Owen (1990) felt that Desio and Orombelli's (1971) Dak Chauki moraine represents moraines left by the Borit Jheel advance and that hummocky moraines on the valley floor represent recessional moraines as the trunk-valley ice retreated, whereas Shroder *et al.* (1989) felt that the Dak Chauki moraines represent the terminus of the Borit Jheel advance.

Scattered till fragments are preserved beneath screes in the lower half of the Hunza valley and can be examined where the screes have recently been excavated. Similar exposures of till are present along the lower half of the Naltar valley several hundreds of m above the present stream level and are considered to be of Borit Jheel age.

Owen (1988a) felt that a number of the sections that Shroder *et al.* (1989) had described as tills belonging to other glaciations were instead related to Borit Jheel ice. For example, in the lower Gilgit valley the tills of the Farhad Bridge outcrop, and the scattered deposits of till on bedrock benches in the valley floor that are overlain by alluvial fan sediments, could be part of this supposed sequence. In addition, Owen (1988a) noted that some of the Jalipur tillites appeared similar to Borit Jheel deposits, being laid down by lodgement processes on the valley floor by main-valley ice. He felt that the Jalipur sequence was therefore of Borit Jheel age and that this ice mass probably did not extend further down valley than

151

Chilas, about where the known Jalipur terminates. One of us (JS) believes that there are no objective physical criteria or age dates, other than supposition, to support this but it remains a working hypothesis.

The upper till and moraine with post-maximal weathering in the Drang section directly below Nanga Parbat (Shroder et al., 1989) is almost certainly of the last glacial and Borit Jheel in age, as well as being derived from Nanga Parbat glaciers. On the north side of the Indus at Gor, Patro and elsewhere in the vicinity, similar cross-valley tills and moraines occur, first noted by Desio and Orombelli (1971), which have the same weathering characteristics. In addition, in a few of these places occur remnants of a moraine slightly older than these; Shroder et al. (1989) equated this oldest of cross-valley moraines, and also possibly Borit Jheel equivalents, to their L1 or earliest last glaciation and another possible Borit Jheel equivalent.

The presence of clear cross-valley moraines at this location in the last glacial or Borit Jheel time is a problem for Owen's (1988a) hypothesized Borit Jheel ice from Hunza that extended nearly to Chilas. He attempted to resolve this by hypothesizing a debris-flow or other origin for several of the deposits near Ginne Gah, Patro and elsewhere, but the presence of pronounced hummocky topography, infilled kettle holes, and sediments backed up behind the moraine dams precludes the idea. For example, at the Patro and Ginne sites, moraine dams from this time or later cross the Indus and produced thick fluvial and lacustrine sediments; at Patro, on the north side of the river above the Shina village, the lake beds are $> 50$ m thick, whereas at Ginne the mixed sediments are $> 100$ m thick.

Though main-valley ice did not extend down the Indus valley as far as the Shatial area in the last glacial stages, tributary-valley ice probably extended into the main Indus valley from the Darel valley. This produced the lake sediments (Burgisser et al., 1982) and large moraines from the Darel valley to the southern side of the Indus valley that Shroder et al. (1989) described as blocking the Indus and producing thick sediments on both sides of the valley for over 50 km up river at Chilas. It also may have produced the younger set of lodgement tills at river level as described earlier. Shroder et al. (1989) dated the lake sediments at 38,000 $\pm$ 2600 TL yr BP. Owen (1988a) indicated that it is reasonable to assign these moraines and the glaciolacustrine sediments to the Borit Jheel glacial stage.

The large south-facing rock embayment north of the Indus near Gor (Shroder, 1989a,c) is infilled with several hundred m of diamictons and moraines confined at the western margin by a large concave asymmetrical slope. These sediments lap up on and partly overlie the older tills that are thought possibly to mark the position of the terminus of the last stade of the middle glaciation. The age of these moraines is unknown but the post-maximal weathering of the surfaces indicates that they were probably deposited during last glacial or Borit Jheel times and perhaps could be an indistinct terminus of the Borit Jheel ice, as Derbyshire et al. (1984) originally mapped. Boulders ($> 5$ m) of the Nanga Parbat gneiss are present within the deposits that appear to have been supplied from the opposite side of the valley across the highly active Raikot fault. Shroder

(1989a,c) and Shroder *et al.* (1989) believed that these clasts were deposited during the late glacial times by ice from the Raikot and Buldar valleys, at the same time that the Indus was blocked by other ice from the south at Patro and Ginne, all of which also appear to be at least partly equivalent to the Ghulkin I stade of Derbyshire *et al.* (1984). The absence of till and the steep and V-shaped lower gorges of the Raikot and Buldar valleys on Nanga Parbat are probably a reflection of the 5.5 mm $yr^{-1}$ of uplift at their base, rather than special pleading for alternate hypotheses to limit glacial growth on the Nanga Parbat massif. Owen (1988a) has suggested, however, that the Nanga Parbat clasts in Gor could possibly have been derived from the main Indus valley ice via a diffluence col north of Gor Gali peak, but this would still require their movement across the entire supraglacial ice stream. Furthermore it is highly unlikely that any separate glaciers developed in the south-facing exposure at Gor at an elevation of only ~ 3000 m in the Nanga Parbat rainshadow.

Upon Borit Jheel ice retreat, many of the tills deposited on the steep valley sides were quickly redeposited as debris flows, forming the main component of the sediment fans in the middle Indus and Gilgit valleys (Owen, 1988a; Derbyshire and Owen, 1990; Owen and Derbyshire, 1988). In addition, the process of ice wasting in the Gilgit area was complex, with ice perhaps retreating up the Hunza and Gilgit valleys simultaneously. Retreat, however, appears to have been punctuated by a further advance of the Hunza valley ice which seems to have blocked the Gilgit valley and allowed a lake to form behind the Hunza ice. Owen (1988a) identified two sets of glaciolacustrine silts along the south side of the Gilgit area that he believed represent a lake which drained and reflooded before the ice eventually retreated up the Hunza valley. Shroder *et al.* (1989) had noted that the normally magnetically polarized lower lake deposits ( > 100,000 TL yr BP) at Dainyor were at exactly the same height as those in Gilgit town, and therefore equated them as the result of a cross-valley ice dam further downstream, but their equivalent altitudes may be only coincidental elevation similarity.

Other lacustrine deposits form the lower limits of sediment fans between Thelichi and Bunji in the Indus valley and may represent dams formed behind cross-valley ice or landslip dams, or differential melting of Indus ice which maintained an ice barrier downstream. Retreat of Bagrot ice or, more likely, a minor advance synchronous with readvance of Hunza ice, formed a small end moraine at the mouth of the Bagrot valley which now forms an inlier within the sediment fan debris-flow sediments.

Owen (1988a) suggested that, in the Haramosh area, ice probably advanced from the Mani Bashar and Phuparash valleys and so supplied ice to the main Indus valley, isolated exposures of till are present along the valley sides at a height of 100–350 m above present river level. The mouth of the Phuparash gorge is incised below the probable former valley floor of Borit Jheel times.

In Baltistan, major valley ice deepened the Shigar valley and entered the Skardu basin, overriding the valley-fill sediments east and north of Blukro rock.

Similarly, ice from the Deosai plateau extended into the Skardu basin from the south, overriding the terraces (Owen, 1988b). Hummocky moraines are present along the valley floors in the Shigar valley; e.g. 3 km north of Shigar village. Lateral moraines are also present along the Shigar valley, though most are deeply dissected. At the mouth of the Baumaharel valley, the glacier blocked the valley and produced a lake which backed up into the valley. The lacustrine silts which accumulated in this lake onlap the lateral moraine. Water flooding from the lake produced an impressive meltwater gorge.

Overall, the Borit Jheel glaciation gave rise to extensive valley glaciation, producing the many till deposits in the Gilgit and middle Indus valley. In the Hunza valley, ice occupied diffluence cols and coalesced to form extensive Hunza ice bodies extending down the valley, certainly to Gilgit and the Dak Chauki moraine, possibly to Gor as an end moraine, and perhaps even as far as Chilas.

## Last glaciation–Ghulkin I stade

Derbyshire *et al.* (1984) showed that this stage represented a phase of limited glacial extent of 'expanded foot' type, restricting ice to the tributary valleys and limited to minor advances into diffluence cols in Hunza. During this time, the glaciers of upper Hunza advanced across the valley and diverted the Hunza river and, although occupying the main Hunza valley, they extended only a few km downstream. In the Hindi embayment (Derbyshire *et al.*, 1984), ice from the glaciers extended up to ~ 1.5 km north of the present Hunza river, thus diverting and ponding drainage. Lacustrine silts intercalated within the tills from the Minipin and Pisan glaciers from Rakaposhi mountain yielded a date of 47,000 ± 2350 TL yr BP (Institute of Geology, Academia Sinica).

Evidence for the Ghulkin I stage in the Gilgit and middle Indus valleys hinges on the interpretation of large regular lobate landforms comprising poorly sorted bouldery polymictic diamictons with a crude subhorizontal stratification, which are deflated of fines on their surfaces. In a few places, such as at Sachen and Mani glaciers, modern examples of these lobate landforms are directly contiguous to existing ice-cored moraines so that some examples at lower elevations have been interpreted by Giardino *et al.* (1988) and Shroder *et al.* (1989) as former ice-cored rock glaciers or moraines, which moved during Borit Jheel time. At the heads of many side valleys and at many other lower elevations, such as Batkor Gah and elsewhere, however, the morphology, sedimentary structures, petrology and microfabrics of these landforms indicate a possible debris-flow origin instead (Derbyshire *et al.*, 1987; Owen, 1988a; Owen and Derbyshire, 1988; Derbyshire and Owen, 1990). They consider that these lobate landforms were emplaced during deglaciation of the Borit Jheel phase and now form inliers within the sediment fans in some places.

Some of the present glaciers in the Gilgit, middle Indus and Haramosh areas, e.g. Hinarche, Raikot and Mani glaciers, appear to show less obvious evidence of this expanded foot stage, except for an obvious downwasting of ice now several hundreds of m below a series of lateral moraines. At Nanga Parbat this seeming lack of evidence for the expanded foot stage is partly illusory because of destruction of much evidence by erosion along the active Raikot fault, but elsewhere the cause is not known. Other glaciers, e.g. the Sachen and Patro from Nanga Parbat, clearly had expanded stages. The Sachen glacier, for example, once extended across the Astor river, and the town of Astor sits astride the prominent moraine. Upstream at Gorikot thick lake beds show that the Sachen ice dam existed for an extended period.

Deformed floodplain sediments and hummocky moraines within the Skardu basin near the mouth of the Satpara valley indicate a possible Ghulkin I expansion of ice from that valley (Owen, 1988b). No evidence for such an advance exists in the Shigar valley, and the Braldu, Biafo and other glaciers apparently did not come down as far as the Shigar valley at that time.

## Last and Holocene glaciations–Ghulkin II, Batura, Pasu I and Pasu II stades

In the upper Hunza valley, Derbyshire *et al.* (1984) differentiated these stades on the basis of weathering criteria and morphostratigraphy, the Ghulkin II stade being considered as a late still-stand in the retreat from the Ghulkin I expansion.

The Batura stade moraines are confined to the tributary glacial troughs and take the form of arcuate moraines and a group of prominent lateral moraines in the upper Hunza valley (Derbyshire *et al.*, 1984). Owen (1988a) recognized similar prominent lateral moraines and felt that some in the areas adjacent to the Mani, Hinarche and Raikot might be equivalent to the Batura moraines. Gardner and Jones (this volume) used tree-rings and pedogenesis to suggest that the Raikot moraines were produced in a Holocene maximum, probably during the Neoglacial or the Little Ice Age.

The Pasu I stade is represented by a set of sharp-crested moraines with a weak yellow rock varnish. Derbyshire *et al.* (1984) suggest that they are probably the result of the Little Ice Age and have yielded radiometric age dates in the range of 800–325 $^{14}$C yr BP (Batura Glacier Investigation Group, 1979, 1980). Moraines apparently equivalent to these occur around other glaciers in the Gilgit and middle Indus areas (Owen, 1988a). It may be that high meltwater discharges from the present-day glaciers have destroyed much of this record. The Pasu II tills have no rock varnish and represent nineteenth and twentieth century advances; moraine crests are sharp and moraines are commonly ice-cored. Equivalent features occur near the fronts of all of the glaciers in the Hunza, Gilgit and middle Indus areas.

## CONCLUSION

Our interpretation of the extents of the glaciations in the Karakoramo and Nanga Parbat Himalaya begins with a pre-Pleistocene planation surface that occurs at an altitude of ~ 5200 m. Three major glaciations during Pleistocene time and at least five minor 'Neoglacial' advances follow. First, it must be stressed, however, that the glacial record in any mountain environment is commonly incomplete (Gibbons et al., 1984) and, because at least 32 glaciations have been identified in the deep sea record for the late Cenozoic (Shackleton and Opdyke, 1973), probably many more glaciations occurred in the western Himalaya than have been recognized at present. Second, the very poor age constraints on the glacial deposits and crude correlations between deposits indicate that many of the sediments which have been assigned to a single glaciation may in fact represent more than one glaciation. Third, the problem of diamict differentiation has been ever present in this work, such that some glacial, mass-movement and other deposits have undoubtedly been incorrectly ascribed to the wrong process. Much further work on all of these problems is indicated.

The recognized first glaciation, the Shanoz, produced extensive ice in large broad open valleys. Deposits from the glaciation are sparse, but glacial surfaces at 4150–4300 m in the Hunza area, above 2600 m in the Gilgit region, above 3200 m in the Haramosh region, above 4000 m on Deosai, and at 3200–3500 m in the Shigar valley area provide much evidence for this glaciation. The Jalipur sequence is both genetically and temporally problematic. The Jalipur tillites are now known to be clearly glacial and some of the unit is downfaulted, down-folded and thereby preserved, but the presence of the deposit at the deepest parts of the valley bottoms and 1500–2000 m below the high-level Shanoz glacial erosion surface remains a problem in need of further fieldwork. Only detailed, large-scale field mapping and a full spectrum of analytical techniques are likely to solve this problem. The reverse magnetism of the Bunthang sequence (Cronin and Johnson, this volume), however, is clear evidence of an oldest glaciation in the Skardu area, and even though the lowermost till is ~ 900–1000 m below the high-level glacial erosion surface of the Shigar valley, the two are possibly related.

The second, middle or Yunz glaciation appears to be the penultimate glaciation. This was an extensive valley glaciation, with ice extending as far as Shatial in the middle Indus valley and ice from the Shigar valley filling the Skardu basin, but whether or not Skardu ice connected past Haramosh with middle Indus ice is as yet unknown. Also as yet unresolved is whether or not there were two phases of ice movement at this time, as JS thinks, or only one, as LO and ED think. The tills of this time period occur at high altitudes (3200–3650 m) in the Hunza valley. Scattered high tills along the Gilgit valley and lodgement tills over 100 m above the present river level in the middle Indus valley are attributed to this glaciation. Some of the thick tills overlain by 2 m of degraded loess on the Shamoskith plains of Deosai probably also relate to this glaciation. Glacially eroded benches in the middle Indus and Gilgit valleys are related to these former Yunz ice levels.

The third (fourth?) or last major glacial advance is best represented by the Borit Jheel glaciation, which produced large glaciers that extended at least as far as Gilgit, probably to Gor, and perhaps to Chilas. It is represented by tills below 3000 m in Hunza, lodgement tills on the valley floors, the major Dak Chauki moraine at Gilgit, and perhaps by scattered outcrops along the valley sides in the Gilgit–Haramosh area. Owen (1988a) at first thought that at least part of the Jalipur sequence fitted here, but we agree that this idea needs further work. Main-valley ice from the Shigar valley also entered the Skardu basin at this time, depositing hummocky tills and lateral moraines in Baltistan. This glacial stage is tentatively correlated with the last Pleistocene glaciation of the Northern Hemisphere but many further dates are needed in order to substantiate adequately this supposition.

Evidence for Neoglacial advance is most evident in the higher regions of the upper Hunza and Astor valleys, but the record is poor in the middle Indus and Gilgit areas. The Ghulkin I stade produced 'expanded foot' type glaciers mainly confined to tributary valleys in the Hunza area. They are probably best represented in the middle Indus area by large lateral moraines below which the present-day glaciers are entrenched several hundred m. The Ghulkin II stade probably represents a still-stand in the Ghulkin I stade and is not significant in the Gilgit and middle Indus areas.

The Batura stade represents a restricted advance that produced large moraine complexes in the Hunza area, but probably had little effect in the Gilgit and Hunza areas, except to help to produce a second set of moraines inset into Ghulkin I moraines below which the present glaciers are entrenched.

The Pasu I stade relates to the Little Ice Age advance that produced moraine sequences in the upper Hunza valley which, however, are missing the Indus–Gilgit area. Moraines of recent centuries, which are deposited near the snouts of the present glaciers, have been grouped together as the Pasu II stade. These appear to have low preservation potential owing to the frequent fluctuations of glacial positions in the Hunza valley.

There is as yet little dating control of these stades and stages of glaciation. As a consequence, the chances of misinterpretation of the glacial sequences in the western Himalaya remain high. The relative role of uplift and glaciation is still not well understood, but it is likely that the critical monsoonal climatic influence was cut off soon after the last Pleistocene glaciation by rapid uplift of the Great Himalaya. In addition, the south-facing rainward side of Nanga Parbat is likely to have had a different pattern of glacial advances than the north, rainshadow side. The rapid uplift of the mountain along the Raikot fault on the north side is also likely to have affected the glaciers there. Inasmuch as exhumation rates are commonly calculated as a function of surface erosion and attendant cooling rate changes as minerals pass through annealing isotherms at depth, the palaeo-configuration of the isothermal surfaces of this region is of significance in neotectonic studies. Establishment of a firm geomorphologic history of the region is therefore critical to this methodology. With present-day relief in the Karakoram and Nanga Parbat Himalaya on the order of 6–7 km, descending cold waters in faulted and fractured glacial valleys, and rising thermal waters along faults, the palaeotopography and directly linked palaeoisothermometry could be of some importance. In addition, the linkages of long-term erosion to isostatic adjustment and the uplift of the western Himalaya are also not well established.

Finally, little work on the geomorphology and Quaternary history has been done in the adjacent areas of the western Himalaya, e.g. eastern Baltistan, Khunjerab (north Karakoram), southern Deosai and northern Kashmir, Chinese Tartary, and Chitral. These areas were obviously glaciated and some helped to supply ice to this study area, so that their role in the Quaternary glacial history of the western Himalaya has still to be assessed. As yet, also, it has not been possible to provide many temporal linkages or other correlations with the prior work in nearby Afghanistan, northern India or Nepal.

# 7

# ALTITUDINAL ORGANIZATION OF KARAKORAM GEOMORPHIC PROCESSES AND DEPOSITIONAL ENVIRONMENTS

*Kenneth Hewitt*

## ABSTRACT

The geomorphology of the Karakoram Himalaya is considered in terms of the altitudinal zonation of climatic–geomorphic conditions and their downslope relations. Factors of greatest importance are the large gradients of precipitation with elevation, temperature regimes including the seasonal migration of frequent freeze–thaw, and the duration of snow on the ground. These factors first result in a major erosional role of the downslope moisture stream from humid, high altitude areas to progressively drier, and eventually arid, areas at low altitudes. Avalanches, glacier flow and seasonal meltwaters each play a large part. Second, local forms and processes exhibit altitudinal zonation, which occurs in four broad belts. Attention is directed to conditions 3000–6000 m in altitude in the central Karakoram and in the Biafo and Barpu–Bualtar glacier basins. Here the landscape is a combination of extensive glacier ablation zones and off-glacier slopes. Moisture supply, including seasonal meltwater production, and freeze–thaw cycles are abundant. It is a landscape dominated, on the one hand, by rock walls and ridges and, on the other, by glacier ablation zones. The best-developed depositional features occur at the interface between glacier margins and off-ice areas. Distinctive systems of deposits follow the margins of the glaciers but depend upon slope, snowmelt, and lacustrine processes, and ice margin activity. They also reflect the variations and relations of these processes as a function of elevation. Both glacier fluctuations and changes in climatic–geomorphic conditions with elevation have relevance to late Pleistocene and Holocene developments and their depositional legacy.

## INTRODUCTION

The main aim of this paper is to examine geomorphic activity in those parts of the central Karakoram lying at 3000–6000 m in altitude where precipitation,

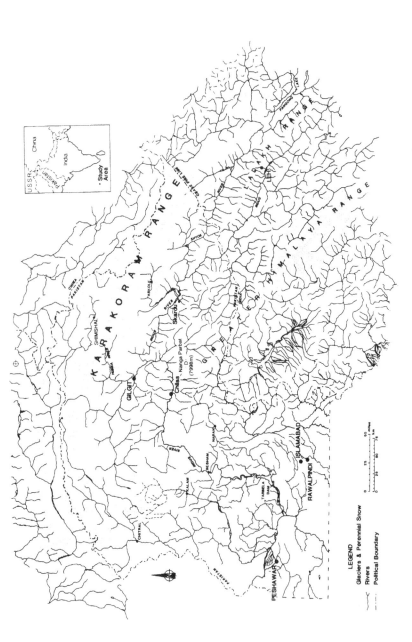

*Figure 7.1* The Karakoram range and its location in southwest Central Asia, showing the extent of perennial snow and ice, and the major river systems.

seasonally released meltwaters and freeze–thaw cycles are relatively abundant.

Glacier ablation zones compose a large fraction of the area (Figure 7.1). Here, and on steep valley walls, erosional processes display exceptional vigour and diversity. Locally, and especially along glacier margins, they give rise to distinctive suites of depositional landforms. This relatively neglected part of the regional geomorphology comprises about two-thirds of the Karakoram Himalaya.

## ALTITUDINAL ORGANIZATION OF THE KARAKORAM ENVIRONMENT

In the Karakoram, because of enormous available relief and deep dissection, altitudinal variations play an overwhelming role in landform patterns and surface processes. The landscape depends fundamentally upon tectonic conditions and varies regionally as a function of lithological conditions (Gansser, 1983). Broad regional gradients of moisture supply, temperature and vegetation cover occur (Paffen et al., 1956; Hewitt, 1968a; Schweinfurth, 1956). Important local variations arise, owing to valley wind systems, aspect, human activity, and sub-basin location within the upper Indus and Yarkand river systems (Dainelli, 1924–35; Miller, 1984). In this paper, however, Karakoram geomorphology as a whole is first considered, followed by its expression in the central parts of the main range.

Topoclimatic conditions, relevant to the geomorphology of all mountain lands, seem to exhibit an exceptional development and control over geomorphic activity in the Karakoram. At this scale, the most general and distinctive features of surface conditions and processes relate to what has been called 'vertical stratification', but also to the downslope linkages in landform patterns and surface processes between 'altitudinal belts' (Isachenko, 1965; Kowalkowski and Starkel, 1984).

## HYPSOMETRY AND REGIONAL GEOMORPHOLOGY

Since von Humboldt (1820) first gave so much attention to hypsometry, it (and the area–altitude relations of conditions that reflect it) has been a common descriptive approach to high-mountain environments (Price, 1981; Klimek and Starkel, 1984). Its importance for the geomorphology of the Karakoram has been recognized only rarely (the von Schlagintweits, 1860–66; Paffen et al., 1956; Hewitt, 1968a).

The hypsometry of the Karakoram may be characterized in terms of three elements (Figure 7.2). The upper and lower elements comprise only a small fraction of the surface area or positive mass. The region below ~ 3000 m actually coincides with most of the 'fluvial zone'. It consists of the gorges and overdeepened valley floors of the major rivers. This makes up less than one-quarter of the total area.

In both area and mass, the bulk of the landscape lies at 3000–6000 m. This

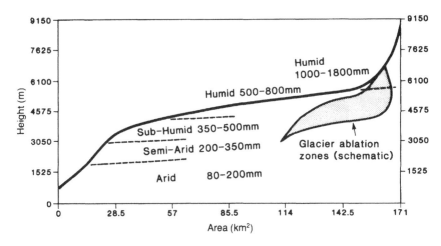

*Figure 7.2* General hypsometric curve for the upper Indus basin above the Tarbela Dam, showing approximate altitudes of precipitation zones. The curve is derived from 2000 ft contour intervals on map sheets of the area at a scale of 1:250,000.

zone contains most of the mountain ridges, the glacierized surface and, on the eastern fringes of the region, some subdued, high-altitude stream basins at the margins of the Tibetan plateau. A third element, comprising perhaps 2 per cent of the area, consists of the very highest ridges and peaks above 6000 m.

Along the main axis of the Karakoram, the predominant altitudinal zone (3000–6000 m) consists largely of glacierized basins. Perennial snow and ice cover some 15,150 km² (von Wissmann, 1959), so that glacier action is of fundamental importance. However, the total of basin areas tributary to ice masses is more than twice as large again. There are also extensive mountain slopes above 3000 m which descend directly to the river gorges. Hence, the geomorphic processes in this zone also reflect large areas that are ice free but have seasonal snow cover. However, much of this slope area has perennial snow and ice at the head, or a glacier margin at the base, or both. Because average slopes at these elevations tend to be steeper than 40°, the true area exposed to surface processes is even greater than for the slopes of the ice cover, which are generally much gentler (Figure 7.3).

Most of the following discussion, based upon the central Karakoram, applies broadly throughout the Southwest Central Asian mountain system. However, it is important to note that altitudinal limits and the importance of particular zones and features change toward the far west and east of the Karakoram range, as well as to the north and south of it. The altitudinal data and detailed observations referred to here relate especially to the basins of Barpu–Bualtar ('Hopar') and Biafo glaciers. The differences between these two basins reflect the range of conditions in the central Karakoram glacier basins as a whole. The huge, open-accumulation areas of the Biafo are absent from the Barpu–Bualtar, which is

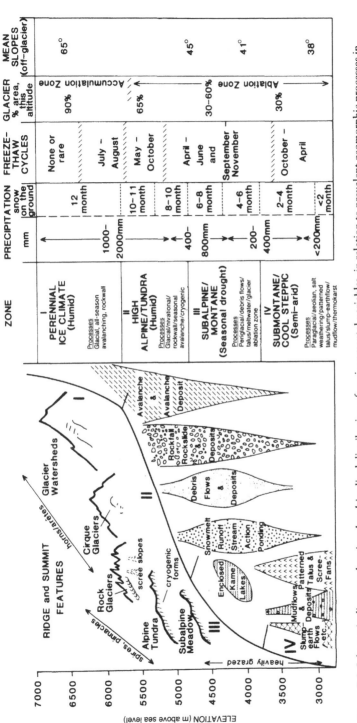

*Figure 7.3* Schematic diagram showing the altitudinal distribution of environmental variables, in relation to prevalent geomorphic processes in the central Karakoram. The hypsometric curve is that of the Barpu–Bualtar basin.

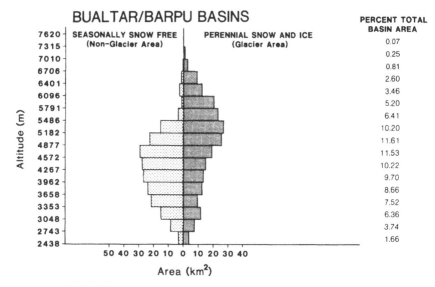

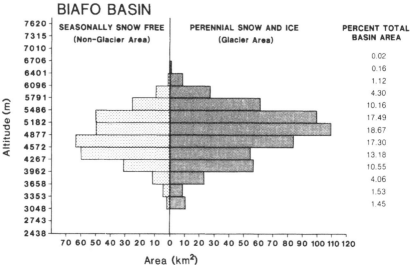

*Figure 7.4* Area–altitude distributions of the glacierized and off-ice surfaces of the Barpu–Bualtar and Biafo glacier basins, central Karakoram.

largely avalanche-nourished (Figure 7.4). While some of the largest glaciers such as Siachen and Rimo resemble the Biafo in this respect, a majority resemble the Barpu–Bualtar.

It should be noted that most of the existing Karakoram literature deals largely with either the lower semi-arid fluvial zone or the uppermost zone of great peaks (Miller, 1984). The important exceptions are works on the larger glacier basins, but few of these deal with geomorphology (Hewitt, 1989a; and the early work of the von Schlagintweits, Oestreich and Dainelli). It thus appears that the bulk of the Karakoram landscape is poorly known and rarely described.

In defining the broad environmental factors likely to control geomorphic processes, attention is directed here to the area–altitude distributions and relations between precipitation, thermal conditions (including freeze–thaw cycles), the duration of snow on the ground, and water yields from the melting of snow and ice. These conditions are subject to altitudinal gradients in two complementary ways. First, the moisture stream from humid high-altitude areas to arid lower ones is of decisive importance to the rate and extent of geomorphic activity everywhere (Hewitt, 1985). Second, the visible (i.e. mappable) landforms suggest a series of altitudinal belts where particular form and process complexes prevail.

## MOISTURE SUPPLY AND MOBILITY

The supply of moisture is dominated by two topoclimatic variables, namely, the vertical distribution of precipitation and the seasonal migration of thermal conditions with altitude. A steep gradient of increasing precipitation with altitude is now well established (Batura Glacier Investigation Group, 1979; Hewitt, 1985). Recent work suggests that maximum precipitation in the central Karakoram occurs at 5000–7000 m altitude. It is ~ 1000–1800 mm water equivalent (Shi and Zhang, 1984; Hewitt, 1986; Wake, 1987a,b). This contrasts with the arid or semi-arid conditions that prevail below ~ 3000 m altitude (Figure 7.2), where precipitation is ~ 100–200 m (Whiteman, 1985; Butz, 1987).

Throughout the Karakoram, the bulk of precipitation is snowfall, and entirely so above ~ 4800 m. Below that, some summer rains occur and occasional rainstorms can have marked, but brief and localized, geomorphic effects. Westerly derived winter precipitation is the main moisture source. Secondary summer precipitation, including occasional monsoonal invasions, accounts for about one-third of snowfall accumulation on central Karakoram glaciers (Mayewski et al., 1980; Wake, 1987a,b). Both the duration and amount of snow on the ground are important indicators of geomorphic conditions (Figure 7.3).

Solar radiation is the main determinant of snow and ice ablation. Nevertheless, temperature is a better indicator of the timing and significance of thermal conditions for geomorphic activity. Obviously, negligible runoff occurs where temperatures remain below freezing for most or all of the day, but important

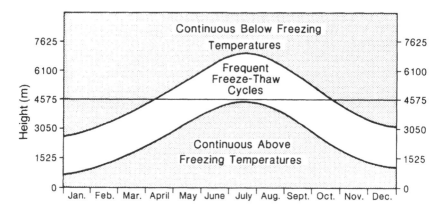

*Figure 7.5* Seasonal migration of temperature zones for the Karakoram range, emphasizing zones of continuous freezing, of above freezing temperatures and of frequent diurnal freeze–thaw cycles (after Hewitt, 1967).

geomorphic work may be performed where temporary melting occurs. A broad, seasonally moving altitudinal belt having frequent, diurnal freeze–thaw cycles is continuously present in the Karakoram (Figure 7.5), although its full impact through cryogenic processes is felt only where it migrates into altitudinal bands which have relatively abundant moisture. In most areas this involves elevations which are > 4000 m and < ~6000 m (Hewitt, 1968b).

The most important factor affecting moisture supply, however, is the way in which the seasonal interaction of thermal and precipitation regimes determines the timing and intensity of the downslope moisture stream. Almost the entire landscape above 5500 m is subject to all-season avalanching or is part of the glacier accumulation zones or is both. A large fraction of all of the snow supply of the region, including that which nourishes glaciers, is avalanched through some hundreds or thousands of m of altitude. The ice stream of the Barpu–Sumaiyar Bar glacier does not extend continuously into the perennial ice zone, the bulk of its nourishment coming from year-round massive avalanching from the watershed of Gannish Chissh ('Golden Peak', 7027 m) and Malubiting (7458 m). It is a typical 'Turkestan' or avalanche-fed glacier (von Klebelsberg, 1925–6). Much of the nourishment of the Miar tributary and Bualtar is also provided by avalanches, but these fall on relatively long and narrow firn basins of the 'firn kettle' or incised reservoir type (Visser and Visser-Hooft, 1935–8; von Wissmann, 1959). The Biafo is unusual in the vast extent of its high-altitude accumulation basins. However, it is comparable to some of the largest Karakoram glaciers, such as the Chiantar, Chogo Lungma, Siachen and Rimo. These are also nourished mainly by direct snowfall on to large, open accumulation basins. They are classified as 'firn-stream' or 'Alpine' in type.

Transportation of moisture from high, cold areas to intermediate altitudes by avalanching and glacier flow provides a huge increase in available moisture at

3500–5000 m altitude. Here, precipitation is moderate, ranging from ~ 250 mm to 800 mm at the higher altitudes. The available heat in the melt season is capable of ablating much more, and the avalanched snow and glacier ice are the larger sources of runoff over much of this zone.

Excluding areas of thick debris cover, annual ablation of ice at 3500–4500 m altitude on the Biafo, Barpu and Bualtar glaciers is of the order of 3000–7000 mm water equivalent (Hewitt, 1986, 1987). The geomorphic impact of this moisture is further enhanced because 75 per cent or more is released in periods of intense melting over 6–10 weeks of the summer. Moreover, it is liberated from only 30–40 per cent of the glacier area and from as little as 20–30 per cent of total basin area, further concentrating its effects. In the glacier ablation zones there is also a progressive downglacier increase in the amount and thickness of the debris cover. On some glaciers such as the Biafo or Baltoro, the lowermost few km or tens of km may be completely masked by debris. This has large and cumulative downglacier effects upon melting, ice-margin conditions and the development of thermokarst forms.

The melting of snow on valley-side slopes is also highly concentrated in time, although most of the snow cover tends to disappear well before the main ice-melt at a given elevation. The downslope redistribution of snow by avalanching and its concentration in warmer areas is again important. Snowmelt from avalanche deposits in gullies and run-out zones tends to lag well behind that of surrounding snow packs. It is also more concentrated and less moisture is lost by evaporation. Runoff from avalanche deposits is thus of great significance for geomorphic processes, occurring after most of the areal snow cover has disappeared. It is then that the thermal conditions promote the most rapid melting. Avalanche snow and the meltwater derived from it are the main sources for the widespread Karakoram debris- and mud-flow activity, and for a majority of those smaller streams that persist after the main snowmelt has passed.

Cryogenic processes are at their most active in the vicinity of the transient snowline, where a moisture supply and bare ground coincide with the zone of frequent frost cycles. Below this zone, the intensity or duration of active erosion and deposition generally becomes progressively more confined along drainage axes. Elsewhere, drought conditions prevail. Even along drainage lines processes tend to become increasingly moisture-starved and debris-rich in a downslope direction. The altitudinal range of the drainage lines also plays an important role. Abundant avalanche, gelifluction, meltwater stream and lacustrine features occur above 4000 m altitude. Debris- and mud-flow deposits, earth pillars, other 'badland' forms, salt weathering and aeolian features predominate below that level.

The valleys below ~ 3000 m altitude not only receive low precipitation but also are subject to extended drought periods, desiccating valley winds and heavy grazing (Schweinfurth, 1956). A few wasted glacier termini descend into this zone. The Minapin glacier briefly reached as low as 2150 m altitude during the past hundred years, the Pisan glacier possibly descending a little lower (Goudie et al.,

1984), but most glaciers terminate above 2800 m. Thus, nearly all of this lower zone has a large moisture deficit (Subrahmanyam, 1956; Whiteman, 1985). Very little runoff derives from the zone, and then only briefly in early spring or during exceptional summer rainstorms. Apart from aeolian action, activity here is even more completely dependent upon snowmelt streams, glacier-derived river flow, and mass movements from higher altitudes.

The greater part of stream flow in the trans-Himalayan upper Indus and Yarkand river systems is derived from glacier basins (Haserodt, 1984; Hewitt, 1985). Most of the main tributaries emerge directly as large rivers from the snouts of some 30–40 of the largest ice masses. Thus the 'fluvial zone' is mainly a zone of transport for moisture and debris from higher altitudes. An important role played by moisture flowing through this zone from above is the reworking and removal of Pleistocene and Holocene lag deposits. Much of the activity here is essentially what Church and Ryder (1972) defined as 'paraglacial' (Li et al., 1984). Long sections of the river gorges, however, are armoured with large boulders and the intermittently cross-bedrock sections. This and the aridity of surrounding slopes and river terraces no doubt reduce the pace of reworking and removal of the abundant lag deposits present in the zone. It also means that their entrainment tends to be episodic in time and space. Catastrophic events such as dam-burst floods play a major role (Hewitt, 1968a, 1982).

Regional erosion rates and sediment supply are complicated by the entrainment of lag deposits in the paraglacial zone (R.I. Ferguson, 1984). They depend, however, largely upon erosion in the higher more humid glacial and alpine zones. The downslope throughflow of moisture and eroded materials determines the overall scale and tempo of erosion. Rates of regional denudation, according to sediment yield data on the major streams, are ~ 1 mm yr$^{-1}$ for the trans-Himalayan, upper Indus basin, and 5–6 mm yr$^{-1}$ for Hunza (Hewitt, 1968a; Goudie et al., 1984).

In sum, the role of surface hydrological processes as agents which redistribute snow, ice and meltwaters downslope is of overriding importance for erosional processes. In most areas it is a far greater influence upon local moisture availability and its geomorphic impact than is direct precipitation.

## ALTITUDINAL ZONING OF PROCESSES AND LOCAL LANDFORMS

Geomorphology is concerned equally with the erosional system as a whole and the sculpting of landforms by locally operating processes. In the Karakoram, the topoclimatic conditions described above also result in particular sets of processes and attendant landforms or deposits being concentrated in altitudinal bands. Their local extent and limits are highly variable and there is no lack of topoclimatic or lithologically determined 'anomalies'. The altitudinal ranges over which particular features are concentrated can vary dramatically with aspect, as will be shown below for the Barpu–Bualtar basin. And, as in all such spatial patterns, it is open to debate whether the phenomena are best described in terms of distinct bands or zones of transition. Nevertheless, a sense of the range and distribution of these local conditions and features is essential to an interpretation of the geomorphology. Four zones are identified below, which summarize conditions in the central Karakoram (Figure 7.3).

### Zone I ( > 5500 m altitude)

The perennial ice belt, with frigid, humid conditions. Cloudiness and snowfall are heavy throughout the year, but especially in winter. Glacionival forms dominate, especially glacier accumulation zones and all-season avalanching on slopes.

### Zone II ( ~ 4000–5500 m altitude)

A cold but seasonally warm, humid belt. It includes the upper and middle ablation zones of glaciers, and a mixture of rockwall, alpine meadow, and tundra conditions. A heavy winter snowfall lies on the ground for 8–11 months. In summer occur a short vigorous melt season, occasional and sometimes large snowfalls, and 2–12 weeks of diurnal freeze–thaw cycles.

### Zone III ( ~ 3000–4000 m altitude)

A cool subhumid belt with warm summers. A moderate winter snowfall lies on the ground for 3–8 months. Several weeks of diurnal freeze–thaw cycles occur in spring and autumn, and drought in summer. The lower ablation zones of glaciers are included, mostly with thick debris covers. Conditions on valley-side slopes are montane. Gentler slopes support a mixture of forest patches or copses and meadows, where not subjected to human activity which has cleared most of the forest area. Heavy summer grazing has helped to give much of the area a dry steppe character. Extensive areas of steep slopes have important seasonal rockwall processes.

### Zone IV ( < 3000 m altitude)

A submontane, cool steppe belt of semi-arid conditions. In winter occurs light ephemeral snowfall and 10–12 weeks of diurnal freeze–thaw cycles. Summers are hot, with extended droughts intensified by katabatic winds and low cloudiness. Except for artificially irrigated settlement areas, gentler slopes are subject to heavy grazing which leaves much bare ground. Most areas are covered by old glacial, glaciofluvial, and slope deposits, and veneers of more recent deposits reflecting moisture-deficient processes. This paraglacial zone involves the reworking and entrainment of sediments by moisture from higher elevations. Aeolian action is widely in evidence too, with sporadic dune fields and frequent dust storms. Other desert features include tafoni, rock varnish and patterned talus forms.

The extent and geomorphic significance of these zones depends heavily upon aspect and local relief. Zone IV conditions can extend ~ 1000 m higher on southern exposures. On steep northern slopes of large altitude range, Zone III conditions can extend below 3000 m.

Because of the importance of the downslope moisture stream, the most characteristic expressions of these belts occur where the tops and bases of slopes are within the given zone. Below ridges and spurs that do not reach much above 3500 m, the desert-like forms of Zone IV achieve their fullest expression. An important but neglected aspect of Karakoram geomorphology concerns those many ridges and lesser peaks that culminate at heights of 4000–7000 m altitude (Figure 7.3). On more rounded summits at the lower elevations in Zone II, well developed cryogenic forms occur. Block fields, stone polygons and gelifluction lobes abound. They rarely occur even on gentler areas below the summits, presumably being overwhelmed by the movement of debris and moisture from above. The great screes of the region also originate mostly from watersheds at the upper limit of Zone II. Most of the rock glaciers of the central Karakoram occur along ridges and peaks at ~ 4500–5500 m. Peaks that reach only ~ 1000 m into Zone I support cirque glaciers and other small ice masses. Valley glaciers such as Biafo and Barpu–Bualtar are derived from higher watersheds.

The enormous area of rockwall and other bedrock forms at all elevations must be noted. Average slope, however, as well as moisture supply increase with altitude so that the relative area of bedrock exposure increases upward. In the highest glacionival zone this is masked by perennial accumulations of snow and ice, even on very steep faces. Here avalanching appears predominant. Rockwall conditions themselves have their most significant expression in Zone II, where the highest frequency of rockfalls and rockslides, and their most extensive deposits, occur (Hewitt, 1968a).

The configuration of the base of slopes or the run-out zones of channelized flows are important in determining the locally developed landforms and deposits. The zone in which a major break of slope, valley floor or glacier margin occurs will determine the kinds of deposits that develop at lower elevations.

## ALTITUDINAL CONTROL OF DEPOSITIONAL ENVIRONMENTS AND FORMS

The occurrence and forms of superficial deposits depend equally upon the altitudinal zoning of geomorphic processes. Zone I is essentially a landscape of bedrock, snow and ice forms. Surface deposits of wind-blown dust, fresh rockfall debris and dirty avalanches are short lived. They are associated with short periods of clear weather in summer and with southerly exposures. By contrast, most of Zone IV is blanketed in superficial deposits, but they are dominated by the legacy of Pleistocene and Holocene events, especially glacial episodes and the last

*Plate 7.1* View down Barpu glacier from the ridge between Miar and Sumaiyar Bar tributaries. It shows the continuous ablation-zone, depositional landform feature on the right flank, and discontinuous one on the left. (Photograph: Hewitt, 29.7.87)

major ice recession. Important exceptions are floodplain deposits of the major rivers, and cones or fans of debris being built where chutes and gullies descend from higher, moister zones.

Once more, the intermediate zones at ~3000–5500 m have vigorous contemporary processes associated with active building of depositional forms. Superficial materials reworked and patterned by cryogenic processes occur on the rare, flat-topped ridges and spurs. Where the rockwalls descend to them, talus, avalanche and other deposits of rapid mass wasting spill across them. These are a small part of the landscape, however, the deposits patchy and thin. In the heavily glacierized central Karakoram, the most distinctive and substantial depositional forms occur along the margins of the glaciers in their ablation zones (Plate 7.1). They consist of an interfingering of ice-margin, kame-terrace, mass-wasting, avalanche, snowmelt, stream and lacustrine deposits.

The purpose here is to identify the role of altitude in controlling these deposits. This is everywhere a transitional environment between the glacier and valley walls. The glaciers act to control the elevation of the depositional wedge that forms here. Secular fluctuations in ice level result in an alteration of ice-margin deposition and erosion. When the ice is higher than the depositional surface between it and the valley wall, lateral moraines are built. Behind the moraines, glaciofluvial and glaciolacustrine deposits are settled in a typical kame-terrace environment. When the ice thins below this level, as in most areas over the last several decades, the deposits at the ice margin tend to erode by gullying, slumping and wind action.

The limited attention given to these features in the past identifies them mainly with glacier activity, but deposition derived from off-ice subaerial processes plays a large role. In many sections it dominates the forms and mass of the deposits. Commonly a broad terrace or depression occurs between the ice margin or a lateral morainic ridge and the valley-side slopes. This feature may stretch unbroken between spurs or at the entrance of tributaries for a km or more. It will generally reach hundreds of m wide. In the depressions or across the terrace, distinctive forms of deposit occur, governed by slope processes and snowmelt streams. The terrace also provides a rare, protected environment of gentle slopes and moist microclimates. The richest vegetation communities in the region develop here and meltwater runoff supports ephemeral streams and lakes. The water may come from the glacier when ice levels are high. In recent decades it has been supplied by snowmelt and avalanching on the surrounding slopes. Evidently, moisture and debris come from the valley slopes regardless of ice levels. Moreover, we observe today how, during episodes of glacier thinning, they continue to build or aggrade the terrace along most of its length, despite erosion at the ice margins.

While research has largely neglected the size and extent of depositional features along the margins of Karakoram glaciers, scientists and travellers to the region have long remarked upon them. The term 'ablation valley' (*Ablations-schlucht*) was first applied to their surface expression along the Chogo Lungma

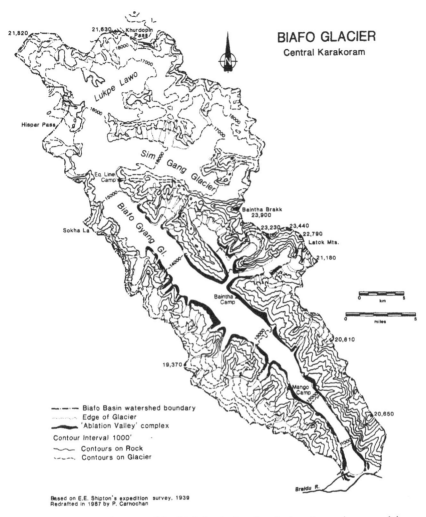

*Figure 7.6* Topographic map of the Biafo basin showing the location and extent of the depositional landform complex along the ablation-zone margins.

glacier by Oestreich (1906). Visser and Visser-Hooft (1935–8) thought that these deposits originated in strong ice-margin ablation during high ice levels. They thought that this would produce a depression which gradually filled with debris, coming predominantly from the glacier. Without doubt, most of the deposition here depends upon the melting of either ice or snow, but there are some serious questions about the proposed origin of these forms and the appropriateness of the term 'ablation valley'. Geomorphologists now seem to prefer to abandon it. Nevertheless, in the literature of the past, the depositional forms to be described

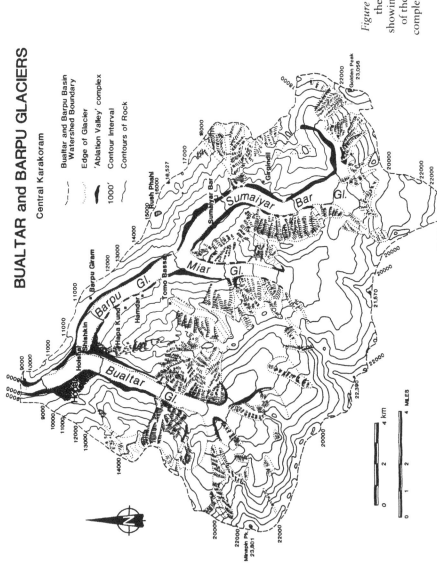

# BUALTAR and BARPU GLACIERS

## Central Karakoram

Bualtar and Barpu Basin
Watershed Boundary

Edge of Glacier

'Ablation Valley' complex

1000'  Contour Interval

Contours of Rock

*Figure 7.7* Topographic map of the Barpu–Bualtar basin showing the location and extent of the depositional landform complex along the ablation-zone margins.

*Plate* 7.2 The glacier margins in Zone III from above the terminus of Barpu glacier. The enormous lateral moraine of the glacier's right flank is evident, as is minimal deposition in the valley-side depression. This plate shows the same area as Plate 7.1, which was taken from the peak in the centre, background. Note that the ice, though 20–35 m below the moraine crest, is higher than the floor of the valley-side depression. (Photograph: Hewitt, 15.5.87)

here are usually identified with the so-called ablation valleys (Mason, 1930). Only their dependence upon altitudinally varying conditions is considered, leaving the complex questions of their origin, scale of development and sedimentology for another occasion.

Along the margins of the Biafo and the Barpu–Bualtar glaciers, a well developed wedge of deposits is present nearly everywhere in Zones II and III. It extends almost continuously through 2000 m or more of elevation, over distances of 10–40 km, and on both flanks of the glaciers (Figures 7.6, 7.7). Although there are many complicated local variations, the general character of these deposits reflects the altitudinal zoning of processes described above and the elevation range of the slopes descending to them.

In the lower sections, near the glacier termini in Zone III, the enormous debris load of the glaciers on the one hand, and moisture-starved slopes on the other, give glacial deposits the larger roles in most areas (Plate 7.2). The lateral moraines, which may rise 30–50 m above the valley-side depression, are generally the dominant forms. Moreover, considerable glacier thinning in the last ~ 100 yr has left a steep cliff of deposits of at least that height above the glacier. Altitudinally varying conditions seem important to resistance to erosion. The processes of lodgement are influential in producing a compact, well indurated mass. That, however, also depends upon the great abundance of debris and much higher proportions of fines in morainic materials of the lower areas of glaciers. Also, the aridity of the climate at this elevation causes rapid surface evaporation, which tends to prevent finer material being washed away. This leads to salt build-up which aids in cementing the deposits (Owen and Derbyshire, 1988). Along the lower 10 km of Bualtar glacier, however, downcutting by the glacier and the lowering of the Hispar river have exposed ~ 300–600 m of the old lateral deposits (Figure 7.7). They are now subject to large-scale, multiple landsliding. It represents a type of 'paraglacial transition' extending into the zone of active glaciation.

Slopes in Zone III are starved of moisture, so that sedimentary delivery at the base of them is proportionally restricted. Unless the valley walls descend from much higher elevations, contemporary deposits are confined to minor spring mudflows and slope wash, patterned talus and minor rockfall debris (Hewitt, 1986). Without doubt, this helps to explain the deep troughs beside the glaciers that, at higher elevations, tend to be filled with > 10 m of slope and kame deposits.

Climate change and secular glacier fluctuations strongly affect the state of these deposits as well. The high lateral moraines seem to reflect activity in the Little Ice Age and perhaps earlier high ice levels. In the depressions and at the base of valley-side slopes is abundant evidence of 'fossil' or degraded deposits of past wetter conditions of more vigorous cryogenic processes, or greater moisture supply at higher ice levels. They include dry-lake deposits, dried-up stream channels, old rockslides, and avalanche boulder tongues.

As a general rule, deposits derived from valley-wall processes, or those that

*Plate 7.3* The ice-margin depression and base of slope deposits at ~ 3500 m altitude along the Barpu glacier. The prominent lateral moraine runs across the centre of the view with the debris-covered Sumaiyar Bar tributary and clear ice of Mair glacier beyond. Note the dense willow woods in the depression behind the moraines, and the ephemeral lake at Phai Phahi. A debris-flow channel and levees descend to the fan in the foreground. (Photograph: Hewitt, 14.5.87)

177

*Plate 7.4* View across the ice-margin depression and base of slope deposits at ~ 4085 m altitude on the left flank of Biafo glacier. It shows a large debris-flow fan, depositional flats and small ponds behind the lateral moraines. Ice-margin processes are complicated by the inflow of Biantha tributary immediately up glacier, and some of the ice is stagnant with thermokarst ponds. (Photograph: Hewitt, 21.7.85)

*Plate 7.5* The depositional apron in altitude Zone II (5000–5300 m altitude) along the margins of 'Golden Peak' tributary, Barpu glacier. Note the remains of avalanches and debris movement from the south-facing valley wall on the right. The ridge above the debris-covered glacier is partly lateral moraines, partly slope-failure debris (see text), and partly avalanche deposited. About 5 km of the ablation valley is shown. The permanent avalanche cones descending to the glacier's left flank and nourishing it occur in the background. They descend from slopes whose relief is > 2000 m and mostly in Zone I. (Photograph: Hewitt, 7.8.87)

form in the depressions and on the terraces between the lateral moraines and valley slopes, become progressively more important with altitude. They may be predominant in the surface area of depositional forms in Zone II and in the upper part of Zone II. Vegetation assumes a special importance, with the communities on the terrace or in the depressions commonly being lush and varied. In terms of deposition this helps to trap and bind sediment. This is evident in the rich vegetation communities at 3500–4200 m altitude along the Biafo and Barpu–Bualtar glaciers. Tall thickets of birch, willow, ash and junipers exist here. They may extend up the flanks of the lateral moraines and over the lower slopes and depositional cones along the valley side. They undergo flooding with water and debris during major episodes of melting. Occasional debris flows and floods carry huge quantities of bouldery sediment into these thickets, where they are trapped (Plate 7.3).

At progressively higher altitudes the quantities of morainic debris delivered by the glacier at its margins tend to diminish, as does the proportion of finer and reworked materials. On the other hand, slopes tend to become steeper, the avalanches and meltwater draining from them more abundant, including progressively greater quantities descending from the perennial snow and ice of Zone I. The glacier and its lateral moraines serve to check and confine much of this surface movement of moisture and debris along the valley sides. Only the larger avalanches or meltwater streams manage to reach the glacier. At ~ 3500–5000 m altitude, the depositional complex mainly reflects abundant avalanching, rockfalls, rockslides, debris flows and summer meltwater movement. Ponding of water in depressions is commonplace. The depressions are commonly formed partly by avalanches, whose thickest deposits of snow lie in them through much of the summer, their meltwater forming ephemeral ponds and lakes. Where avalanches do not reach the terrace, their meltwater tends to carry large amounts of debris from steep chutes or canyons, including rare large debris flows. These form conspicuous depositional cones across the terraces (Plate 7.4). The debris-flow deposits may also check or dam water and materials channelled down the depressions behind the lateral moraines.

In the upper part of Zone II, slope processes may completely dominate the formation of the depositional wedge along the glacier (Plate 7.5). An area of the upper Barpu glacier at 4500–5200 m altitude, for example, is swept by and buried under avalanched snow for 10–11 months. In the brief but vigorous summer melt, frequent rockfalls sweep across the compact avalanche snow deposits or come to rest on them. Slushers and debris flows move over them or around their edges. Many of the slope-process events have the momentum to reach the glacier. Slope processes tend to inhibit lateral moraine development and it is difficult to distinguish moraines from avalanche and rockfall-rampart deposits that merge with them. In late July and August, however, large quantities of meltwater are delivered from the slopes and melting avalanche deposits. These rework the surface of the terrace, tending to create a broad depression covered with braided channels.

The amounts of meltwater liberated at these altitudes are too great to be ponded and they flow to the glacier. In the Barpu example, however, the stream channels are confined to the terrace surface or behind avalanche ramparts and lateral moraines over a distance of 3.5 km before cutting through to the glacier.

The area is also one of several sections where a substantial part of the depositional mass consists of remnants of one or more catastrophic rock slides. They were of the type sometimes referred to as 'rock avalanches' or 'sturzstroms' (Hsu, 1975; Eisbacher and Clague, 1984). Exposed sections of the landslide deposited here are 5–35 m thick and 150–200 m wide. They occur over 7.5 km along the glacier. They compose ~ 0.75 km$^2$ of the surface of the depositional terrace, with the exposed parts consisting of boulders 5–45 m in diameter. This is easily mistaken for moraine. While it occurs along the north flank of the glacier, it was carried there by a prehistoric landslide derived from an area near the summit of Gannish Chissh 'Golden Peak' (7027 m) on the southern flank. The total volume would have been $> 100 \times 10^6 m^3$ and must have completely buried the glacier at that time. To judge from a series of similar but smaller rockslides that buried 4.2 km$^2$ of Bualtar glacier in 1986, the massive slope failure of Gannish Chissh probably caused the glacier to surge (Hewitt, 1988). This is not just an extreme and atypical observation. In five different areas, catastrophic landslide deposits form glacier margin deposits of comparable thickness, though of lesser extent, along the Barpu–Bualtar glaciers. Lateral moraines for several km down glacier from them have lithological characteristics showing that they derive from the reworking of landslides. In general, the evidence from Zone II is that most of the marginal 'moraine' derives from rockfalls, rockslides and avalanched debris.

The upper Barpu example also illustrates that, in Zone II especially, aspect comes to have a decisive influence upon the processes that occur. The above description applies to only the south-facing flank of the glacier. No depositional wedge and almost no deposits occur on the opposite flank (Plate 7.5). Such is the case over the uppermost 10 km and through ~ 1000 m of altitude change, where deposits do form at the base of the southerly slope. Instead, massive avalanching in all months of the year from the $> 2500$ m wall of the Gannish Chissh interfluve sustains permanent large cones of snow. In fact, this area is entirely below the regional snowline and firn limits on glaciers and is nourished by the coalescing of these cones of avalanched snow along the left side of the glacier.

The general differentiation of the altitudinal range of processes and deposits as a function of aspect should be stressed again here, especially in Zones II and III. Permanent snow cover reaches as far as 1000 m lower in elevation on the north flanks of the steepest, east–west trending valley walls in the central Karakoram, compared to south-facing slopes. On the Biafo and Barpu–Bualtar, avalanches and summer meltwater streams that cut through to the glacier are more numerous and occur at much lower elevations on the north-facing flanks. The lowest reaching contemporary avalanches observed on the Barpu glacier were 4200 m on the south flanks but 3100 m on the north. Streams that cut through to the glacier occurred 800 m lower down on the north flank. That does not reflect greater elevation of the latter in the middle and lower glacier areas. In general, northerly slopes also remain humid and better vegetated to lower altitudes, although that may be masked by the extent of grazing and the clearance of trees and bushes. Below ~ 3200 m altitude in the areas surveyed, aspect differences are less pronounced because the system of desiccating katabatic valley winds tends to dry out the lower parts of the northerly slopes as well.

## CONCLUSION

The main concern of this paper has been to draw attention to the dominant role of altitudinally varying conditions in the broad patterns and contemporary processes of Karakoram geomorphology. This also has, however, a bearing upon interpretations of the historical development of the landscape during the Pleistocene and Holocene, a subject that creates great interest among earth scientists. On the one hand, most of the evidence used to reconstruct past developments comes from lower elevations, essentially termed Zone IV here. On the other, it is generally thought that the dominant pattern of change in earth-surface conditions and processes, at least since the mid-Pleistocene, has been expanding and contracting glaciation accompanied by vertical shifts in snowlines. The suites of old moraines and the enormous river terraces cut into old glacial, glaciofluvial and glaciolacustrine sediments in Zone IV obviously record dramatic changes in earth-surface processes at lower elevations. But they, in turn, reflect changes in past as well as present climatic–geomorphic conditions at higher elevations. These have usually been assumed to involve altitudinal shifts in temperature and precipitation zones, and the area of perennial ice climate (Godwin-Austin, 1864; Dainelli, 1924–35; de Terra and Paterson, 1939; von Wissmann, 1959; Li et al., 1984). It seems reasonable to suppose that the climatic–geomorphic zones recognized today have shifted vertically in concert with the glaciers and perennial snow cover. This is certainly complicated by the probable high rate of uplift of the entire region since the late Tertiary, and large-scale orographic modification of weather systems, notably the penetration of the monsoon (Mayewski et al., 1980). The zones identified here cannot be expected to shift vertically with no modification in character or altitudinal width. Broadly, however, the lag deposits occurring along valley sides, from times of thicker and more extensive glacier systems, seem likely to reflect processes and depositional environments found today only at higher elevations.

The best evidence of this also occurs in contemporary Zones II and III. First, we have noted that the deposits occurring in Zone III, at least, include some that record secular climatic changes at the given altitude. Specifically they include deposits of processes now observed only at higher elevations. This suggests that the character of deposition will be controlled by altitudinal changes in climate. Second, in many places in Zone III, and above the present glacier surfaces, ancient deposits occur, commonly in the form of a depositional 'bench', and areas of earth pillars and 'hoodoos' (Hewitt, 1968a). The materials and their locations record ice levels hundreds and sometimes > 1000 m higher than today's. To date, these higher altitude deposits have not been studied sufficiently by earth scientists, although Shroder *et al.* (1989) initiated work on such features in the lower valleys.

In conclusion, this paper has sought to show how large altitudinal range and gradients of conditions serve as decisive controls upon earth-surface processes in the Karakoram Himalaya. These earth-surface processes reflect especially the relation between gradients in both moisture supply and temperature conditions with altitude. It must be stressed that the central Karakoram is characterized by a predominance of steep slopes and enormous local relief. As a result, while locally operating processes reflect altitude, more or less rapid and large-scale movements of moisture and materials occur between altitude zones. Linked sequences of processes between zones of very different climate assume major importance. That is obvious in the glacier entrainment of debris from altitudes > 7000 m down to < 3000 m. It is also commonplace, however, for off-glacier processes. In many areas, for example, the avalanching of snow from high altitudes in Zones I and II primes debris- and mud-flow processes in gullies and on slopes in Zones III or IV. In this way, extensive gully and chute systems, linking two or more of the altitudinal zones identified above, give rise to seasonally as well as altitudinally phased sequences of geomorphic processes, within the general downslope streaming of moisture and debris (Hewitt, 1968a).

The geomorphic features and events most commonly studied by earth scientists are those of Zone IV, or what historically has been identified with the 'semi-arid Himalaya' (Goudie *et al.* 1984). Here occur the debris- and mudflows and their depositional fans so often reported (Bonney, 1902; Rabot, 1905; Rickmers, 1913; Hewitt, 1988). Here is also the enormous and richest legacy of Pleistocene and Holocene deposition. Although some dune fields occur, however, few of these past or contemporary deposits derive from arid-land geomorphic processes. Rather they derive from the large-scale downslope movements of moisture from higher, humid altitudes and the debris entrained, which is subsequently remobilized by local strong winds.

# 8

# SEDIMENT TRANSPORT AND YIELD AT THE RAIKOT GLACIER, NANGA PARBAT, PUNJAB HIMALAYA

*James S. Gardner and Norman K. Jones*

## ABSTRACT

The Raikot (Rakhiot) glacier is one of the largest of the 69 separate glacier systems on the north slope of the Nanga Parbat massif (8125 m). Although this glacier apparently once descended into the Indus river, its present terminus is 15 km from and 2000 m above the valley. Raikot glacier is 14 km long, with its highest point at ~ 7700 m and its terminus at 3150 m. Glacier mass is fed by frequent snow and ice avalanches and high snowfall. The equilibrium-line altitude (ELA) is ~ 4800 m with an uncharacteristically low accumulation/ablation area ratio of ~ 0.60, perhaps caused by the high rates of accumulation and rapid transfer of large volumes of ice ( ~ 70 × 10^6 m^3 yr^{-1}) through the ELA to maintain the extensive ablation zone.

Raikot glacier is nested within prominent and well wooded lateral moraines with advanced pedogenesis which, together with tree-ring dates, suggest a Neoglacial age, probably Little Ice Age. Since 1934, total terminus recession has been 250 m (4.9 m yr^{-1}), but there was rapid recession of 450 m during 1934–54; an advance of at least 200 m from mid-1960s to early 1980s; and a recession of 5 m yr^{-1} in the mid-1980s. In 1985 and 1986, re-establishment of old stations allowed determination of velocities of 18–83 cm d^{-1} 2 km from the terminus; 21–47 cm d^{-1} 1 km from the terminus; and 7–26 cm d^{-1} near the terminus. Recovery of old mountain climbing gear shows an average annual downglacier velocity of 350 m yr^{-1}, or a mass transport time of ~ 40 yr from accumulation area to terminus.

A total sediment load of 4.45 × 10^6 m^3 was calculated by mapping and measuring variable distribution and thickness of surface debris and sediment concentrations within the ice. Measured ablation values of water equivalent are 53.1–86.9 × 10^6 m^3. Adding in a reasonably estimated 10.4 × 10^6 m^3 of basal water loss gives a mean annual discharge of ~ 2.1–3.1 m^3 s^{-1}. Measured mean sediment concentrations were an extremely high 5270 mg l^{-1}. Net denudation rates thus work out to 4.6–6.9 mm yr^{-1} for the glacier area and 1.4–2.1 mm yr^{-1} for the basin area.

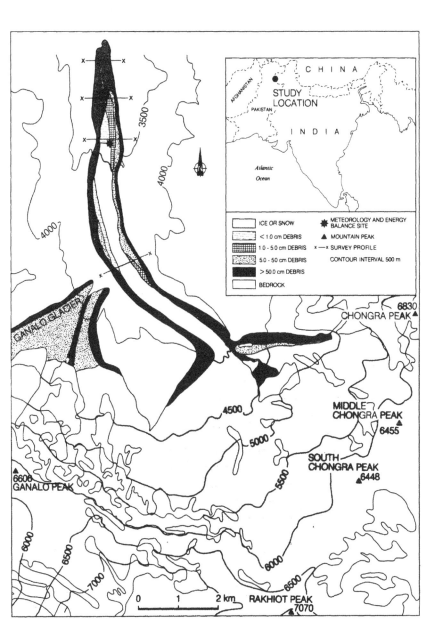

*Figure 8.1* The study area location and the Raikot (Rakhiot) glacier.

# INTRODUCTION

The Nanga Parbat (Diamir) massif is a significant topographic feature in the upper Indus basin of northern Pakistan. The main summit rises to 8125 m, while a number of subsidiary peaks, such as Chongra, Raikot (Rakhiot), Ganalo and Mazeno all rise to 7000 m (Figure 8.1). The massif is one of the most heavily glacierized in the western Himalaya. Being topographically isolated from other major mountain groups, it is influenced by both westerly and monsoonal meteorological effects. The resulting year-round snowfall above 4000 m gives rise to 69 separate glacier systems covering 302 km$^2$ and having an estimated volume of 25 km$^3$ (Kick, 1980). The Raikot (Rakhiot) glacier, on the north slope, is one of the larger glaciers on Nanga Parbat (Plate 8.1).

This paper describes the morphology, dynamics and sediment transport characteristics of the Raikot glacier. Shroder *et al.* (1989) have indicated that the Raikot glacier and the neighbouring Buldar and Patro glaciers together played significant roles in the ice-damming of and the provision of sediments to the Indus river and its valley in the late Pleistocene. This paper focuses on the modern Raikot, the terminus of which is situated 15 km from, and 2000 m above, the Indus. It may be reasonably typical of high speed, erosive western Himalayan glaciers which are driven by high accumulation rates, steep topographic gradients and high rates of water loss in their ablation areas, and which yield high concentrations of sediment in their meltwaters. Such glaciers are thought to be active agents and linkages in the regional denudation system.

The Nanga Parbat region shares a long tradition of scientific observation with the rest of the western Himalaya and Karakoram (Hewitt, 1989a). Owing to its topographic dominance adjacent to a favoured route from southern Kashmir to the Karakoram, Nanga Parbat attracted the attention of many early travellers such as Vigne and H.A. von Schlagintweit. Topographic and geodetic surveys were carried out in the area early in the twentieth century (Kick, 1975). Beginning in 1934, a number of the Nanga Parbat glaciers, including the Raikot, were surveyed and resurveyed. The 1934 German Nanga Parbat expedition attempted to climb the mountain via the Raikot glacier. A scientific team was included in the expedition and it produced a detailed 1:50,000 topographic map of Nanga Parbat. Its glaciological surveys provide the basis for all subsequent glaciological research in the region (Finsterwalder, 1935, 1937). Subsequent observations at the Raikot glacier were made by Troll (1938), Pillewizer (1956) and Gardner (1986). The observations reported in this paper are based on research carried out in a ten-day period in June 1985 and a two-month period in June, July and August 1986 in association with the joint Pakistan–Canada Snow and Ice Hydrology Project.

# GLACIER MORPHOLOGY

The Raikot glacier is a complex, high-gradient glacier which, together with the Diamir, Patro and Buldar glaciers, drains the steep north slopes of Nanga Parbat.

*Plate 8.1* The terminus and lower portion of the Raikot glacier in July 1986 with the north face of Nanga Parbat in the background. The photograph was taken from Finsterwalder's Fairy Meadow Station 1B. (Photograph: Gardner, 1986)

Frequent snow and ice avalanches and periods of extremely high rates of snowfall (Wien, 1936) make the Raikot glacier among the most infamous in the world, having claimed over 30 lives in 50 years. The glacier is 14 km in length over an altitude range of ~ 4600 m, with the terminus at an elevation of 3150 m and the highest point at ~ 7700 m on the summit plateau of Nanga Parbat. In long profile (Figure 8.2), the glacier is composed of three morphological units: above 5500 m is the Raikot firn, a basin surrounded by very steep slopes which rise to the summit areas of Chongra, Raikot and Nanga Parbat peaks (Figure 8.1); between 4100 m and 5500 m is a high-gradient icefall; and from 4100 m to the terminus is a relatively low-gradient glacier tongue (Plate 8.1). Glacier surface morphology and characteristics reflect this zonation. The icefall and upper tongue are extremely broken by crevassing into large ice blocks (Plate 8.2). The

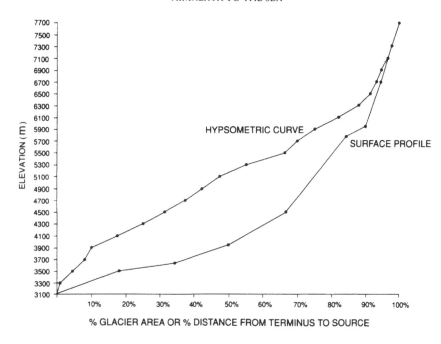

*Figure 8.2* The longitudinal surface profile of the Raikot glacier and the hypsometric (area–altitude) distribution for the glacier-covered area.

tongue is progressively less broken and more debris-covered toward the terminus (Figure 8.1; Plate 8.1).

The total connected glacier area of the Raikot is about 32 km². However, numerous unconnected hanging glaciers, such as those on the north face of Nanga Parbat (Plate 8.1) and small proximous glaciers (Figure 8.1) increase the total glacierized area of the Raikot basin to 56 km² or about 60 per cent of the total basin area above the terminus of the Raikot glacier. A number of these apparently unconnected glaciers contribute mass to the Raikot glacier through large and frequent snow and ice avalanches (average of 8 d⁻¹ over 8 days in 1985). This adds significantly to the functional accumulation area of the Raikot glacier. In addition, avalanches add mass and alter the surface albedo of parts of the ablation area at the base of the north face of Nanga Parbat (Buhl, 1986).

The altitudinal distribution of connected and unconnected glacier ice in the Raikot system is portrayed in Figure 8.2. The equilibrium-line altitude is at ~ 4800 m, giving an ablation/accumulation area ratio of about 0.60, somewhat lower than the usual (0.70). Implications of a low AAR, among others, are that high rates of accumulation and rapid transfer of large volumes of ice across the equilibrium line (70 × 10⁶ m³ yr⁻¹ w.e.) are able to sustain a relatively large ablation zone (Finsterwalder, 1937). This rapid transfer of glacier mass together

*Plate 8.2*  The upper ablation zone, icefalls and part of the accumulation area of the Raikot glacier. Shown is the heavy surface debris load on the glacier's left margin, derived largely from the remobilization of lateral moraine material. The upper accumulation area and the Raikot firn are obscured by cloud. (Photograph: Gardner, 1986)

with the avalanching activity may be significant factors in the sediment transport characteristics of the Raikot glacier.

## GLACIER DYNAMICS

Mass balance considerations and topographic conditions provide the context for glacier dynamics at several time scales: the immediate or contemporary, the Neoglacial; the Holocene; and the Pleistocene. Here we consider the first and part of the second time scales. Morainic and other deposits in the Indus valley in the vicinity of Raikot Bridge and derived from the Raikot glacier are described by Shroder *et al.* (1989) and have been tied to the period of the 'last glaciation' of the Indus valley (max. 140,000 yr BP). Up the Raikot valley from the Indus, in the vicinity of Tato and Fairy Meadow, are limited remnants of large lateral moraines probably associated with these late Pleistocene glaciations. The present Raikot glacier is nested within lateral moraines which are well wooded and show advanced pedogenesis (Plate 8.1). Tree-ring dating of the older trees on these moraines give a maximum age of 500 ± 50 yr. It is suggested that these moraines define the maximum advance during the Holocene. Their small downvalley extent from the existing glacier (1000 m) and the 'freshness' of their proximal slopes suggests that they have been occupied, if not formed, during the Neoglacial period and quite likely within the Little Ice Age. Down valley from the present terminus there are no end or terminal moraines to define recent advances.

The 1934 German Nanga Parbat Expedition survey of the Raikot glacier set the basis for our understanding of the very recent glacier behaviour. Subsequent surveys by Pillewizer (1956), photographs by travellers in the area, and the 1985 and 1986 surveys (Gardner, 1986) permit the following summation of recent behaviour of the Raikot glacier. Since 1934 terminus recession has amounted to 250 m (4.9 m yr$^{-1}$). However, within this are three phases: rapid recession of 450 m (23 m yr$^{-1}$) during 1934–54; an advance of at least 200 m between the mid-1960s and the early 1980s; and recession of about 5 m yr$^{-1}$ in the mid-1980s. These patterns are generally consistent with recent results from studies of Karakoram and western Himalayan glaciers (Mayewski and Jeschke, 1979; Zhang Xiang-song et al., 1981) and add credence to the proposal that glaciers in the region show similar, though variable, rates of response to generalized environmental changes (Mayewski et al., 1980).

Finsterwalder (1935, 1937) produced the first flow-velocity data for Nanga Parbat glaciers. The surveys on the Raikot glacier revealed flow characteristics not described before. In the first instance, high velocities, in the order of 900 m yr$^{-1}$ were found during midsummer near the base of the Raikot icefall. Down glacier, velocities decreased to a value of 125 m yr$^{-1}$ at a cross-sectional profile ~ 4 km from the terminus. Also observed in the icefall and upper glacier tongue was the seemingly independent movement of large blocks of ice within the glacier and sustained high velocities right to the glacier margin. The high velocity and block-like flow led Finsterwalder to the descriptor: 'blockschollen' flow. This type of flow has clear implications for sediment incorporation, transport and yield.

Pillewizer (1956) found similar patterns of flow in 1954 on the Raikot glacier. His survey did indicate flow velocities as much as 20 per cent greater than those measured in 1934.

The surveys in 1985 and 1986 re-established Profile 5 of Finsterwalder and Pillewizer, using large prominent boulders as surface markers. In addition, three cross-sectional profiles, located at 2 km, 1 km and 0.5 km from the terminus, were established and monitored over a six-week period in 1986. Photographic comparisons between 1934, 1954 and 1985–6 revealed few morphological changes, including virtually no surface lowering up glacier from Profile 5, suggesting similar flow patterns and velocities. At Profile 5, surface velocities were 35–60 cm$^{-1}$, with the lower values located toward the glacier margins. These midsummer values could translate into annual values comparable to those estimated in previous surveys, making an assumption for a winter reduction in flow velocity.

The downglacier profiles were established to provide information pertinent to sediment transport in the most heavily sediment-laden portion of the Raikot glacier (Figure 8.1; Plate 8.1). The 1986 surveys produced velocity values of 18–83 cm d$^{-1}$ 2 km from the terminus, 21–47 cm d$^{-1}$ 1 km from the terminus, and 7–26 cm d$^{-1}$ near the terminus, a progressive downglacier decrease suggestive of compressive flow in this zone of the glacier. At all cross-sections, highest velocities were found near the glacier centre line and the lowest were found toward the margins, the latter generally being 40–50 per cent of the former. Finsterwalder (1950) used a mean velocity to width ratio of 0.16 as the lower threshold for defining blockschollen flow. On the lowest 2 km of the Raikot glacier, where surface evidence of blockschollen flow is absent, this value is 0.06, well below the suggested threshold.

In 1986, garbage from a mountaineering camp emerged on the glacier surface. The material included a collection of parts of tin cans, of metal tent pegs and of a Primus stove. The material was collected and returned to Canada for identification. Photographs of the items were then sent to Germany for further identification and it was suggested that the combination of items would make them consistent with the 1953 Nanga Parbat expedition (Kick, pers. comm., 1987). Locations of camps during that expedition suggest that Camp 3 at 5900 m in the lower Raikot firn was the lowest, most likely point of origin. If this was the point of origin, an average annual downglacier velocity of 350 m yr$^{-1}$ would result, giving a mass transport time of about 40 years from accumulation area to terminus.

## GLACIER SEDIMENT LOAD

Two aspects of the sediment load of the Raikot glacier were examined in the field: the distribution and thickness of surface debris; and the sediment concentrations in the ice. The extent of surface debris was mapped using transects across the surface in the lower 4 km of the glacier and visually from the margins up glacier. Debris thicknesses were estimated from spot measurements across transects in the lower glacier. Pits dug in the surface debris at the glacier margins revealed thicknesses of 1.5 m. Upglacier debris thicknesses were estimated by extrapolating patterns found on the lower glacier and by surface appearance as compared with that down glacier.

The distribution of surficial debris has changed little since 1934 and 1954, suggesting that input, transport and output patterns are unchanged in at least a century. Figure 8.1 gives an approximation of debris cover and thickness over the glacier surface. In the terminus area and up glacier a distance of 1.6 km, the glacier surface is completely debris covered, though with varying thickness. Greatest debris thicknesses ( > 1.5 m) were found along the glacier margins, with decreasing thicknesses toward the centre line of the glacier. Debris coverage extends up glacier as lateral strips to near the base of the icefall (Plate 8.2). The left-bank lateral strip is more extensive and contains more debris than that on the right bank.

Complete debris cover in the terminus region is not uncommon on western Himalaya and Karakoram glaciers. It is common, in varying degree, to all glaciers of Nanga Parbat. As such, the debris cover is an important component in any mass balance consideration since it acts as a significant ablation retardant at thicknesses in excess of 3 cm (Mattson and Gardner, 1989). This condition permits extension of glacier ice and associated geomorphic processes to lower elevations than might otherwise be the case.

Using the data presented in Figure 8.1, it is possible to produce an approximation of total surficial sediment load. Debris volumes are calculated on the basis of the areal extent and thickness in four zones (75 cm, 20 cm, 2.5 cm and 0.5 cm) and a correction factor for density of the unconsolidated debris. This calculation yields a total surface volume of $2.02 \times 10^6$ m$^3$.

The surficial debris on the glacier ranges from very large blocks in excess of 3 m in median diameter to fine clastics in the fine sand–silt range. In the lower 1 km of the glacier three transects were run across the glacier surface, with randomly spaced sample stations at each of which 25 clasts were measured for size and roundness. This sampling indicated a slight increase in mean clast size toward the terminus (11.3 cm $\pm$ b axis to 6.1 cm $\pm$ b axis) and an increasing tendency from subangularity to subroundness.

The character and volume of englacial sediment load is difficult to estimate. A preliminary approach to this was taken using samples of glacier ice which were melted and filtered for sediment. This approach underestimates the total load by not sampling or by ignoring large clasts which are contained in the ice. Our estimate is based on thirteen such ice samples taken from various locations on the surface, taking care not to include ice which had undergone ablation and regulation and thus debris concentration. In addition, we used a basal ice sample taken from inside the stream portal at the Raikot terminus (Plate 8.3). Examination of the ice within the portal revealed a further problem in estimating englacial sediment load: that of the sediment banding within the ice which is very strongly pronounced in the Raikot glacier. It is thought that such banding is the result of the annual ablation/sediment concentration/accumulation cycle which occurs in the lower accumulation–upper ablation area of the glacier. These concentrations of sediment and the incorporation of occasional large clasts make total load estimation merely approximate. Based on a total of 13 samples of ice, an estimated englacial sediment concentration of 2125.04 mg l$^{-1}$ resulted. Using a conservative estimate of ice volume in the ablation zone, based on an assumed mean ice depth of 200 m, a total internal sediment load of $2.430 \times 10^6$ m$^3$ resulted.

Several sources contribute sediment to the Raikot glacier system. Basal

*Plate 8.3* Basal conditions at the Raikot glacier as exposed within the meltwater stream portal. Illustrated are examples of the banding of fine sediment within the ice and the ice-ground interface which shows the glacier to be detached and lacking in a distinctive basal sediment concentration. (Photograph: Jones, 1986)

contributions probably are a factor in the cold-based sections of the glacier in the accumulation and upper ablation areas. Material is added through the snow and ice avalanching, particularly that from the north face of Nanga Parbat. Above the equilibrium line, this material is eventually incorporated in the englacial load. Below the equilibrium line the material becomes part of the surficial load. Our field observations in the upper ablation zone suggest that the most significant contribution of surficial debris comes from the collapse of left-bank morainic material on to the glacier margins (Plate 8.2). On the right bank, similar though less voluminous collapse occurs while greater contributions occur via mass wasting of talus and other debris from the lower slopes of Buldar Peak (Plate 8.2).

The terminus debris cover accumulates through three mechanisms: the concentration of the laterally transported surficial material through the decrease in velocities of transport toward the terminus; the surficial concentration of englacial debris resulting from ablation processes; and shearing upwards of englacial material in the compressive flow zone of the lower ablation area and terminus.

These data indicate that the Raikot glacier is effective in incorporating and transporting sediment. To be a viable link in the basin and regional denudation system, the sediment must be transferred from the glacier to the proglacial area and, ultimately, beyond the basin. Huge deposits of glacial sediments stored in the Indus valley downstream from Raikot Bridge indicate that the late-Quaternary Raikot glacier was most effective in removing material from the basin (Shroder *et al.*, 1989). In its depleted post-Little Ice Age form, the Raikot is obviously less effective in this way. Indeed, the relative paucity of large morainic deposits in the present proglacial area (Plate 8.1), despite the apparently large glacial sediment load, suggests that the direct physical transfer from glacier to proglacial area is a relatively unimportant aspect of the sediment yield, leading to the supposition that yield via meltwater is significantly more important.

## MASS BALANCE CONSIDERATIONS

Glacier mass balance provides one approach to estimating transfer of mass, and thus sediment, within the glacier system. The ablation component, conditioned by assumptions regarding non-melt losses such as evaporation, provides a means for estimating water yield. Coupled with data on meltwater sediment concentrations, water yield estimates can be used to estimate glacier sediment yield by this route.

Glacier surface ablation was measured over a six-week period from June–August 1986 at 25 locations in the lower ablation zone of the Raikot glacier. The locations were chosen to give a cross-section of surface conditions including differing ice face exposures. Each location used white stakes against which ablation was measured. The number of individual measurements at each site over the six-week period ranged from 12 to 43. The time period during which the monitoring took place included a full spectrum of ablation season weather types in the area, ranging from monsoonal rain with complete overcast to clear sky conditions under low humidity, high pressure air mass types. This sampling provides a reasonable basis for approximating ablation rates over the glacier surface. Unfortunately, the data do not include the complete altitudinal range found within the ablation zone. This shortcoming is tempered somewhat by the fact that 90 per cent of the ablation zone is located within an altitudinal range of $\sim 800$ m situated well below the equilibrium line (Figure 8.2; Plate 8.2). We suspect, therefore, a small ablation gradient over most of the ablation zone with a sharp ablation gradient restricted to the icefall zone below 4800 m.

Measured ablation values during the monitoring period ranged from zero under debris thicknesses of $> 0.5$ m to 13.3 cm d$^{-1}$ under clear sky conditions on clean ice or ice with a dusting of sediment. The mean ablation value for all measurements of all sites was 7.52 cm d$^{-1}$. Based on very scanty knowledge of the seasonality of climate in the area (founded on reading of expedition accounts), an ablation period of 90 days was used to calculate total ablation. This duration at 7.5 cm d$^{-1}$ would produce an average ice loss of 6.75 m. To provide a range in estimate, an ablation value of 4.5 cm d$^{-1}$ was also used. It produces a total ice loss of 4.2 m over the hypothetical ablation season. Using an ablation zone area of 14.3 km$^2$, these estimates produce total ice loss values of 60.06 × 10$^6$ m$^3$ to 96.53 × 10$^6$ m$^3$. Translated to water equivalent these values are: 53.1 × 10$^6$ m$^3$ to 86.9 × 10$^6$ m$^3$.

Basal ice loss owing to geothermal energy and frictional energy from sliding will also produce a meltwater yield. Using the 12 per cent of surface ice loss as an estimate and assuming the base of the glacier is a flat as opposed to a curved surface and assuming uniform loss over the whole basal area, produces an estimated 10.4 × 10$^6$ m$^3$ annual water loss for the larger ablation estimate. Added to the ablation values, this results in an annual water yield of 63.5 × 10$^6$ m$^3$ to 97.3 × 10$^6$ m$^3$. Stated in terms of mean annual discharge, the water yield is in the range of 2.1–3.1 m$^3$ s$^{-1}$.

By necessity these are very rough approximations, based as they are on a sampling of measured ablation rates and calculated using a number of assumptions about the surface and subsurface of the glacier. We have little independent information from which to validate the estimates, apart from the estimate of ice discharge across the equilibrium line which was calculated using measurements of surface velocity and the Legally formula by Finsterwalder (1937). This value of 70 × 10$^6$ m$^3$ yr$^{-1}$ suggests that our estimates may be reasonable. Surface velocity increases of as much as 20 per cent, observed by Pillewizer (1956) during a recessional stage in 1954, imply somewhat higher rates of ice discharge and water yield which would be consistent with the higher values generated by our ablation measurements. Comparative data on terminus position in 1985 and 1986 indicate a current recession of ~ 5 m yr$^{-1}$, which would account for ~ 1–1.5 per cent of the total annual water yield. We believe that the lower value of total ablation water yield is conservative given that the surface has been treated as flat, resulting in a significant underestimate of total surface area. This may be significant because of the very rough and broken surface of the Raikot glacier. The estimate of subglacial water yield also is conservative, again because the surface is treated as flat and because other sources of energy such as inflow of water from surrounding terrain and tributary glaciers, both of which are significant on the Raikot, are not treated.

The water yield estimates presented here permit some estimation of the accumulation component of the glacier mass balance. Assuming equilibrium, accumulation would amount to 3.6–5.6 m yr$^{-1}$ water equivalent, somewhat less than 6–8 m yr$^{-1}$ water equivalent suggested by Finsterwalder (1937). Under negative mass balance conditions, accumulation values could be slightly less than these estimates.

## TOTAL SEDIMENT YIELD

Sediment leaves the glacier system and basin by physical transfer at the terminus and via meltwater discharge. Sediment which is deposited basally and laterally up glacier from the terminus is thought to remain within the glacier system. The data on surface sediment load, coupled with that on the flow velocity at or near the terminus, permit an estimate of the physical transfer. The calculation uses surface area which moves to the terminus and a mean depth of debris (0.75 m) over that surface. The volume is corrected for density to produce a 'solid' value of $12.4 \times 10^3$ m$^3$ yr$^{-1}$.

To estimate sediment yield via meltwater, water samples from the portal of the Raikot glacier stream were taken over a six-week period in 1986, over a range of weather conditions and time of day. During the ablation season, we found little fluctuation in stage (discharge) or sediment concentration on a diurnal basis or any that could be correlated with day-to-day weather variation. Using 20 samples, sediment concentrations were estimated using a field filtering procedure. Sixty ml of the sampled water was air-forced through a preweighed milipore filter, the filter was placed in a sealed container and transported to Canada where it was again weighed on a precision balance. Weight increase was used to calculate sediment concentrations. Individual sample concentrations ranged from 2652 mg l$^{-1}$ to 10,054 mg l$^{-1}$, with a mean value of 5270 mg l$^{-1}$, an extremely high sediment load.

The mean sediment concentration when applied to the estimates of total water yield produces an approximation of total sediment yield. Four values are produced: two representing the two water yield estimates, and two representing the Raikot glacier area and the basin area upstream from the glacier terminus. These values are given as a volume of solid rock per surface area per year. We have assumed a specific gravity of 2.5 in the translation of measured concentrations to solid rock. For the lower water yield estimate, sediment yields are $4.2 \times 10^3$ m$^3$ km$^{-2}$ yr$^{-1}$ for the glacier area and $1.3 \times 10^3$ m$^3$ km$^{-2}$ yr$^{-1}$ for the total basin. The higher water yield estimate produced values of $6.5 \times 10^3$ m$^3$ km$^{-2}$ yr$^{-1}$ and $2.0 \times 10^3$ m$^3$ km$^{-2}$ yr$^{-1}$ respectively. The physical transfer of sediment at the glacier terminus produces yield values of $0.39 \times 10^3$ m$^3$ km$^{-2}$ yr$^{-1}$ and $0.12 \times 10^3$ m$^3$ km$^{-2}$ yr$^{-1}$ respectively or, in other words, 8 per cent and 5.75 per cent of the yield estimates via meltwater. Therefore, total sediment yields are: $4.6-6.9 \times 10^3$ m$^3$ km$^{-2}$ yr$^{-1}$ for the glacier area; and $1.4-2.1 \times 10^3$ m$^3$ km$^{-2}$ yr$^{-1}$ for the basin area. These translate into denudation rates of 4.6–6.9 mm yr$^{-1}$ for the glacier area and 1.4–2.1 mm yr$^{-1}$ for the basin area.

## CONCLUSION

The material presented in this paper provides a description of the Raikot glacier as a sediment transport system and geomorphological agent in the western Himalayan context. The estimates of sediment yield and denudation must be treated as quite preliminary approximations which are based on many assumptions and which are not easily validated by independently derived estimates either in the same area or in nearby areas with similar conditions and size. Of interest is the observation of Quaternary uplift in the Nanga Parbat area on the order of 5 mm yr$^{-1}$ (Zeitler, 1985), a value which is in the range of denudation values derived for the basin of the Raikot glacier. In addition, we know that denudation rates over much of the upper Indus basin are high (R.I. Ferguson, 1984).

The field observations and some measurements indicate that the Raikot glacier and its meltwaters are the most vigorous elements in the erosional, and thus denudation, system in the immediate basin. For example, sediment loads were monitored on two non-glacier streams in the Raikot basin during 1986, one a small spring-fed stream and one a snowmelt stream. Sediment concentrations on both were generally an order of magnitude less than those from the glacier meltwater stream. The glacier entrains or mobilizes sediment and moves it at a rate of $\sim 350$ m yr$^{-1}$. Some of this sediment is derived from newly weathered and/or eroded surfaces. Much of the sediment on the surface of the Raikot glacier is remobilized morainic material from periods of more vigorous and extensive glaciation. Essentially, the glacier modifies the repositions the sediment in such a way that the glacial meltwater can transport it out of the basin. Thus, without the mechanical and hydrological role played by the glacier, extra-basin sediment transfer would be much reduced.

# 9

# POSITION OF THE PALAEOINDUS AS REVEALED BY THE MAGNETIC STRATIGRAPHY OF THE SHINGHAR AND SURGHAR RANGES, PAKISTAN

*M. Javed Khan and Neil D. Opdyke*

## ABSTRACT

Four stratigraphic sequences were measured and sampled from the Shinghar and Surghar ranges, Pakistan, for detailed magnetostratigraphic studies of the fluviatile Neogene and Quaternary Siwalik Group. The samples were subjected to blanket thermal demagnetization at 500 °C. Some of the samples, particularly those from the pale brown siltstone/claystone units, were further demagnetized to 600 °C to isolate the stable characteristic component of magnetization which was used to identify the magnetic polarity. The observed magnetic polarity reversal sequence is correlated with the standard Magnetic Polarity Time Scale (Berggren *et al.*, 1985a), based on the magnetic pattern and the vertebrate fossils collected from the nearby Daud Khel area (Hussain *et al.*, 1977) and from the Shinghar range during palaeomagnetic sampling. This correlation suggests that deposition of the Silwalik Group in this area started during the basal chron C5A time (11.8 myr) and continued till the later Matuyama chron (0.85 myr). A very thick sandstone is present in the section up to 3.5 km in thickness. This sandstone was probably deposited along the axis of the ancient Indus from 8 myr to 2.7 myr. At that time the axis of the Indus apparently shifted to the east.

## INTRODUCTION

The Siwalik Group of the Surghar–Shinghar ranges is composed of Neogene and Quaternary fluviatile sedimentary rocks exposed along the northern and western flanks of the arcuate and asymmetric Makarwal anticline (Figure 9.1), which forms the northwestern part of the Trans-Indus Salt range situated between the Indus and the Kurram rivers. The Siwalik Group overlies, with angular unconformity, the Sakessar Limestone of early Eocene age. However, in the southern part of the Makarwal anticline, the Siwalik sediments are thought to overlie the Mitha Khattak formation conformably (Danilchik and Shah, 1976).

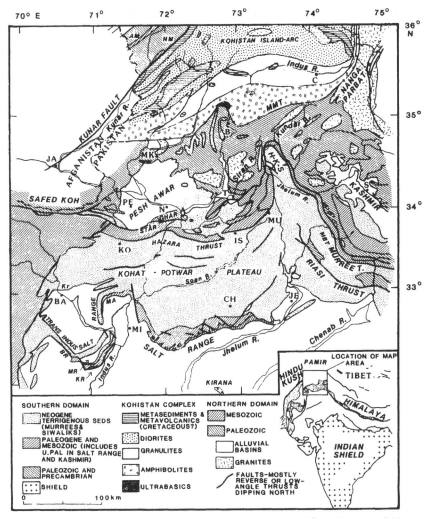

*Figure 9.1* Geologic sketch-map of northern Pakistan (adapted from Gansser, 1964; Desio, 1964; Calkins *et al.*, 1975; Tahirkheli and Jan, 1979; Tahirkheli *et al.*, 1979).

A Attock;
AM Asiatic mass;
B Besham;
BA Bannu;
BR Bhittani range;
C Chilas;
CH Chakwal;
H–K–S Hazara–Kashmir Syntaxis;
IS Islamabad;

JA Jalalabad;
JE Jhelum;
KR Khassor range;
KO Kohat;
KR Kurram river;
MBT Main Boundary Thrust;
MA Makarwal anticline;
MK Malakand;

MR Marwat range;
MI Mianwali;
MU Murree;
NM Northern Megashear;
N Nowshera;
PE Peshawar;
S Srinagar;
T Tarbela.

199

Throughout the extent of the Makarwal anticline, the Siwalik Group consists of three distinct lithologic units. From base to top these units are as follows.

## Red bed zone

This part of the stratigraphic sequence consists of dark reddish-brown claystone/siltstone units, each 10–60 m thick, alternating with light-brown and greyish-brown sandstone units, each 2–20 m thick. The thickness of sandstones increases whereas that of the claystone decreases towards the upper part of the red bed zone. Caliche zones are abundant in the overbank deposits of the lower part, but less common in the upper part. Two sections were measured and samples from this zone were taken from the Khora Baroch and Chichali sections; their locations are shown in Figure 9.2. It is this basal red bed zone of the Siwalik Group which gives the name to the Surghar range ('red mountain', in the local language).

## Sandstone zone

The red bed sequence is overlain by one of the thickest uninterrupted sandstone sequences known to the authors. The massive light-grey Makarwal Sandstone ranges up to 3.5 km in thickness and, unlike the usual molasse sequence sandstones observed in Pakistan, is uninterrupted by any overbank mudstone or siltstone units. In the upper part of the sequence, however, sandstones 100–200 m thick alternate with conglomerate units up to 100 m thick. These conglomerate units are present only in the northwestern part of the Makarwal anticline and are absent in the southern part of this anticline. Owing to the lack of overbank mudstones and siltstones in this part of the stratigraphic sequence, it was not possible to obtain samples for palaeomagnetic studies.

## Siltstone/sandstone/conglomerate

Makarwal Sandstone is conformably overlain by a sequence of alternating light-grey sandstones and pale-brown and yellowish-orange siltstone/claystone beds. The thickness of sandstone and siltstone units ranges between 5 m and 100 m, but in the upper part of the sequence the amount of sand and the thickness of the sandstone units increase. The uppermost 200–300 m of the stratigraphic sequence is composed mainly of light-brownish-grey coarse-grained sandstone overlain by conglomerate. It is the sandstones in this part of alternating sandstones and siltstones succession which collectively constitute the Shinghar range ('white/green mountain', in the local language).

## PALAEOMAGNETIC SAMPLING

The sampling strategy in the region is constrained by the fact that the sedi-

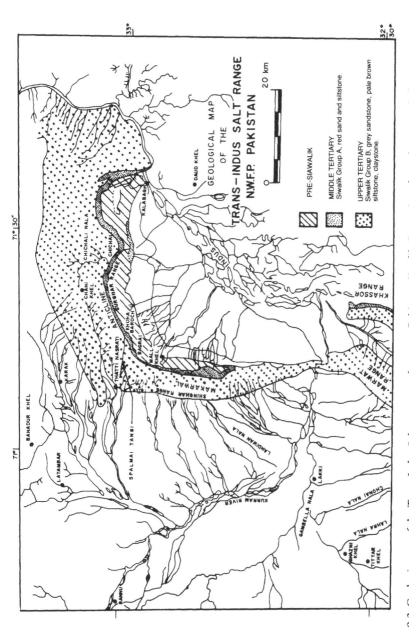

*Figure 9.2* Geologic map of the Trans-Indus Salt range, northwestern Pakistan. Siwalik Group A, red standstones and siltstones, are also exposed in the axial part of the Bhittani range but are not shown in this map because of their small thickness.

mentary section, although very long, does not contain suitable lithologies for palaeomagnetic study in the middle of the section because of the presence of the Makarwal Sandstone. Samples were, therefore, taken from two sections above this sandstone at Spalmai Tangi and Chani Khel and from below it at Khora Baroch and Chichali (Figure 9.2). This sampling strategy was designed in an effort to date the inception of sedimentation, and both the age and duration of the Makarwal Sandstone as well as the basal unconformity.

Three to five oriented samples were collected from each site using a hand rasp and Brunton compass. Forty-one sites were sampled from the Spalmai Tangi section spanning 1500 m and 21 sites were sampled ( > 325 m) at the Khora Baroch section, 11 sites were sampled from the 200 m thick section at Chani Khel and 53 sites were sampled from a 900 m thick stratigraphic sequence from the Chichali section. The strata generally strike ENE–WSW in the northern part of the Makarwal anticline with dips ranging from 25° to 35° to the northwest. In the southern part of the anticline the beds strike N–S dipping to the west between 25° and 35°.

## LABORATORY PROCEDURES

Samples were returned from the field and fashioned into rough 2.5 cm cubes. The direction and intensity of the natural remanent magnetization (NRM) of these samples were then measured on a fluxgate spinner magnetometer or on an SCT superconducting magnetometer.

## THERMAL DEMAGNETIZATION

A pilot group of representative samples was then selected for stepwise thermal demagnetization at temperatures ranging from 100 °C to 660 °C, with intervals of 25–100 °C. A representative orthogonal plot is shown in Figure 9.3. The magnetic behaviour of these samples is quite straightforward, with a low temperature overprint present in many samples which unblocks at temperatures below 400 °C. The magnetization vector decays linearly toward the origin from 400 °C to 660 °C. This suggested that a temperature of at least 400 °C is required to isolate the primary component of magnetization. The remaining samples were subjected to a single step demagnetization at 500 °C and some of these were then heated to 600 °C to verify the stability of the directions of magnetization. The results of thermal demagnetization are given in Table 9.1.

## TESTS OF DATA RELIABILITY

The characteristic directions of magnetization revealed through thermal demagnetization were used to calculate site-mean directions. The sites were classified as class A, B or C in order of decreasing reliability following Khan *et al.* (1988). Thirty-seven of 41 sites from the Spalmai Tangi section are class A sites and 4 are

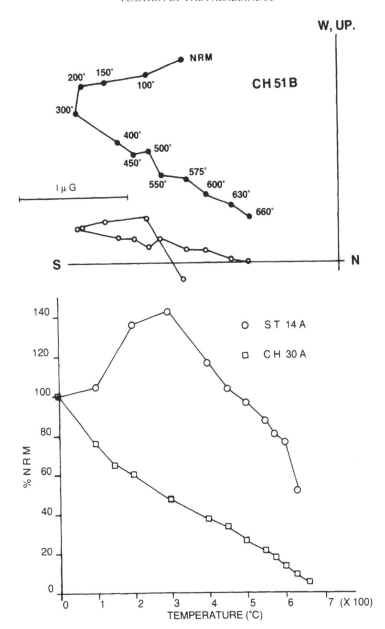

*Figure 9.3* An example of partial thermal demagnetization: orthogonal diagrams (Zijderveld diagrams) based on successive end points of partial thermal demagnetization vectors. Solid (open) circles are plotted on a horizontal (vertical) plane. Short bars indicate the scale for remanent magnetization intensity. All of these plots are based on bedding-corrected directions of magnetizations.

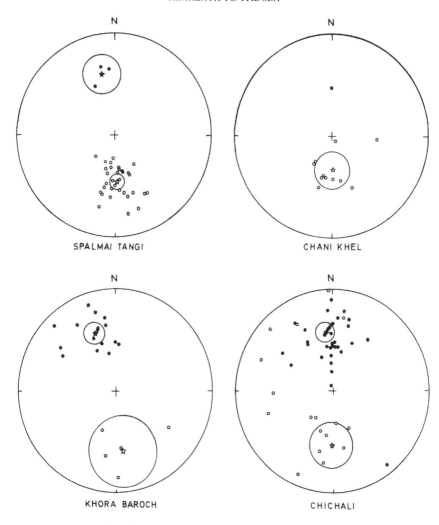

*Figure 9.4* Equal-angle stereographic projections of bedding-corrected site-mean directions obtained after thermal demagnetization at 500 °C and 600 °C. Solid (open) circles are plots on the lower (upper) hemisphere. Solid (open) stars are the mean of normal (reversed) polarity sites, and the large circles represent limits of 95 per cent confidence level.

class B. Sixteen of 21 sites from the Khora Baroch section are class A and 5 are class B. All eleven sites from the Chani Khel section are class A; however, the data from the Chichali section are not as good and only 36 of 53 of the sites are class A, 14 are class B, and 3 are class C. The data of class C sites were not used in further analysis. No fold test is possible from this area; however, the mean directions of normal and reversely magnetized sites from the Spalmai Tangi,

Khora Baroch and Chichali sections are almost anti-parallel, indicating reasonable stability (Figure 9.4; Table 9.1). The data from the Chani Khel section are all reversely magnetized, so that a reversal test is not possible; however, the direction is anti-parallel to the present-day earth's magnetic field at this location. Although it is not possible to obtain a fold test on these sediments, a companion study on sediments of the same age in the next range to the south (Marwat range) successfully pass the fold test (Khan, 1983).

## MAGNETIC POLARITY STRATIGRAPHY

The latitudes of the virtual geomagnetic pole (VGP) positions were calculated from the characteristic magnetizations at each site. The VGP latitudes were

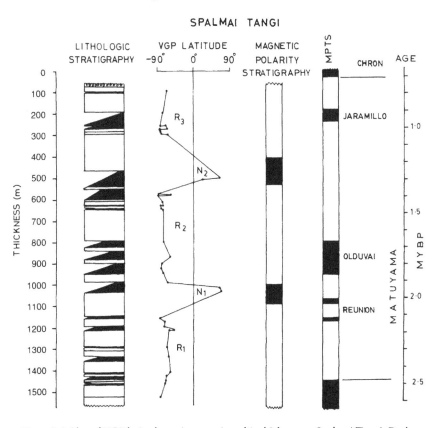

*Figure 9.5* Plot of VGP latitude against stratigraphic thickness at Spalmai Tangi. Dark (white) blocks in the lithologic stratigraphic column represent vertical accretion siltstone/claystone (lateral accretion sandstone) deposits of the molasse sequence. Dark (white) columns of MPS represent normal (reverse) magnetic polarities. The boundaries of magnetic polarity reversals are placed at intermediate positions between two successive sites having opposite polarities.

205

plotted against stratigraphic thickness which resulted in a magnetic polarity reversal sequence for each section, described individually as follows. The sections above the Makarwal Sandstone at Spalmai Tangi and Khora Baroch will be presented first.

## Spalmai Tangi

A plot of VGP latitudes versus stratigraphic thickness (Figure 9.5) reveals five magnetozones marked by four magnetic polarity reversal boundaries. This section is dominantly marked by reversed magnetic polarity and only two short normal polarity magnetozones (N1 and N2) are observed. A diamictite unit correlated to the Bain diamictite of the Bhittani range and Pezu Pass known to be 2 myr in age (Khan *et al.*, 1985, 1988) is present ~ 200 m below magnetozone N1. Because of the presence of multi-storied sandstone units in this section, the boundaries of N1 and N2 magnetozones may actually be slightly different from those shown in Figure 9.5, but significant difference is most unlikely. However, as each polarity zone is marked by at least two sites, the polarity zones are considered to be reliable.

During the course of a palaeomagnetic sampling an *Elephas* sp. jaw was collected from the lower part of the measured section. This indicates the presence of Pinjor-stage fauna. It has previously been demonstrated by Opdyke *et al.* (1979) in the Potwar plateau that Pinjor age faunas containing *Elephas* are younger than 3 myr. An age younger than 3 myr is plausible because the section represents the top of the Siwalik sequence in this area. It is also known that Siwalik sediments in the Bhittani, Marwat and Khassor ranges immediately to the south are of Plio–Pleistocene age (Khan *et al.*, 1988). Therefore, the dominance of reversed magnetic polarity and the presence of the Bain diamictite constrain the correlation of the magnetic stratigraphy to the Matuyama chron. The short polarity zones N1 and N2, therefore, probably correlate to the Olduvai and Jaramillo subchrons. This interpretation suggests that the section ranges in age from ~ 2.3 myr at the base to younger than 800,000 yr at the top. Sediments younger than this are present and form the top of stratigraphic sequence. However, these younger sediments consist predominantly of coarse-grained sandstones and conglomerate and thus are not suitable for palaeomagnetic studies. It is unlikely that sedimentation in the area continued into Brunhes time.

## Chani Khel

A second short section was sampled above the Makarwal Sandstone in the upper part of the sequence. This section was reversely magnetized in entirety (Figure 9.6). Considering the proximity of this section to Spalmai Tangi and the position of this section at the top of the sedimentary sequence, it is most likely that this reversed section is Upper Matuyama in age.

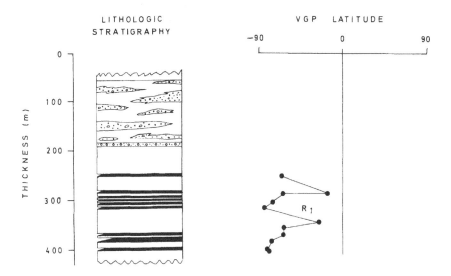

*Figure* 9.6 Plot of VGP latitude versus stratigraphic thickness at Chani Khel. Plotting conventions are the same as for Figure 9.5.

## Chichali

The Chichali section lies directly beneath the Makarwal Sandstone and directly overlies the basal unconformity. The magnetic polarity sequence shows 17 magnetozones. It is unfortunate that some of these magnetozones are defined by only one site, owing largely to the amount of sand in the section as the Makarwal Sandstone is approached at the top of the sequence. It is important to constrain the age of this sequence because of the tectonic and sedimentological significance of the region.

The age of the sediments overlying the Makarwal Sandstone are known to be Matuyama chron in age. The sedimentation rate can be extrapolated at the Spalmai Tangi section, using the pick of the upper Olduvai and lower Jaramillo transitions as the best defined part of the reversal sequence. The bottom of the Makarwal then would be ~ 7.3 myr. To expect rates of sedimentation to remain constant over such a long interval is, of course, unrealistic; however, it is most reasonable that the base of the Makarwal Sandstone could be 7–8 myr or of upper Miocene age. The Chichali section contains 16 reversals. Johnson and McGee (1983) have provided a means of estimating the length of time involved in a magnetic reversal sequence if the number of sites and reversals is known and assuming that the reversals of the earth's magnetic field have a Poisson distribution. If we assume that the reversal frequency of the earth's magnetic field is reasonably well known, it is possible to calculate the probable length of time

207

involved in the sequence. The Johnson/McGee method yields 4.3 myr for the time involved in the Chichali section. Therefore the base of the section could be expected to be on the order of 11 myr.

Unfortunately, no fossils are directly related to the magnetic sequence at Chichali. However, to the west of Chichali, near the Makarwal coal field, one of the authors (Opdyke) found a *Hipparion* sp. tooth at several hundred m above the local base of the Siwalik sediments. *Hipparion* is known to appear in the Siwaliks within the lower part of chron 11 (C5N) at 10 myr (Barry *et al.*, 1985) immediately east of the Indus river, ~ 100 km to the southeast of Chichali. Hussain *et al.* (1977) have described a fauna which is located in lower chron 11 in sediments similar in lithology to those at Chichali. Therefore, a correlation of the sediments at Chichali to the Magnetic Polarity Time Scale (MPTS) in the 7–10 myr range is a reasonable hypothesis.

Recently, Cerveny *et al.* (1988) have studied the age distribution of detrital zircons in the Indus river system and in an elegant study they have shown that ages of detrital zircons lag the depositional age of the sediments by ~ 3–5 myr. Detrital zircons from directly above the basal unconformity at the Chani Khel section gives an initial peak at 17 myr. We place the age of this unconformity at 11.75 myr, giving a 5 myr lag time for the zircon ages. This is in reasonable agreement with lag times observed in the Chingi section further to the east. We therefore believe that the placement of the beginning of sedimentation in the area in lower chron C5AN is consistent with all known facts.

It can be seen by examination of Figure 9.7 that the central part of the stratigraphic section between 750 mm and 350 mm is dominated by normal polarity (N3–N7) punctuated by short reversed intervals. A long normal zone has been identified in both the Khaur area and along the northern flank of the Salt range in the Potwar plateau area to the east. This long normal zone is correlated with chron 11 in this region, based on fission-track dates and superposition (N.M. Johnson *et al.*, 1982; Opdyke *et al.*, 1982; Tauxe and Opdyke, 1982). It seems most probable that the zone of normal polarity observed in the Chichali section is correlative to chron 11. The most probable correlation of this section to the MPTS is given in Figure 9.7. This proposed correlation would correlate the top of the sequence to chron 7 and the base to chron C5A. The correlation given is based upon the constraints mentioned previously. The correlation from chron 7 to chron 11 is convincing; however, the proposed correlation at the base of the sequence would require a slowing of the rate of sedimentation toward the basal unconformity.

The magnetic stratigraphy of the Khora Baroch section (Figure 9.7) reveals three magnetozones marked by two magnetic polarity reversals. All magnetozones are present in more than one site and are considered to represent magnetic field behaviour. A correlation to the section at Chichali is possible (Figure 9.7) in which the R1 magnetozone at Khora Baroch is clearly correlative to magnetozone R2 at Chichali. This correlation requires a simultaneous onset of sedimentation at both localities.

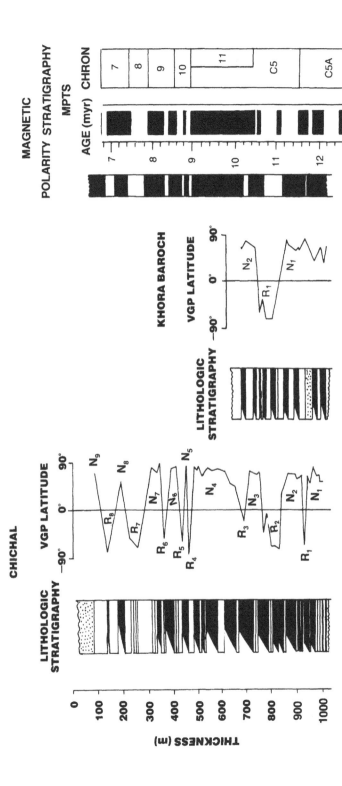

*Figure 9.7* Plot of VGP latitude versus stratigraphic thickness at Chichali and Khora Baroch. Plotting conventions are the same as for Figure 9.5.

# DISCUSSION

Undoubtedly the most fascinating aspect of the stratigraphy and sedimentology of the Shinghar–Surghar range is the enormous sandstone, up to 3.5 km in thickness, which dominates the middle of the stratigraphic section. The unbroken continuous nature of this sand body makes it clear that its origin differs from that of sandstones of the same age on the Potwar plateau.

The Chichali and Spalmai Tangi sections serve to date the beginning and termination of this large sand unit. At Spalmai Tangi the deposition of the sand body ends at 2.3 myr in upper Pliocene time. However, sedimentation continues in the region until the middle Pleistocene probably terminating in the Brunhes. In prior studies the sedimentary sequence measured at Spalmai Tangi, which is Plio–Pleistocene in age, has been correlated with the Dhok Pathan formation of the Potwar plateau, based primarily on lithologic similarities (Fatmi, 1974; Danilchik and Shah, 1976; Shah, 1977). It is now known that the Nagri formation, in the sense used by the previous writers, is older than the section studied here, ranging in age from chron 10 (8.7 myr) to chron C5R Spalmai Tangi (11 myr; G.D. Johnson et al., 1982). However, the section at Spalmai Tangi is similar in age to sediments in the Bhittani and Marwat ranges to the south (Khan et al., 1988), and to sections on the eastern end of the Salt range (Opdyke et al., 1979). The rate of subsidence (i.e. sedimentation rate) derived from the Spalmai Tangi section would be $\sim 69$ cm 1000 yr$^{-1}$, a rate similar to that seen in the Marwat range to the south.

The time of the beginning of deposition of the sand body is constrained by the section at Chichali where the preferred correlation to the MPTS would place the onset of the sand deposition at 7 myr. This is close to the estimate obtained by extrapolating the rate of sedimentation from Spalmai Tangi downward (7.3 myr). This thick sandstone unit was therefore deposited in $\sim 5.7$ myr spanning upper Miocene and Pliocene time.

A sand body of this magnitude was most likely to have been deposited by a major river system, considering the sedimentary context of the region as a whole (Burbank and Johnson, 1982). The floodplain of the Indus river lies today only $\sim 20$ km to the east. We believe that the most likely source for this enormous sand sheet is the palaeoIndus river which probably flowed through the area occupied by the present Makarwal anticline during 7–2.3 myr. The Indus was probably displaced to the east at 2.3 myr, which is close in age to the tectonic activity in the region associated with the folding of the Soan anticline (Raynolds and Johnson, 1985). Sedimentation continued at the site of the Surghar range until continuing tectonism uplifted the area in the upper Pleistocene and formed the main Salt range as well (Johnson et al., 1986). It has been previously suggested by several authors (Behrensmeyer and Tauxe, 1982; Johnson et al., 1986; Burbank et al., 1986) that the palaeoIndus flowed from west to east through the Siwalik basins during the deposition of the Nagri facies, with its thick multi-storied sandstones. The beginning of the deposition of Nagri type

Table 9.1 Directions of remnant magnetization for the four sites studied in the Shinghar and Surghar ranges.

| | N | D | I | K | ∞95 | $N_e$ | $D_c$ | $I_c$ | $K_c$ | $∞95_c$ |
|---|---|---|---|---|---|---|---|---|---|---|
| Spalmai Tangi | 4 | 342.1 | 38.4 | 4.9 | 46.3 | 4 | 334.3 | 24.0 | 8.9 | 32.6 |
| | | | | | | | (347.7)* | (26.9)* | (61.8)* | (15.8)* |
| | 37 | 201.1 | −41.5 | 18.2 | 5.7 | 37 | 177.0 | −39.5 | 17.7 | 5.8 |
| Khora Baroch | 16 | 1.8 | 52.9 | 16.5 | 9.3 | 16 | 340.7 | 27.5 | 18.7 | 8.7 |
| | 5 | 209.2 | −50.7 | 5.1 | 37.5 | 5 | 174.4 | 29.1 | 7.6 | 29.4 |
| Chichali | 39 | 5.6 | 51.2 | 6.0 | 10.2 | 37 | 354.3 | 30.9 | 8.4 | 8.7 |
| | 11 | 214.9 | −53.0 | 3.3 | 29.7 | 13 | 195.9 | −38.0 | 5.7 | 19.0 |
| Chani Khel | 2 | 37.7 | −61.0 | 6.9 | >90 | 0 | – | – | – | – |
| | 9 | 219.2 | −69.2 | 33.8 | 8.9 | 11 | 179.1 | −51.5 | 10.4 | 14.8 |

N   No. of sites
D   Declination
I   Inclination
K   Precision parameter

∞95   95% confidence level (Fisher 1953), all sites thermally demagnetized
$N_c$   No. of sites after bedding connection
$D_c$   Declination corrected for bedding tilt
$I_c$   Inclination corrected for bedding tilt

* Directions obtained after one site with intermediate directions is excluded.

facies begins in the Surghar range in chron 9 at ~ 8 myr. This undoubtedly represents the shift of the Indus from flowing west–east to a route similar to the present north–south axis. Apparently the palaeoIndus became partially or completely blocked from flowing to the east, perhaps by the initial movements which eventually led to the Pir Panjal range.

At Chichali, sedimentation began at ~ 12 myr. This contrasts with the inception of Siwalik sedimentation to the east in the Potwar plateau at 18 myr (Johnson *et al.*, 1986). We, therefore, see a general younging of the onset of sedimentation toward the south and west, probably resulting from progressive loading of the northern margins of the Indian plate by the southward progression of the Himalayan front.

# 10

# PALAEOECOLOGIC RECONSTRUCTION OF FLOODPLAIN ENVIRONMENTS USING PALAEOSOLS FROM UPPER SIWALIK GROUP SEDIMENTS, NORTHERN PAKISTAN

*Jay Quade, Thure E. Cerling, John R. Bowman and M. Asif Jah*

## ABSTRACT

Upper Siwalik sediments in northern Pakistan contain a well exposed record of palaeosols spanning most of the Pleistocene. Three sections at Mirpur, Rohtas, and in the Pabbi Hills on the margin of the Potwar plateau were studied. Palaeosols at all three sequences contain abundant pedogenic carbonate and soil organic matter (palaeohumus). The stable carbon isotopic composition of both of these phases indicates that nearly pure grasslands covered floodplains throughout the Pleistocene. Neutral to slightly alkaline values of soil pH, and leached zones dominated by Ca-smectite, chlorite and illite are also consistent with soils formed under grassland (mollisols). Stable oxygen isotopes in palaeosol carbonates are similar to those of modern soil carbonate in the area. The combined evidence strongly suggests that a monsoonal climate with moderate annual rainfall such as characterizes the eastern Potwar region today was in place throughout the Pleistocene. The carbon isotope chemistry of a post-Siwalik ( < 400,000 yr) palaeosol and modern soils reflects the mixed thorn-scrub/grass cover typical of the broken hill country that developed after uplift and erosion of the Siwaliks.

## SECTION LOCATIONS AND AGES

Plio–Pleistocene age fluvial sediments, loosely known as the 'upper Siwaliks', reach up to several km thickness in many areas of northern Pakistan. Pleistocene palaeosols were studied in three sections where the palaeomagnetic and fission-track ages had been previously established. At Mirpur (Figure 10.1), sediments covering the period 3.4 myr to ~ 1.4 myr are exposed along gentle dips in the

213

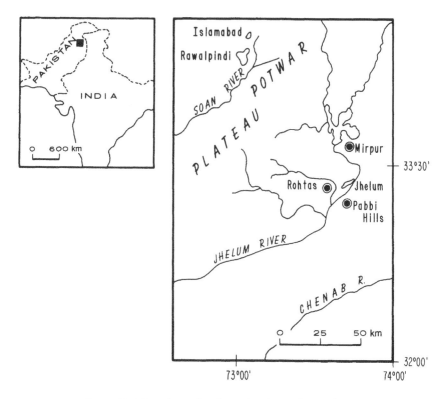

*Figure 10.1* Locations of study sections in northern Pakistan.

northern limb of the Mangla–Samwal anticline (Figure 10.2). The study section follows the same wash that Johnson *et al.* (1979) sampled for palaeomagnetics, which passes the village of Pothi and abuts the south side of New Mirpur town. In the Pabbi Hills to the south, palaeosols falling beetween ~ 1.8 and 0.4 myr were studied. The northwest limb of the Pabbi Hills anticline, along the east–west railroad cut that transects the hills, was sampled; this is the same section that was sampled by Keller *et al.* (1977) for palaeomagnetics. Work on the Pleistocene at nearby Rohtas was much less detailed than the other two sections. Samples came from along Matial Kas, the next major wash north of Basawa Kas where Johnson *et al.* (1979) conducted their palaeomagnetic studies. At Rohtas all of the Pleistocene up to the Brunhes–Matuyama geomagnetic boundary (1.8–0.7 myr) was covered.

## METHODS

An analysis was made of the carbon and oxygen isotopic composition of a single nodule from each palaeosol that was collected. Nodules were roasted under

214

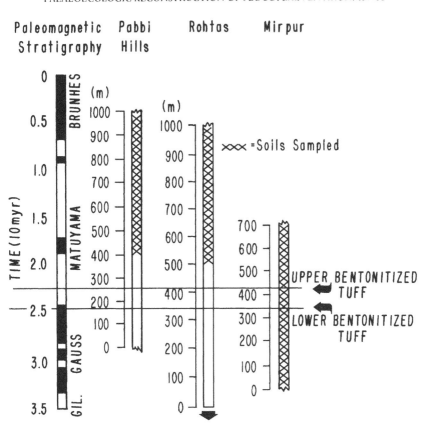

*Figure 10.2* Ages of the study sections and approximate intervals in which palaeosols were sampled for isotopic analysis.

vacuum at 450 °C prior to conversion to $CO_2$ with phosphoric acid. Soil organic matter was picked and screened for modern rootlets, which were present only in a few samples. It was then digested for 45 minutes in 6 N HCl, washed with distilled water, and finally combusted at 900 °C in the presence of cupric oxide and silver foil. Isotopic results for both carbon and oxygen are presented in the usual $\delta$ notation as the permil deviation of the sample $CO_2$ from the PDB standard, where $R = {}^{13}C/{}^{12}C$ or ${}^{18}O/{}^{16}O$, and $\delta = \{(R_{sample}/R_{standard}) - 1\}\ 1000$.

The procedure of Jackson (1979) for measuring soil pH of leached zones was followed. Distilled $H_2O$ and soil were mixed at about 1:1 by weight prior to submersion of the probe. Reproducibility on the pH measurements was about $\pm$ 0.2 pH units.

The <2 micron fraction was separated from leached zones for X-ray diffraction analysis of clays. Separate scans were performed on untreated, glycolated and heated (200 °C and 500 °C) portions of each sample.

215

## GEOLOGIC SETTING AND PALAEOSOL RECOGNITION

Sediments in the Pabbi Hills, Rohtas and Mirpur sections were deposited from the Pliocene to mid-Pleistocene by large river systems ancestral to the modern Jhelum river. Deposition changed to uplift and erosion in the mid- to late Pleistocene. Moderately cemented sandstones, usually < 20 m thick, cap the many ridges in the area. Intraformational conglomerates, largely composed of reworked soil nodules, often accompany the sandstones. The sandstones and conglomerate alternate with weakly indurated, floodplain sands and silts. Where unmodified by pedogenesis, the silts are finely laminated and contain 10–20 per cent detrital carbonate. Most often the silts are oxidized to a pale pink colour. No organic matter is usually present in this facies. Locally, however, beds are purplish-grey, moderately to strongly bioturbated and weakly calcareous. Well preserved aquatic molluscs sometimes occur in these settings.

Palaeosols display a suite of features that make field recognition straightforward. Most obvious is the yellowish or more rarely orangish colour (Figure 10.3). These colours go with the B horizons of the palaeosols. The B horizons are thoroughly bioturbated and usually are leached of detrital carbonate, especially toward the top. Clay skins and a fine filigree pattern of rootlet tubes are often also apparent. Palaeosols can be developed on weakly consolidated thin sands, such as are common in the Pabbi Hills, or on siltier floodplain sediments, as often the case at Mirpur.

Carbonate nodules nearly always accompany soils, generally at the base of the leached zone (B horizon) and sometimes extending into sediment below (Figure 10.3). The nodules are usually a centimetre or less in diameter and are irregularly shaped. They are composed of 75–90 per cent micritic calcite. The concentration of nodules varies from soil to soil, but no continuous 'calcrete' ledges were observed in any section.

Many but not all soils are capped by organic-rich A horizons or palaeohumus. They appear as pale olive grey (5 Y 5/2 dry) zones several tens of cm or, in exceptional cases, up to 1 m in thickness. Detrital carbonate has usually been leached out; complete lack of bedding and abundant fine rootlet traces are common. At first glance it is tempting to call these horizons swamp or pond deposits. But the absence of aquatic fauna, the leaching of carbonate, and the common association with yellow B horizons and carbonate nodules point to a pedogenic origin in nearly every case. This conclusion is reinforced by isotopic evidence discussed later.

One potential complication of these criteria is that palaeosols often occur in superimposed clusters or 'composite soils' (Morrison, 1964). For example, one 30 m section from Mirpur at about the level of the Olduvai event is composed entirely of overlapping palaeosols. From afar it is a yellowish zone sandwiched between pinkish floodplain sediments. In detail, numerous individual humic, leached B, and nodular zones are visible, often one superimposed on the next. Thus, where carbonate leaching is a diagnostic feature of the upper portions of

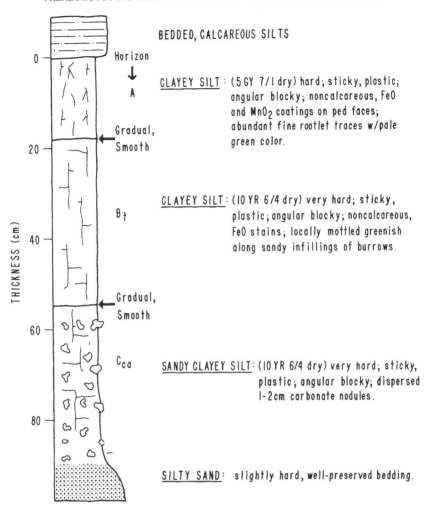

*Figure 10.3* Profile of a palaeosol from about the 2.8 myr level at Mirpur. A greyish A horizon and yellow Bt horizon, both leached of carbonate, make recognition of soils such as this straightforward. The greyish colour results from the presence of small quantities (< 1 per cent) organic matter. Bt denotes an argillic or clay-enriched horizon into which soil clays have been translocated from above. Depth of carbonate leaching to ~ 50 cm and a dispersed nodular zone below also typify Plio–Pleistocene age palaeosols throughout the region.

individual soils, in these composite situations carbonate from overlying soils often cements A and B horizons of underlying soils. This must be kept in mind when estimating the average depth of carbonate leaching in palaeosols for this period; it is best to confine oneself to simple, isolated soil occurrences.

217

## SOIL CHEMISTRY

Three geochemical aspects of the palaeosols were examined: the stable isotopic composition of soil humus and nodules, clay mineralogy of B horizons, and soil pH.

### Stable isotope geochemistry

The stable isotopic composition of palaeosol humus or soil carbonate, both common in the upper Siwaliks, can be used to reconstruct some aspects of palaeovegetation once present during pedogenesis (Cerling, 1984; Cerling et al., 1989). Plants metabolize and in the process fractionate carbon isotopes along three distinct pathways. $C_3$ plants include nearly all trees and shrubs, and those grasses favouring a cool growing season. They average $-27 \pm 6$ ml$^{-1}$ worldwide, the deviation being the result of variations in light intensity, moisture stress, plant longevity, and other factors. $C_4$ plants include some shrubs, particularly in the families Chenopodiaceae and Euphorbiaceae, and grasses favouring warm growing seasons. $C_4$ plants average about $-13$ ml$^{-1}$. Even accounting for deviations, $C_3$ and $C_4$ plants do not overlap isotopically. CAM plants, which include many of the desert succulents, occur most commonly in settings drier than northern Pakistan. They generally constitute only a small fraction of the biomass of any ecosystem.

The most direct measure of the relative contribution of $C_4$ and $C_3$ plants to former floodplain biomasses is to measure the carbon isotopic content of palaeohumus. This is commonly available in palaeosols in the upper Siwaliks where A horizons have 0.5–1.0 per cent organic matter. However, soil carbonate is even more abundant. Three recent studies of modern soils (Cerling, 1984; Cerling et al., 1989; Quade et al., 1989) indicate that there is a direct link between the carbon isotopic composition of coexisting organic matter and carbonate in soils where respiration rates are high. The reason for this is that plants, as they respire and decay, produce soil $CO_2$ whose carbon isotopic composition directly reflects the relative ratio of $C_4$ to $C_3$ plants in the local biomass. In modern soils it has been demonstrated that soil carbonate precipitates in equilibrium with this soil $CO_2$ (Quade et al., 1989; Cerling et al., 1989). The observed carbon isotopic difference between coexisting organic matter and carbonate is 14–16 ml$^{-1}$, depending on temperature. This systematic difference arises from equilibrium fractionation and gas diffusive effects (Cerling, 1984).

Soil organic matter from all five analysed humus horizons range from $-16.3$ to $-12.8$ ml$^{-1}$ (Table 10.1; Figure 10.4), indicating that the original biomass was composed entirely or almost entirely of $C_4$ plants. All of these samples come from Mirpur and range $\sim 3.1$–1.5 myr. Soil nodules from fifty palaeosols in the three sections were analysed. Between 3.4 and 1.5 myr at Mirpur, the $\delta^{13}C$ (PDB) of soil carbonate ranges between $-1.9$ and $+2.6$ ml$^{-1}$, and averages $+0.8$ overall. Carbon isotope values fall between $-2.4$ and 2.8 ml$^{-1}$ (Figure 10.4) in

*Table 10.1* Stable carbon and oxygen isotope analyses from upper Siwalik Group and modern soils, northern Pakistan.

| Sample no. | Area | $\delta^{13}C$ (PDB) | $\delta^{18}O$ (PDB) | Age (myr) |
|---|---|---|---|---|
| Nodular carbonate | | | | |
| 7905 | Pabbi Hills | 1.8 | −4.7 | 1.67 |
| 9189 | Pabbi Hills | 1.1 | −4.8 | 1.67 |
| 9190 | Pabbi Hills | 2.3 | −5.5 | 1.67 |
| 9191 | Pabbi Hills | 2.8 | −4.6 | 1.67 |
| 9192 | Pabbi Hills | 1.3 | −5.8 | 1.55 |
| 9193 | Pabbi Hills | 0.8 | −5.6 | 1.48 |
| 9194 | Pabbi Hills | 2.1 | −5.8 | 1.45 |
| 9196 | Pabbi Hills | 1.9 | −6.6 | 1.40 |
| 9197 | Pabbi Hills | 0.2 | −6.3 | 1.34 |
| 9200 | Pabbi Hills | −0.9 | −5.9 | 1.32 |
| 9201 | Pabbi Hills | −0.8 | −6.0 | 1.28 |
| 9202 | Pabbi Hills | 1.2 | −8.1 | 1.27 |
| 9203 | Pabbi Hills | −0.1 | −7.9 | 1.17 |
| 9204 | Pabbi Hills | 1.4 | −6.4 | 1.13 |
| 9205 | Pabbi Hills | 1.3 | −5.2 | 1.10 |
| 9206 | Pabbi Hills | 0.3 | −4.4 | 1.08 |
| 9207 | Pabbi Hills | 1.3 | −5.2 | 1.08 |
| 9208 | Pabbi Hills | 0.9 | −5.6 | 1.07 |
| 9209 | Pabbi Hills | 2.5 | −6.6 | 1.06 |
| 9210 | Pabbi Hills | 2.3 | −6.8 | 1.03 |
| 9211 | Pabbi Hills | −0.1 | −6.1 | 1.01 |
| 9212 | Pabbi Hills | −2.1 | −8.5 | 0.96 |
| 9216 | Pabbi Hills | 1.9 | −6.5 | 0.72 |
| 9217 | Pabbi Hills | 0.9 | −6.8 | 0.69 |
| 9718 | Pabbi Hills | 1.6 | −6.1 | 0.67 |
| 9720 | Pabbi Hills | 0.5 | −6.7 | 0.58 |
| 9721 | Pabbi Hills | −2.4 | −9.0 | 0.54 |
| 9722 | Pabbi Hills | 2.1 | −5.7 | 0.48 |
| 8007 | Mirpur | 0.6 | −6.7 | 3.18 |
| 8012 | Mirpur | 0.9 | −5.5 | 3.13 |
| 8014 | Mirpur | 1.2 | −6.9 | 3.10 |
| 8016 | Mirpur | 0.7 | −7.8 | 3.05 |
| 8024 | Mirpur | 0.1 | −7.2 | 2.97 |
| 8037 | Mirpur | 2.0 | −7.2 | 2.78 |
| 8043 | Mirpur | 1.7 | −7.5 | 2.58 |
| 8051 | Mirpur | 1.0 | −6.2 | 2.49 |
| 8056 | Mirpur | −1.9 | −6.8 | 2.36 |
| 8065 | Mirpur | 0.4 | −6.2 | 2.05 |
| 8075 | Mirpur | 2.6 | −6.0 | 1.54 |
| 8076 | Mirpur | 0.8 | −5.0 | 1.52 |
| 8082 | Mirpur | −5.1 | −7.3 | 0.05 |
| 8005a | Mirpur | −7.2 | −8.9 | modern |
| 8005c | Mirpur | −5.6 | −6.2 | modern |
| R-4 | Rohtas | 0.6 | −5.0 | 2.3 |
| R-8 | Rohtas | 0.5 | −7.4 | 2.2 |

219

*Table 10.1 continued.*

| Sample no. | Area | $\delta^{13}C$ (PDB) | $\delta^{18}O$ (PDB) | Age (myr) |
|---|---|---|---|---|
| R-16 | Rohtas | 0.5 | −5.6 | 1.8 |
| R-20 | Rohtas | −0.6 | −6.5 | 1.0 |
| R-26 | Rohtas | 1.8 | −5.2 | 0.6 |
| Floodplain sediment | | | | |
| 8009 | Mirpur | −3.3 | −10.7 | 3.15 |
| 8015 | Mirpur | −2.5 | −10.7 | 3.10 |
| 9032 | Mirpur | −4.1 | −9.6 | 1.67 |
| 9035 | Mirpur | −3.9 | −10.8 | 3.00 |
| 9188 | Pabbi Hills | −10.3 | −17.1 | 1.67 |
| 9199 | Pabbi Hills | −17.7 | −19.9 | 1.32 |
| Sandstone nodules | | | | |
| 8006 | Mirpur | −2.8 | −10.3 | 3.30 |
| 8010a | Mirpur | −2.7 | −11.1 | 3.18 |
| 8010b | Mirpur | −3.0 | −10.9 | 3.18 |
| 8010c | Mirpur | −2.9 | −10.0 | 3.18 |
| 8048 | Mirpur | −3.8 | −10.0 | 2.50 |
| 8077 | Mirpur | −3.3 | −8.7 | 1.45 |
| Organic matter | | | | |
| 8032 | Mirpur | −16.3 | — | 3.00 |
| 8046 | Mirpur | −14.9 | — | 2.58 |
| 8055 | Mirpur | −15.7 | — | 2.36 |
| 8071 | Mirpur | −12.8 | — | 1.67 |
| Modern grasses | | | | |
| 9601 | — | −13.0 | — | — |
| 9602 | — | −11.4 | — | — |
| 9603 | — | −12.6 | — | — |
| 9604 | — | −10.5 | — | — |

the Pabbi Hills between 1.7 and 0.4 myr, with an average also of +0.8 ml⁻¹. At Rohtas, the range is −0.6–1.8 ml⁻¹ over the span 2.2–0.8 myr. Taking an overall average of +0.8 ml⁻¹, carbonate is then 15.7 ml⁻¹ heavier than the average organic matter, well within the range of difference observed in modern soils. Both phases then yield a consistent picture of a $C_4$ dominated biomass on floodplains throughout the Pleistocene.

The $\delta^{18}O$ (PDB) of soil carbonate for all three upper Siwalik sections falls between −8.9 and −4.4 ml⁻¹ (Table 10.1; Figure 10.5), but by far the majority fall between −7 and −5 ml⁻¹. This range is similar to that displayed by modern soils.

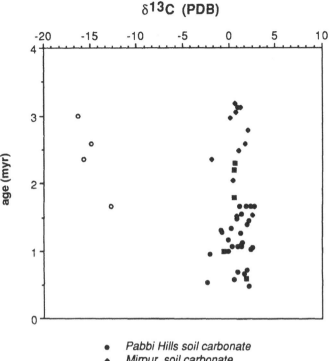

*Figure 10.4* The stable carbon isotope contents of soil carbonate and organic matter through time. The average difference between the two phases is 15.7 ml⁻¹ within the same narrow range of difference (14–16 ml⁻¹) displayed by coexisting phases in modern soils. This is strong evidence against diagenetic alteration of either phase, notwithstanding up to 1 km of burial. Soil organic matter averages −14.7 ml⁻¹, thus indicating the presence of a pure or nearly pure $C_4$ biomass on floodplains throughout the Plio–Pleistocene. This biomass was almost certainly a grassland.

## Clay mineralogy and soil pH

Analysis was made of the clay content of three B horizons and two unaltered floodplain silts with X-ray diffraction, confining the study to the <2 micron fraction. Ca-smectite and chlorite dominate both the soil clays and the unaltered floodplain clays under examination. A 10 A mica (illite?), vermiculite (?), and pyrophyllite were also present in smaller amounts.

Soil pH measured on six different leached zones (B horizons) from Mirpur came out at 7.0 to 7.9 ± 0.2 in all horizons. Soil pH is directly related to base saturation; this range indicates base saturation of 75–100 per cent (Brady, 1984).

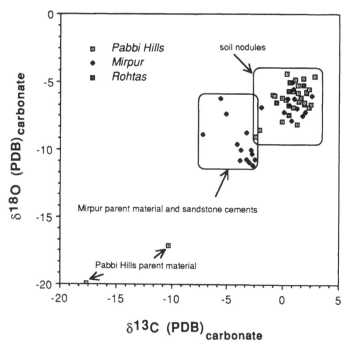

*Figure 10.5* The stable carbon and oxygen content of soil nodules, sandstone nodules, and silty parent material in upper Siwalik beds. Local groundwaters are likely to be buffered at depth by the isotopic composition of the calcareous parent material, which forms the largest carbonate reservoir in the upper Siwaliks. Sandstone cements overlap with the parent material field, suggesting that their isotopic composition is determined by the parent material. Soil nodules lie almost entirely outside this field, indicating little or no isotopic exchange with parent material after burial.

## POST-BURIAL CEMENTS

Even though soil carbonate nodules are highly visible in the sediments, it turns out that sandstone cements and unaltered floodplain sediments constitute about 80–90 per cent of the total carbonate reservoir in the upper Siwaliks. Their geochemistry was examined in order to assess their potential role in diagenetic modification of the isotopic composition of soil carbonate.

Unaltered floodplain sediment runs ~ 15–25 per cent carbonate. Most of this is probably detrital as there seems to be little post-burial movement of carbonate in the fine-grained facies, as evidenced by the fact that B horizons in soils retain their original leached condition. At Mirpur the $\delta^{13}C$ (PDB) of sediment runs −2.5 to −3.9 ml⁻¹, and $\delta^{18}O$ (PDB) is −9.6 to −10.8. In the Pabbi Hills these values are −10.3 to −17.7, and −17.1 to −19.9, respectively (Table 10.1; Figure 10.5). Why the ranges differ so much in areas so close together is not yet clear.

The important point is that there is no overlap by either parent material with the isotopic range displayed by soil carbonates.

Ellipsoidal nodules averaging 30–50 cm in length are common in sandstones throughout the upper Siwaliks. Being better cemented, the nodules stand out in relief on many outcrops. The 'sandstone nodules' at Mirpur and Rohtas run ~ 35–50 per cent pure sparry calcite cement. This texture is in marked contrast to the dirty ( ~ 75 per cent calcite), micritic appearance of soil nodules, which make up ~ 10 per cent of the grains in the coarser sands. No limestone or marble fragments were counted. Thus, calcite cement makes up most of the total carbonate present. At Mirpur and Rohtas the range for the $\delta^{13}C$ (PDB) and $\delta^{18}O$ (PDB) of bulk carbonate in sandstones is −2.8 to −3.3 and −8.7 to −10.9 ml$^{-1}$, respectively (Table 10.1; Figure 10.5). These ranges overlap closely with the values for floodplain sediment at Mirpur. There are no results on sandstones from the Pabbi Hills.

## POST-SIWALIK SOILS

The Siwaliks first began to experience uplift and thereafter erosion at different times for the three sections (Johnson *et al.*, 1979). The top of the Mirpur section dates to ~ 1.4 myr when erosion and later deposition of the Mirpur Gravels occurred. At both Rohtas and the Pabbi Hills, stable floodplains continued to receive fine-grained sediment to somewhat above the Brunhes–Matuyama geomagnetic boundary (0.7 myr).

The oldest post-Siwalik soils at Mirpur and Rohtas are patchy remnants of reddish soil capping the topographically highest outcrops of Siwalik sediments. The soil is sharply unconformable with the underlying beds. The associated B horizon is quite red (5YR 4/6 dry). A calcareous horizon is also present. Younger soils can be found at several levels inset into the eroded Siwalik topography.

The $\delta^{13}C$ (PDB) contents of the capping soil (Table 10.1, sample 8082) and an immature (Holocene) soil (samples 8005a, b) from an inset surface are −5.1 to −7.2 ml$^{-1}$, notably lighter than results from upper Siwalik soils. The $\delta^{18}O$ (PDB) of carbonates is −6.2 to −8.9 ml$^{-1}$, comparable to the older upper Siwalik carbonates.

## DISCUSSION

A major question for any geochemist interested in palaeoenvironments in the Siwaliks is the potential role of diagenesis. The bases of all the sections studied were buried under 1 km or more of sediment. The geothermal gradient in nearby oilfields has been measured at 21–22 °C km$^{-1}$ (Khan *et al.*, 1986). Combined with modern mean annual temperature this equates to burial temperatures of ~ 50–60 °C at the base of sections. Depth to water is usually < 10 m where village wells are located in wash bottoms. Sandstones seem to be the main aquifers. In all, modest temperatures plus local saturation make the potential for post-burial alteration a concern.

## Diagenesis assessed

As noted previously, the average $\delta^{13}C$ (PDB) of soil carbonate and coexisting organic matter differs by 15.7 ml$^{-1}$ within the range of difference observed in modern soils (Cerling et al., 1989; Figure 10.4). Preservation of this difference, had diagenesis occurred, seems very unlikely in light of the differing chemical natures, organic and inorganic, of the two phases. The absolute values for both phases are also consistent with observed modern values. These two lines of evidence are the most compelling argument against diagenetic resetting of the original isotopic composition of soil organic matter and carbonate.

The evidence from post-burial cements supports this interpretation. At Mirpur for example, the largest reservoir of carbonate in the upper Siwaliks, floodplain sediment is consistently 3–4 ml$^{-1}$ less negative in $\delta^{13}C$ (PDB) than the soil carbonate (Figure 10.5). It is probable that at depth the carbon species in groundwater are in isotopic equilibrium with this reservoir. This is reflected in the identical carbon isotopic composition of calcite cement in ellipsoidal sandstone nodules. Being much more permeable, the sandstones are likely to be the main groundwater aquifers. Partial equilibration of the soil nodules with this reservoir would have caused a negative shift. At +0.8 ml$^{-1}$, the nodules are already near the positive endmember composition observed for modern carbonate, and thus any such diagenetic effect is improbable. The same arguments hold for the Pabbi Hills where the $\delta^{13}C$ (PDB) of the parent material is even more negative.

Oxygen in sandstone cements is 3–5 ml$^{-1}$ more negative than that in soil carbonate (Figure 10.5). Either this reflects a different parent water from the soils or it means higher temperatures of precipitation for the cements in the presence of similar waters. A 3–5 ml$^{-1}$ difference implies ~ 0.5 km of burial, assuming the present geothermal gradient. But thin sections of the ellipsoidal nodules reveal little or no evidence of compaction: mica grains and other platey minerals are undistorted. Moreover, the cements in the nodules run ~ 35–50 per cent, implying cementation very early, before much compaction. Davies (1967) came to the same conclusions regarding these types of nodules from the Devonian in England.

The most likely origin, therefore, of the light $\delta^{18}O$ (PDB) results from the cements is precipitation at shallow depths in the presence of isotopically light groundwater. Groundwater underlying floodplains may well have been dominated by recharge travelling entirely underground from the high mountains, or fed locally by adjacent major rivers, or both.

## Upper Siwalik palaeoenvironments

Carbon isotope evidence from both soil carbonate and organic matter show that $C_4$ plants dominated floodplain biomasses throughout the Pleistocene. This means that grasslands dominated floodplains because the only $C_4$ biomasses of large extent in this region are grasslands. The few more negative numbers that were obtained probably originate from the local presence of $C_3$ plants in, for example, a riparian setting. These types of plants would come and go in the soil isotopic record as river and floodplain configurations randomly changed, or perhaps because climate periodically shifted in the Pleistocene. Since the two processes cannot be distinguished, the structure observed in the isotopic record cannot be interpreted as climatically driven.

A further important implication of the presence of $C_4$ plants throughout the Pleistocene is that the climate over that period was monsoonal and comparatively dry in character, as it is today. The basis for this is that $C_4$ grasses favour high growing season temperatures, although the exact temperature limits appear to depend on latitude (Rundel, 1980). For example, in the mid-west of the USA, $C_4$ plants will dominate where daytime maximum and minimum temperatures exceed 32 °C and 8 °C respectively (Teeri and Stowe, 1976; Ehleringer, 1978). The same figures for tropical grasses are ~ 21 °C and 9 °C (Rundel, 1980). They will dominate a year-round biomass only when such warm season conditions coincide with the main rainy season. Grasslands in general depend on moderate, seasonally distributed rainfall. Pronounced dry seasons, particularly where accompanied by frequent lightning and consequent brush fires, may be particularly critical to inhibiting encroachment of forest/shrub cover.

Such conditions basically describe the monsoonal climate of the area today. Seventy per cent of the mean annual rainfall of 40 cm falls in the warm summer months. Mean daytime maximum growing season temperatures are 30–35 °C, the range favoured by the $C_4$ grasses. Lightning activity often accompanies the sporadic pre-monsoonal (May–June) rains. Modern vegetation has been seriously disturbed by the grazing of goats and cows, and by cultivation. However, vegetation lists for nearby western India (Sankhla *et al.*, 1975) and measurements of extant modern grasses (Table 10.1) suggest that modern grasses in the Potwar region are largely $C_4$. A few areas along the active margins of the Jhelum river (Figure 10.1) appear to be undisturbed. A very large (> 3 m) $C_4$ grass (Table 10.1, sample 9601) comprises nearly the entire floodplain biomass. All indications are that the biomass on floodplains of the large river systems throughout the Pleistocene, up until disturbance by humans in the past few millennia, was a $C_4$ grassland.

Most of the prairie region of the mid-western USA is underlain by a class of soils which the US Soil Conservation Service calls mollisols. A number of diagnostic features of this type of soil compares favourably with the upper Siwalik palaeosols described above. Foremost, mollisols are distinguished by a prominent organic horizon, termed a 'mollic epipedon'. It is typically tens of cm thick, as are the palaeohumus zones in the upper Siwalik palaeosols. Modern mollisols can run as high as 8–9 per cent organic matter, much higher than for the palaeosols. But oxidative loss of organic matter during burial should not be surprising.

Mollisols are typified by > 50 per cent base saturation, as are the Pakistani palaeosols. This points to moderate leaching and therefore moderate rainfall levels. Although the possibility of diagenetic alteration of base saturation indices cannot be precluded, the moderate burial temperatures and the fact that carbonate at least is not being reintroduced into leached zones after burial encourages the view that soil pH has remained pristine.

Ca-smectite, illite, chlorite and vermiculite are the dominant clay minerals in mollisols (Brady, 1984, pp. 167–9), the same suite found in the B horizons in upper Siwalik palaeosols. Kaolinite is not typical of mollisols, and it is conspicuously absent in upper Siwalik palaeosols. Sub-100 °C burial temperatures and short burial times should not have led to major clay mineral alteration. It is interesting that the clay mineralogy of both soils and parent material is the same. The implication is that recycling of fine-grained sediment in floodplains was extensive. Certainly erosion of floodplain soils was the principal source of material for conglomerates in the upper Siwaliks, which are composed largely of abraded soil nodules.

Mollisols are usually accompanied by nodular calcareous horizons. Dispersed nodules rather than calcretes, irrespective of soil maturity, are common for mollisols and for the palaeosols. As noted previously, depth of carbonate leaching is difficult to quantify in many palaeosols because they are often composite profiles. In the isolated profiles that were measured (e.g. Figure 10.3), leaching thickness did not generally exceed 50 cm. Accounting for minor compaction, this sort of value suggests only moderate average rainfall throughout the Pleistocene, on the order of 40–60 cm yr$^{-1}$ (Jenny, 1941). While very qualitative, this value suggests a semi-arid to slightly wetter climate throughout the Pleistocene.

Oxygen isotopes for the upper Siwalik palaeosols fall in the same range as modern values for soils of the region (Table 10.1; Quade, unpub.). The oxygen isotopic composition of soil carbonate correlates with that of local rainfall (Cerling, 1984). The similarity of modern and ancient values suggests that, taken over the span of the entire Pleistocene, weather patterns and temperatures of carbonate precipitation have been fairly constant. This is consistent with the fairly uniform palaeovegetational picture based on carbon isotopes.

The $\delta^{13}C$ (PDB) of carbonate in post-Pleistocene soils is much more negative ($-5$ to $-7$ ml$^{-1}$) than in upper Siwalik soils. The cause is probably tectonic rather than climatic. The younger soils are located on broken topography typical for erosion of Siwalik beds. The modern vegetation is a mix of trees and shrubs, which are $C_3$ plants, and grasses, which are $C_4$ plants. Such a vegetational mix is reflected in the carbon isotope results. It is not known why the hills are partially forested, while the floodplains are grasslands. Damage owing to natural fires is probably less pervasive in the broken hill terrane, a factor which would encourage tree growth. Soil stoniness may be another factor. The results from the post-Pleistocene soils do serve to underscore the limitations of this palaeovegetational reconstruction. The chemistry and morphology of upper Siwalik palaeosols tell only about the presence of grasslands on the floodplains. It is quite possible that lowland hill country, for which there is almost no sedimentologic record, was partially wooded throughout the Pleistocene, as it is today.

# 11

# THE PALAEOCLIMATIC SIGNIFICANCE OF THE LOESS DEPOSITS OF NORTHERN PAKISTAN

## H.M. Rendell

## ABSTRACT

The evidence of glacial/interglacial cycles within the Quaternary sedimentary record of the Potwar plateau and adjacent areas of northern Pakistan is reviewed. Thermoluminescence (TL) dating of the loess deposits of the Potwar provides a chronology for the late Pleistocene and enables comparison with the record from deep-sea cores from the Indian Ocean and Arabian Sea. The record of loess deposition and preservation is used to infer changes in the degree of penetration of the southwest monsoon into the northern part of the subcontinent during the late Quaternary.

## INTRODUCTION

The last 2 myr of the geological record in northern Pakistan appear to be dominated by tectonic, rather than climatic, events. Although some evidence as to the number and the extent of Pleistocene glaciations of the Pir Panjal and Karakoram ranges does exist, evidence of the actual timing of these events is practically non-existent. In the unglaciated areas to the south of the mountains, evidence of glacial/interglacial cycles is both indirect and limited. Although potentially datable sequences of Pleistocene deposits can be found in the Peshawar intermontane basin, the main foredeep area, which now forms the Potwar plateau, is largely devoid of deposits that could be considered palaeoclimaticly significant. One of the consequences of the sequential nature of the tectonic disruption of the foredeep area has been widespread erosion of the eastern Potwar during the last 500,000 yr (Burbank and Raynolds, 1984).

The recent development of thermoluminescence (TL) dating techniques for sediments has facilitated the dating of the most recent portion of the Quaternary sequence and allowed correlation of the fragmentary terrestrial record with the continuous one from deep-sea cores. The results of a dating programme carried out on the loess deposits of the Rawalpindi and adjacent areas (Figure 11.1) are discussed in this paper.

*Figure 11.1* Location of study area in northern Pakistan. The Potwar plateau on both sides of the Soan river was emphasized in this study.

## EVIDENCE OF CLIMATICALLY RELATED DEPOSITS IN THE POTWAR AND ADJACENT AREAS

The establishment of a record of Quaternary glaciations has focused on the recognition of points of maximum glacial advance, the number of glacial advances, and glacial or glacially related deposits. Most early workers estimated that glaciation had extended to an altitude of ~ 2440–1800 m. Middlemiss (1896) first recognized evidence of glaciation in the Hazara area where he noted the presence of moraines at 1830 m. Subsequently, Norin (1925) identified terminal moraines at 2070 m in Kashmir and at 2440 m in Baltistan. Coulson (1938) concluded that glaciation extended to 1370 m in Hazara. Porter (1970)

*Table 11.1* Evidence of climatically related deposits in the Potwar area.

| Deposit | Proposed by | Current interpretation |
|---|---|---|
| 1 Presence of glacial erratics | de Terra and Teilhard de Chardin (1936) | Catastrophic flooding by Indus river Wynne (1879, 1881), Cotter (1933), Rendell *et al.*, (1989b). |
| 2 Bain boulder bed (englacial material) | Morris (1938) | Volcanic lahar Khan *et al.* (1985). |
| 3 Presence of terrace sequence in Soan valley related to glacial/interglacial conditions and representing three glacial/interglacial cycles | de Terra and Paterson (1939) | No complete system of terraces exists, the De Terra sequence is wrong in terms of both relative and absolute chronology Gill (1952), Rendell *et al.* (1989b). |
| 4 Loess interpreted as a lacustrine silt | Theobald (1877) | Loess interpreted as aeolian silt with some reworking in places Wadia (1928), Cotter (1933), de Terra and Paterson (1939), Gill (1952). |

identified a lower limit for Pleistocene glacial advance in Swat Kohistan at 1525 m and found traces of only three main periods of glacial advance in Swat Kohistan. Recently, Shroder and Khan (1988) have reported evidence of three glacial advances in the Nanga Parbat area. Deposits identified by earlier workers as moraines have been re-evaluated. Some of the Kashmir basin moraines identified by de Terra and Paterson (1939) have been reinterpreted as tectonically controlled fanglomerates (Burbank, 1983a,b; Burbank and Johnson, 1983). Holmes (1988) recently undertook a complete re-evaluation of the glacial record in Kashmir and concluded that the extent of Quaternary glaciation in Kashmir was less than previously supposed, with no tills identified below 2200 m.

Because moraines were restricted to the higher parts of the Pir Panjal and Swat Kohistan, there is no direct evidence of glacial advance in the Peshawar basin and Potwar plateau areas. Interest has consequently focused on a number of deposits potentially associated with glacial activity or with particular climatic conditions. These deposits are discussed fully by Rendell *et al.* (1989) and information concerning them is summarized in Table 11.1.

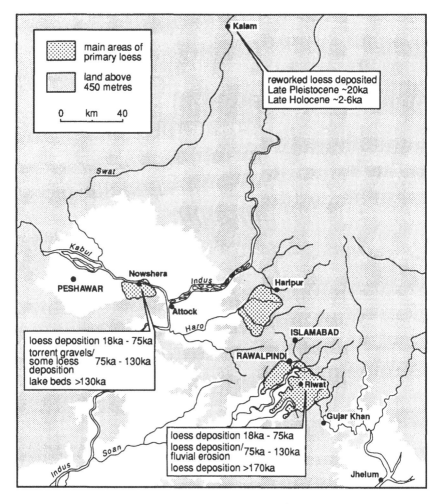

*Figure 11.2* Detailed map of study area and thermoluminescence (TL) dates of the deposit studied. The Potwar plateau occurs south of the highlands south of the Haro river, and on both sides of the Soan river, and east of the Indus river to Jhelum on this map. The term 'ka' indicates TL dates in thousands of years or × 10⁵ yr.

## THE LOESS SEQUENCE OF THE POTWAR AND ADJACENT AREAS

'Potwar loessic silt' is a term used (de Terra and Teilhard de Chardin, 1936; de Terra and Paterson, 1939) to describe the deposits of silt, sand and, in some areas, fine gravel which cover the eroded surfaces of the Siwalik strata and of the post-Siwalik Lei Conglomerate, within the middle Soan valley. Thickness of silt > 10 m are exposed in the Haro valley to the north and west of Hasan Abdal and in the Soan valley to the southeast and southwest of Rawalpindi. De Terra and Paterson (1939) thought that the main phase of loess deposition in the Potwar occurred during the Riss glaciation.

Loess is widely distributed in the northern part of the Potwar plateau, from the Haro valley in the west to the Gujar Khan area in the east (Figure 11.2). In the middle Soan valley and in the Haripur and Taxila areas of the Haro valley, the loess is a loess, in the strict sense of an aeolian deposit composed of fine quartz-rich silt. Fluvial agencies appear to have reworked the silts in other parts of the western Potwar.

Loess is an unusual deposit in the context of the sedimentary history of the Potwar area. It is an aeolian deposit in an otherwise fluvial sequence. Loess is traditionally regarded as a glacially related deposit. This is partly because of the continuing debate over the source of silt-sized quartz material (with glacial grinding providing one possible mechanism) and partly because, at a global level, loess is essentially a Pleistocene deposit.

It is generally recognized that the following conditions must be satisfied for loess to accumulate:

(a)  There must be a source of silt-sized quartz particles;
(b)  These particles must be exposed for entrainment by strong turbulent winds;
(c)  Suitable surfaces for deposition and preservation of loess must exist.

Much of the loess debate has focused on the source of silt-sized quartz particles, with the timing of loess accumulation associated with the periodic availability of a supply of suitable material (Smalley and Smalley, 1983). In northern Pakistan, the supply of silt-sized material is not a problem. Although the nearest source of 'desert loess' lies 700 km to the south of the Potwar, it is unnecessary to invoke 'desert' loess in order to explain the loess deposits in the Potwar. Processes of weathering and erosion in the rising Himalaya have ensured a continuous supply of fine-grained material to the plains; indeed, with the exception of relatively small pockets of loess or silt accumulation in intermontane basins, silts are effectively flushed out of the Himalayan system and on to the plains. The critical questions for loess formation concern the conditions under which aeolian entrainment of the silt-sized material and subsequent accumulation and preservation are favoured.

Silt-sized material could be exposed at a number of points along the mountain front, on braid bars and on the surfaces of alluvial fans. Entrainment would depend on the degree of exposure (lack of vegetation) and on the presence of strong turbulent winds, both climatically controlled factors.

# TL CHRONOLOGY OF LOESS DEPOSITION IN THE POTWAR AND ADJACENT AREAS

## TL dating of loess

The need to establish a chronology for loess deposition in northern Pakistan involved determining an appropriate technique. The loess contains nothing that could be dated by radiocarbon techniques, because it lacks both charcoal and humic material. Nor do the associated archaeological remains include hearths or other materials amenable to conventional thermoluminescence dating. Palaeomagnetic measurements of the loess merely indicated an age of <700,000 yr, but no other dates were available. In this event, the only option was to use a TL dating of sediment grains to date the deposition of the loess.

TL dating of sediments is based on the assumption that the exposure of sediment grains to light, prior to burial, has reduced the measurable TL signal to a value close to zero. In order to obtain a TL age estimate the following are required:

(a) Measurement of the Natural TL signal (TL accumulated since burial);
(b) Measurement of the TL sensitivity to radiation dose;
(c) An estimate of the annual radiation dose since burial. The flux of alpha, beta and gamma radiation is calculated on the basis of the uranium, thorium and potassium content of the sediment.

A crucial stage in the dating process involves a decision about how to calculate the Equivalent Dose (ED): the laboratory radiation dose that is equivalent to the radiation dose that the sample has received since burial. Detailed laboratory analysis revealed that, for the Pakistan loess material at least, the so-called 'regeneration method' was inappropriate (Rendell and Townsend, 1988). The TL ages reported in this study are therefore based on the 'Additive Dose method'.

## Laboratory procedures

All TL measurements were made on the 2–10 micron fractions of the loess. Carbonates and clay minerals were removed during pretreatment, leaving a mixture of quartz and feldspar grains. Batches of 40 discs were made up from each sample by pipetting aliquots of the 2–10 micron fraction, suspended in acetone, on to 10 mm diameter aluminium discs, and allowed to evaporate to dryness. TL measurements were made at an argon atmosphere at a heating rate of 150 °C min$^{-1}$, with a 1 min preheat at 230 °C. TL emissions were monitored through Schott UG11 (ultra-violet) and Chance-Pilkington HA3 (heat-absorbing) filters with an EMI19635Q photo-multiplier tube. Temperature and luminescence data were recorded using an Intertec microcomputer. Equivalent doses were calculated for glow curve temperatures of 270–320 °C.

## Dosimetry

Uranium and thorium contents were determined by either thick-source alpha counting or neutron activation. Potassium was determined by atomic absorption spectrophotometry. On-site dosimeters were used to measure annual gamma and cosmic radiation doses.

## Results of the TL dating programme

The results of a programme of TL analysis of over 80 samples indicate that the most recent phase of loess deposition in northern Pakistan dates to between 18,000 and 75,000 yr BP (Rendell and Dennell, 1987; Rendell and Townsend, 1988; Rendell, 1988, 1989). During the period 75,000–135,000 yr BP, loess was preserved in isolated pockets in the Soan area and was affected by contemporary weathering processes. Torrent gravels were deposited in the Peshawar basin during the same period. A much earlier phase of loess deposition, dating to > 170,000 yr BP, was also identified. It is concluded that:

(a) The end of the most recent phase of loess deposition in northern Pakistan coincided with the last glacial maximum in northwest Europe.

(b) Loess supply was more or less continuous from 135,000 ± 12,800 yr BP (i.e. Stage 5e of the oceanic record) until 18,000 yr BP but, before 65,000 yr BP, preservation was poor, with fluvial activity dominating the Potwar and Peshawar areas.

(c) The main phase of loess deposition was 75,000–18,000 yr BP. This loess is unweathered, pointing to a phase of increased aridity.

(d) A much earlier phase of loess deposition occurred before 170,000 yr BP.

A small study of Himalayan loess was also undertaken in addition to the main emphasis on the analysis of loess from the Potwar area. A limited number of samples of silts from Swat Kohistan, identified by Porter (1970) as loesses, were analysed. The silts overlying the Gabral II outwash terrace date to 18,000–20,000 yr BP, while those for the Kalam I terrace are Holocene, dating to 3000–6000 yr BP. Unlike the Potwar loess, the Swat Kohistan material appears to be essentially colluvial, albeit with a primary aeolian component. Clearly these dates cast some doubt on Porter's (1970) chronology, since the Kalam I material is late Holocene rather than Pleistocene in age.

## DISCUSSION: PALAEOCLIMATIC INTERPRETATION OF THE LOESS RECORD

The establishment of the dynamic causal mechanism for the southwest monsoon is the key to understanding recent climatic change in the northern subcontinent. The rates of uplift associated with the Pleistocene phase of the Himalayan orogeny could have had a significant effect on the climate of the foredeep area, the intermontane basins and the Tibetan plateau. Evidence that the Great

Himalayan range was at the same level as the Tibetan plateau 2 myr ago (Wang *et al.* 1982) implies that the southwesterly monsoon could have penetrated to Central Asia at that time. Subsequent uplift of the Great Himalayan range has blocked this penetration. This change can be expected to have had a marked effect on the albedo of the Tibetan plateau. Similarly, the late Pleistocene uplift of the Pir Panjal would have cut off the Kashmir basin from the influence of the southwest monsoon. The inferred rates of uplift in the Himalayan region recently have been challenged by Molnar and England (1990), who argue that they are more apparent than real. Nevertheless, the role of the Tibetan plateau (a land area of 2,400,000 km$^2$ at a height of 5 km) in forcing climatic patterns is not disputed. Duplessy (1982) points to current conditions in which the presence of a snow-free Tibetan plateau actually accelerates the onset and increases the intensity of the Asian monsoon, owing to the plateau's rapid heating during the spring and the resultant pressure difference between the high Asian continental mass and the Indian Ocean. Manabe and Hahn (1977) modelled a significant weakening in the southwest monsoon in response to an increase in surface albedo over an ice-covered Tibetan plateau.

The loess deposits provide a terrestrial record of continental aridity for the late Pleistocene; a record that can be interpreted in terms of the relative strength of the southwest monsoon. Climate controls the entrainment deposition and preservation of loess both directly and indirectly. The presence of vegetation cover in potential source areas for silt deflation will limit exposure and therefore entrainment. Some silt will fall in areas where reworking will be immediate, so that accumulation and preservation are possible only in areas that are not the focus of fluvial activity. In Pakistan, the loess deposits are preserved on high plateau surfaces and terraces above the present level of fluvial activity.

The picture of the late Pleistocene that emerges from the present analysis can be summarized as follows:

18,000 yr BP to present day    Re-establishment of northward penetration of the southwest monsoon. Fluvial activity is dominant in the Potwar and Peshawar areas.

75,000–18,000 yr BP    A weakened southwest monsoon leads to continental aridity (during oxygen isotope stages 2–4 of the oceanic record). Loess deposition and preservation are widespread.

135,000–75,000 yr BP    There is some monsoonal activity. Some loess continues to be deposited throughout oxygen isotope stage 5. This loess sequence is preserved only in isolated pockets. Much of the Potwar and Peshawar areas are dominated by renewed fluvial activity, with torrent gravels deposited in the Peshawar Basin.

The earlier part of the record is fragmentary but is dominated by fluvial activity with only one earlier ( > 170,000 yr BP) loess deposit preserved.

Evidence of climatic change on the Indian subcontinent has also been deduced from the analysis of deep-sea cores from the Indian Ocean and the Arabian Sea, and from cores taken from lakes in Rajasthan. This evidence is expressed in terms of either continental aridity or the relative strength of the southwest monsoon.

Data for the last 22,000 yr derived from palynological analysis of sediment cores from off the southwestern coast of India show an arid phase (22,000–18,000 yr BP) with a very weak southwesterly airflow. This was followed by a humid phase with progressive northward penetration of monsoon rainfall culminating in a maximum at 11,000 yr BP (Van Campo, 1986). Other workers, however, contend that aridity was still dominant at 11,000 yr BP (Hashimi and Nair, 1986). Data on monsoon strength, based on evidence of relative abundance of the Foraminifera *Globigerina bulloides* associated with upwelling along the Arabian coast, also point to a weakening of the monsoon at 18,000 yr BP (Prell *et al.*, 1980; Prell, 1984). Other deep-sea data confirm this monsoonal weakening at 18,000 yr BP, while pointing to an increase in the dry northeast monsoon (Duplessy, 1982), while data from Rajasthan confirm the existence of an arid phase ~ 18,000 yr BP (Bryson and Swain, 1981).

More recently, much longer term records have emerged from cores taken off the Arabian coast, and, in particular, from the Owen Ridge, during ODP Leg 117 (Clemens and Prell, 1989, 1990). The analysis of the terrigenous components within the cores has yielded data on both the aridity of the Arabian land mass and the relative strength of monsoonal winds (Clemens and Prell, 1989). These data do not, however, provide any direct indication of the extent of the northward penetration of the southwest monsoon into the Indian subcontinent, although measurement of monsoonal strength is a useful surrogate.

The loess record presented here is in reasonably good agreement with the results of the analysis of deep-sea cores off the Indian and Arabian coasts, as outlined above. The sudden cessation of loess deposition soon after 18,000 yr BP appears to be out of step with global trends of loess deposition but in step with a strengthening of the southwest monsoon. The Pakistan loess is unusual in terms of both its latitude (35°N) and its not exhibiting a typical 'loess/palaeosol' sequence. The sequence is instead punctuated by periods of erosion.

## SUMMARY

Loess is a climatically significant Quaternary deposit in northern Pakistan. The dating of the loess sequence has enabled correlation with the existing record from deep-sea cores. The timing of loess deposition and preservation have been used to infer changes in the degree of northward penetration of the southwest monsoon. The evidence of climatic change preserved in the fluvial sedimentary record of the Upper Siwalik and post-Siwalik deposits of northern Pakistan is more subtle and more difficult to decode than that of the loess. A great deal of recent work on the late Cenozoic of Pakistan has concentrated on tectonics, to the virtual exclusion of climatic change. It is hoped that future research will go some way toward redressing the balance.

# 12

# DECREASED TECTONISM OF THE SUI DOME

## D. Papastamatiou and C. Vita-Finzi

## ABSTRACT

The Sui Dome, which contains Pakistan's major gas field, lies south of a mountain loop linking the Sulaiman range in the east with the Chaman fault and Kirthar range in the west. The structure includes beds of the Plio-Pleistocene Siwalik Group. Linear fissures recently exposed by excavation have been interpreted as seismic fractures but they are more probably joints which have been enlarged by subsurface erosion to produce pipes. True faults are few and small in throw. The field evidence accords with the seismic records. This suggests that the structure is now inactive, perhaps because strike-slip movement on the Chaman fault can cope with north–south shortening in this part of the axial belt.

## INTRODUCTION

The Sui Dome, Pakistan's largest gas field, lies south of a mountain loop which links the Sulaiman range with the Kirthar range (Figure 12.1). These are the principal features of a major shear belt between the Makran ranges in the west and the Himalayas to the east and north (Abdel-Gawad, 1971). The dome, which can readily be picked out on satellite imagery (Plate 12.1), is a doubly plunging fold measuring ~ 30 km × 15 km at the surface. It consists of clastic beds of the lower, middle and upper Siwalik Group, which are generally thought to date from Plio–Pleistocene times (Figure 12.2).

At first glance the dome could be taken to represent an embryonic fold (Hobbs et al., 1976). The area is thought to experience the highest level of seismic activity in Pakistan (Quittmeyer et al., 1979; Figure 12.3), an assessment given added urgency by reports of neotectonic features on the dome and in particular of faults and cracks cutting through late Quaternary deposits. The present paper reports on fieldwork designed to evaluate the ground evidence for recent movement coupled with a review of the seismic record. In view of the considerable errors of location that often characterize teleseismic data, and the deficiencies in the historical catalogue of areas as sparsely populated as the Sui region, a geomorphological study seemed more than usually appropriate (cf. Allen, 1975).

236

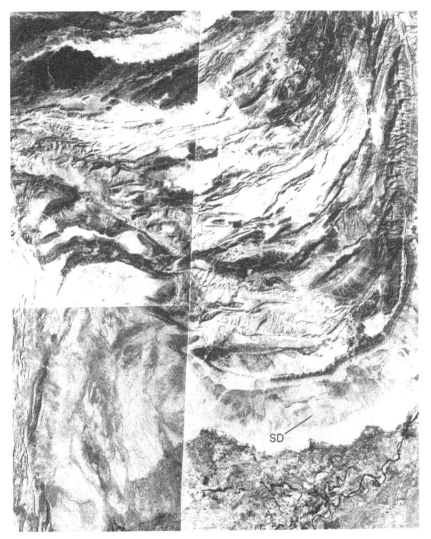

*Plate 12.1* Landsat mosaic of the region. The river is the Indus, SD marks the position of the Sui Dome. (Photograph: US Geological Survey, EROS Data Center, Sioux Falls, South Dakota)

## SEISMOTECTONICS

Figure 12.3 shows the location of epicentres in southern Pakistan for the period 1914–81. The epicentral plot inside the Quetta Transverse Zone shows a southern arcuate band of seismicity within 50 km of Sui and a northern band of east–west trending activity. The two zones show a strong correlation with

237

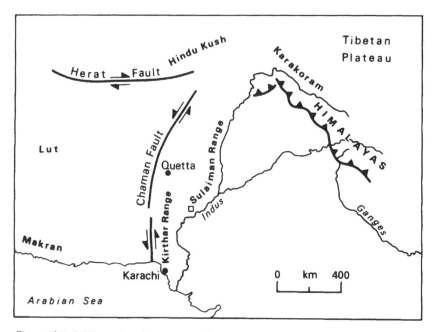

*Figure 12.1* Sui Dome (small square) in relation to major structural units in northwestern Pakistan.

geomorphological trends (Plate 12.1). The northern zone follows the trend of faults which can be mapped at the surface. This broadly east–west alignment of seismicity is associated with thrusting of the Indian plate as it moves north. The southern arcuate zone may be related to movements at depth which are expressed at the surface as a festoon of *en échelon* folding (Quittmeyer *et al.*, 1979).

Seeber and Armbruster (1979) concluded that the seismic data support the correlation of re-entrants along the fold belts with strike-slip faulting in the basement, notably in the Bannu and the Sibi re-entrants respectively east and west of the Sulaiman Arc. They also suggest that similar fault movement along the western Himalayan syntaxis (east of the Salt range) may now be fossil and that, in any case, active belts affected by decollement are likely to display complicated relationships between superficial structures and seismicity. Their interpretation of the Sibi re-entrant is supported by the dynamic manifestations of the 1909 Shahpur earthquake, west of Sui. This earthquake (M = 7.2) generated a meizoseismal area whose elongated shape suggests strike-slip motion in the basement which is not reflected in the geomorphology.

To judge from the satellite imagery (Plate 12.1) the presently active structural form is echoed in the broader inactive structures further north. The repetition of structure suggests that activity advances towards the south. Though localized,

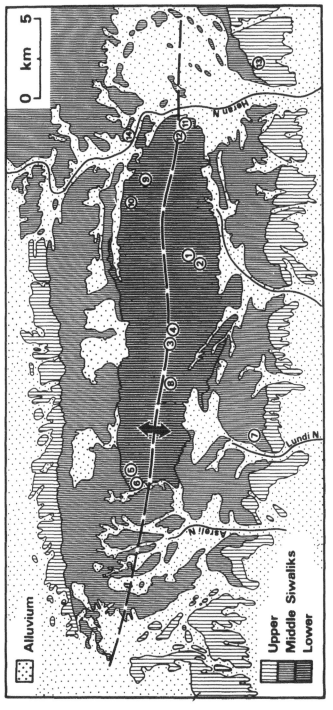

*Figure 12.2* Major subdivision of Siwalik Group within the Sui Dome and location of sites mentioned in text.

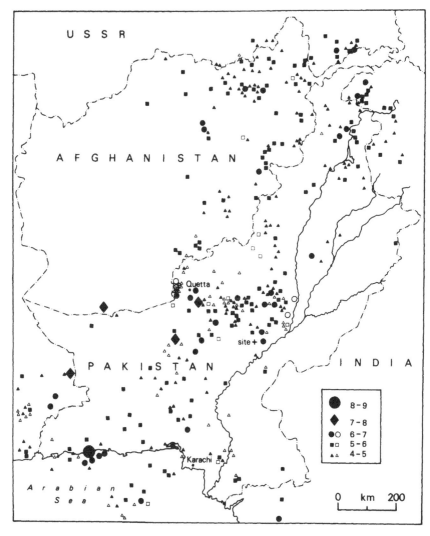

*Figure 12.3* Distribution of teleseismic events 1914–15 (solid symbols) and 1975–81 (hollow symbols) after Quittmeyer *et al.* (1979) and other sources.

this study may thus help to reveal whether any such southward advance is still in progress.

## Stratigraphy and structure of the Sui Dome

The strata exposed by erosion of the dome consist of gravels, sands and clays predominantly fluviatile in origin. Faunal evidence places the earliest beds in the

240

*Plate 12.2* View northward from Site 3 showing structural benches developed by relatively resistant strata. The concordance of heights, inviting explanations by phases of uplift separated by periods of stability, is illusory. (Photograph: Vita-Finzi, 1985)

Siwalik sequence in the Vindobonian (Pilbeam, 1972) dating from ~ 20 myr ago. The upper Siwaliks are commonly placed in the lower Pleistocene. Magnetic stratigraphy in the Salt range and in Kashmir suggests that deposition of the upper Siwaliks locally continued until 400,000 yr ago (Johnson *et al.*, 1979). Artefacts of middle Palaeolithic type were found on the surface of the upper Siwalik gravels near Sui (Site 13). They were unrolled and sufficiently numerous to indicate undisturbed surface sites.

It has long been accepted that the Siwaliks are composed of material supplied by uplift of the Himalayas (Gill, 1952). The assumption is that the basins in which the sediments accumulated were uplifted after the close of deposition, but there is some evidence for intrabasinal uplift while the upper Siwaliks were being laid down (Tandon and Kumar, 1984). In the Sui Dome, dips are lower for the younger strata than for the older but the differences are insufficiently well documented for syndepositional uplift to be estimated. Again, geologists of Pakistan Petroleum Ltd (PPL) claim that the Miocene Gaj/Nari formation shows some thinning beneath the dome, the implication being that an anticlinal axis was already in existence in mid-Tertiary times. This may be compared with the conclusion reached by Johnson *et al.* (1979) that the Pabbi structure, which they regard as the youngest element of the Himalayan fold belt in the Jhelum area, attained surface expression ~ 400,000 yr ago.

*Plate 12.3* Fill terrace (middle ground and left; car gives scale) and bench produced by hard bed (skyline) in Asreli valley (Site 5). (Photograph: Vita-Finzi, 1985)

The distinction between upper, middle and lower Siwalik units on the dome is based primarily on lithology because macrofossils, on which the Siwaliks have been subdivided elsewhere are lacking in the beds. The stratigraphical scheme used by the PPL geologists was drawn up in the Sibi re-entrant. The Jhutti nala section may be summarized as follows (PPL *pers. comm.*, 1985):

Upper Siwaliks     Pebbly conglomerates and gritty sandstone;
Middle Siwaliks    Calcareous sandstone and clays;
Lower Siwaliks     Calcareous grits and sandstones.

Dips are generally 2–3⁰, the highest recorded value being 4°25' where the lower Siwaliks plunge east near the anticlinal axis. Well developed stepped topography appears to reflect successive phases of erosion separated by episodes of uplift (Plate 12.2), but closer inspection shows that the benches are purely local and the products of relatively resistant horizons within the sands of the lower and middle Siwaliks. Likewise, what appear to be erosional terraces within the larger watercourses, as at Site 5 (Plate 12.3) and Site 6, are in fact either structural benches again or fill terraces representing slight changes in the distribution of annual rainfall.

## NEOTECTONICS

Reports of widespread evidence for recent extension near the axial part of the dome told of normal faults cutting late Quaternary gravels and picked out mainly by north–south wadis, and also of predominantly east–west cracks not far beneath the surface and indicative of seismogenic deformation. Analogy with other anticlinal or domed structures in late Cenozoic deposits (King and Vita-Finzi, 1980) would lead one to consider the possibility that the dome is the surface expression of a fault at depth and hence prone to seismic deformation. To judge from the case of El Asnam in Algeria, the most prominent faulting would be secondary and, if normal, not necessarily an indication that extension was the driving process.

The association between surface deformation and subsurface structures in areas where the seismic record is defective and aftershock data are lacking, needs to be supported by firm dating of the structures in question, especially if strike-slip movement is suspected.

No faults or folds at Sui could be detected in the upper Siwaliks and any minor structures that might be present were obscured by the relatively incoherent and coarse-grained nature of the gravels even at Site 13, where they are exposed to a depth of 20 m. The only interruption in the crude cross-bedding that was locally detectable consisted of minor joints along which some calcium carbonate had

*Plate 12.4* Normal fault in lower Siwaliks near Site 1. Tape measures 1 m. (Photograph: Vita-Finzi, 1985)

243

*Plate 12.5* Slickensides in probable fault gouge near Site 1. (Photograph: Vita-Finzi, 1985)

migrated. Similarly, no faults were found within the middle Siwalik exposures examined, whereas some east–west jointing was found notably at Site 7.

The lower Siwaliks sands alone displayed unambiguous faults but their throws did not exceed 25 cm. At Site 1, for example, a normal fault striking 350° dips at 55° with a slip of 20 cm (Plate 12.4). At Site 9 a normal fault with a similar orientation has a slip of 10 cm. At Site 10 a set of normal faults runs roughly north–south and has produced horsts and grabens with throws of 10–25 cm. One of these faults can be traced over a distance of 200 m. All of them are manifested at the ground surface by slickensided, cemented sands representing fault gouge consolidated by percolating or seismically pumped groundwater (Plate 12.5). In the normal faults at Sites 11 and 12 (where the outcome is a graben striking 80°), the cement is at least partly gypsum. It does not follow that faulting followed erosion: induration by calcium carbonate or gypsum is insuffi-cient for differential erosion to pick out the fault traces even if movement occurred when the middle Siwalik beds were locally still in place.

Although the small number of faults sampled precludes any statistical analysis of fault orientation, there is some accordance between the predominant north–south and east–west orientations favoured by the joints and the fault patterns summarized above, with the former trend most in evidence. In addition some strike-slip faults, both sinistral and dextral and with slips of 1–2 cm, were recorded at Site 14 (Plate 12.6) with a strike of 40°. The pattern of faulting as a

*Plate 12.6* Strike-slip fault in lower Siwalik sands at Site 14 (sinistral sense of motion; notebook gives scale). (Photograph: Vita-Finzi, 1985)

whole is thus entirely consistent with doming of the Siwalik series.

Elsewhere, undercutting by subsurface flow and channel erosion within ravines leads to slumping, and the resulting block rotation can be mistaken for normal faulting (Figure 12.4). The feature shown in Plate 12.7 appears to indicate a throw of 3 m (Site 1) but the absence of slickensiding and the rapid reduction in the 'throw' downstream (Site 2), though explicable by scissoring, is more simply accounted for by local sliding. The cemented gravels that provide the reference level for computing displacement at Site 1 are characteristic of the lower and middle Siwaliks, so that their disruption could date from any time since the beds were laid down. Their resistance to erosion explains why they

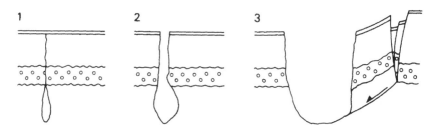

*Figure 12.4* Diagram showing how undercutting gives rise to linear topography.

survive along the wadi sides and hence give the illusion of being Quaternary terraces.

Further information on the current rate of deformation was provided by PPL engineers from an operation of the Sui gas field, which called for the lowering of boring equipment down existing drill holes to depths of ~ 1.5 km. The boring equipment is run through 250 m of limestone from the Sul formation overlying the gas reservoir within the very strict specifications of a tilt allowance of 1.5 per cent and a diameter allowance of 3.2 mm. This particular operation has been conducted at intervals of up to 20 years for the last 50 years without any failures and suggests that current uplift of the limestone does not exceed 0.15 mm $yr^{-1}$.

*Plate 12.7* Vertically displaced cemented gravels in lower Siwaliks (upper right; lower left) produced either by normal faulting or more probably by block collapse after undercutting by wadi flow. (Photograph: Vita-Finzi, 1985)

## Joints and pipes

Much more widespread than the genuine faults are east–west and north–south lineaments picked out by the drainage network and exposed here and there by erosion or excavation. Their origin becomes obvious only if one chances upon a wadi section at the appropriate location. They are the product of jointing in the lower and middle Siwaliks and the subterranean process of erosion, termed piping. Pipes are created by erosion, predominantly mechanical in nature, and result in tubular conduits. They can develop in a wide range of materials and climates. The one prerequisite is a steep hydraulic gradient near a steep free face. The process operates most readily where high infiltration capacities occur in

*Plate 12.8* Pipe (middle ground) leading to enlarged joint in bank of ravine. Note capping of encrusted deposits. Near Site 3. (Photograph: Vita-Finzi, 1985)

247

*Plate 12.9* Discontinuous stepped gully developed from partially collapsed pipe.
(Photograph: Vita-Finzi, 1985)

materials of low intrinsic permeability, e.g. where desiccation cracking is promoted by marked seasonal variation in rainfall. In addition, 'in arid areas major vertical pipes can also be found, suggesting tectonic joint control .... The largest pipes are usually associated with an indurated layer or cap-rock giving some protection against collapse and assuring more prolonged development' (Bryan and Yair, 1982). A well documented system of pipes in a badland area of southern Spain reveals some association between the conduits and gypsum veins; tension cracks on the walls of erosion gullies are also favourable to pipe development (Harvey, 1982).

Analysis of a sample of sand at Site 3, at a point where piping was well developed, showed it to consist of fairly well sorted, medium to fine sand of which 75 per cent comprised subangular quartz and the rest subrounded carbonate fragments. The water-soluble Na content was low (13.9 mg/100 g) but, in view of the smallness of the clay content fraction (13 per cent), this probably reflects leaching. In short, the deposit is well suited to the kind of mechanical erosion postulated above.

Surface crusting by gypsum and other salts conspires with the presence of cemented horizons within the Siwaliks to supply a protective capping to joints enlarged by piping, as at Site 4. The inlets to pipe networks (Plate 12.8) are also prone to enlargement by surface erosion and, in due course, break through the cap-rocks to produce discontinuous, stepped gullies which in turn give way to

*Plate 12.10* Close-up of partially collapsed pipe of Plate 12.9. (Photograph: Vita-Finzi, 1985)

fully integrated watercourses; hence the north–south wadi alignments whose resemblance to grabens is fostered by bank collapse along secondary pipes or joints (Plates 12.9, 12.10).

## DISCUSSION

Granted that, as in other contexts, absence of evidence is not evidence of absence, both the lack of unambiguous indications of faulting in the upper Siwaliks and the limited extent of genuine faults in the older constituents of the Sui Dome suggest that seismogenic surface rupture resulting from folding has become slight or negligible. This does not preclude continuing movement on subsurface faults. On the other hand, there is observational evidence for an abrupt drop in seismicity north of Sui.

Marked variations in tectonic activity within the axial zone have been noted by other workers. Auden (1974), for example, contrasted the highly seismic Quetta re-entrant with the Hazara re-entrant, where 70 years of record indicated feeble seismicity. Rowlands (1978) reported strike-slip movement on the Kingri Fault in the southern Sulaiman range during the Holocene, whereas Yeats *et al.* (1984) found no deformation of Holocene alluvium in a part of the Salt range where the upper Siwalik and later strata are strongly folded and faulted.

Yeats *et al.* (1984) explained the discrepancy by a faulting history with recurrence intervals thousands of years in length. The present authors feel that the Sui record represents abortive rather than incipient fold development. In other words, southward migration of tectonic deformation continued to the point where the southern segment of the Sulaiman Arc met the Sibi re-entrant, but any further regional shortening was then taken up by strike-slip movement of the Chaman fault complex.

# 13

# SURFACE SOILS AND INDUS RIVER SEDIMENTS

*Mirza Arshad Ali Beg*

## ABSTRACT

Surface soils of Pakistan contain predominantly transported materials, mainly loess, old alluvial deposits, mountain outwash, and recent stream-valley deposits. In general, three ages of deposition are recognized: Pleistocene, subrecent or older Holocene and the recent or younger Holocene. The piedmont plain of the Himalayan foothills and the Indus and its tributary floodplain deposits are the major areas discussed in this paper.

The alluvial sediments are rich in weatherable minerals contributed by crystalline rocks of the high Himalaya and the limestones, sandstones, shales and clays in the lower and middle Himalaya. Small tributaries from the Potwar and Salt range in the north and Kohistan, Sulaiman and Kirthar ranges in the south contribute additional material. Sorting of sediments, giving rise to different textures, took place at the time of parent material deposition. Climate, relief, groundwater, vegetation and human activities have controlled soil-formation processes in parent material.

Soils of the active floodplains are laminated silt loams, sands and silty clays. Their composition depends on the catchment area drained by the river in question. Their homogenization or destruction of stratification is limited to a few cm of the surface. Laminated but homogenized silty clays are the major constituents of the younger floodplains, while a wide variety in texture, with predominance of silty clays, silty clay loams, clay loams, loams and silty loams, characterizes older floodplains. The soils, moderately to strongly calcareous, have pH values in the range 7.8–8.4 in non-alkali soils, and of the order of 10.0 in alkali soils. Their organic matter content is usually < 0.75 per cent.

One of the most significant human activities has been river-damming which has changed sediment load downstream. Strong environmental damage is underway through accelerated soil erosion, water logging and salinization.

## INTRODUCTION

Extensive thick marine sediments were deposited on the edges of slowly subsiding geosynclines on the edges of the Indian and Eurasian plates from the beginning of the Palaeozoic era. The convergence of the plates occurred toward the end of the Eocene. This created the first land bridge across the Tethys Sea north of the Indian plate and restricted the flow of water into the sea (Crawford, 1979; Powell, 1979; Farah *et al.*, 1984b). The main collision of the Indian plate with the Eurasian plate took place toward late Oligocene, resulting in the uplift of the Himalaya to a significant level. Continued major uplift occurred during the mid- to late Miocene (Coumes and Kolla, 1984), and the Tethys Sea was closed further. The Indus river came into existence in about this time as well (Shroder, this volume). These events mark the formation of the land mass that became Pakistan.

The alluvial plain of Pakistan was formed in two stages: first by the deposition of alluvium in the Tethys Sea and later by tectonic uplift and further filling of the sea bed and formation of floodplains. This is suggested by the similarity and uniformity of the deposition of detritus by the ancient rivers into the sea trough (Nazir, 1974). Formation of land from sea seems to have followed the same pattern as at present; i.e. by transportation and deposition of alluvial matter during floods similar to the one that occurred in 1988 in the Indus and its tributaries (Tables 13.1, 13.2).

Tectonic movements, coupled with climate change and geomorphic adjustments in the Himalaya and its foothills, have continued during the Quaternary. Broad uplift of the Himalaya, local faulting and folding, periods of increased stream flow or aridity, increased periglacial and glacial activity, and stream changes through capture, avulsion, and other geomorphic adjustments have changed much of the drainage system of northern India (see Shroder, this volume, for drainage descriptions and site locations). Surface soils were formed during the filling of the various sedimentary basins with late Cenozoic sediments (see Quade *et al.*, this volume), and alluvium of 7–8 km thickness was deposited in the Indo–Gangetic basin at the foot of the Himalaya. Continued erosion of Himalayan uplands in late Cenozoic has led to deposition of gravels, sands and saprolitic clays (Lawrence and Shroder, 1985). Loess was deposited during the glacial periods (see Rendell, this volume).

The Indus plain is a system constituting the Indus and six major tributaries: one western, the Kabul river; and five eastern, the Jhelum, Chenab, Ravi, Sutlej and Beas, plus a number of other minor streams or channels. Their drainage basin extends over China, Afghanistan, Kashmir, Pakistan and India. The Indus system contributes the sediment and forms the vast plain. This plain, stretching from the south of the Himalaya and the Salt range to the Arabian Sea, is a land of several confluences called doabs: the Bari between Beas and Ravi, the Rechna between the Ravi and Chenab, the Chaj between the Chenab and Jhelum, and the Sind Sagar or Thai between the Jhelum–Chenab and the Indus.

The Indus and Sutlej have a trans-Himalayan origin while the Jhelum, Chenab, Ravi and Beas rise from the southern side of the Himalayas. The flow in the rivers results from snowmelt as well as intense, heavy and widespread

Table 13.1 Major characteristics of the Indus and its tributaries.

| River | Length | | No. of | | | | | Highest floods | | |
| | Total | in Pakistan | No. of tributaries | Structures across the river | Barrages | Dams | Year | Discharge (CFS) | 1988 discharge (CFS) |
|---|---|---|---|---|---|---|---|---|---|
| Sutlej | 900 | 336 | 8 | 14 | 5[a] | 4[a] | 1955 | 596,870 | 499,000 (gange brown) |
| Ravi | 410 | 370 | 5 | 11 | 3[b] | – | 1955 | 680,000 275,000 (Baloki) | 389,000 |
| Chenab | 440 | 352 | 12 | 13 | 5 | – | 1957 1973 | 1,100,000 854,391 (Qaderabad) | 892,299 |
| Jhelum | 430 | 319 | 10 | 6 | 1 | 1 | 1929 | 1,100,000 | |
| Upper Indus | 1830 | 430 | 27 | 9 | 3 | 1 | 1942 | 950,000 | |

[a] Three barrages and four dams in India.
[b] One barrage in India.

Table 13.2 Area affected as a result of 1988 floods.

| River | Section | Inundated area (km²) | | | |
|---|---|---|---|---|---|
| | | 12.9.88 | 6.10.88 | 7.10.88 | 8.10.88 |
| Sutlej | Minchinabad and Pakpattan to Hasilpur | 1620 | 3770 | 3600 | 3260 |
| | Hasilpur to Panjnad Headworks | 460 | 1250 | 1200 | 1800 |
| Ravi | Wagah to Sidhnai | 1890 | 3840 | 3690 | 3530 |
| Chenab | Akhnur to Sidhnai | 2240 | 3840 | 3600 | 3100 |
| | Sidhnai to Panjnad Headworks | 800 | 1990 | 1700 | 1590 |
| | Panjnad Headworks to Panjnad | 600 | 1900 | 1740 | 1700 |
| Sind | Panjnad to Sukkur Barrage | 3150 | 3640 | 3780 | 3780 |
| Total | | 10760 | 20230 | 19310 | 18760 |

Source: Data from Landsat/SPOT remote sensing satellite, kindly supplied by SUPARCO.

precipitation over the Siwaliks, the Himalayas and adjoining plains, causing flood flows between July and late September. The entire upper catchment area commonly is flooded because the flow volume increases beyond the drainage capacity.

The eastern tributaries of the Punjab join to form a single channel called the Panjnad below the multiple confluences. The Panjnad ultimately joins the Indus at Mithankot. From here the single Indus channel flows on low gradient to the sea (see Jorgensen et al., this volume). From Thatta the land constitutes the delta, marked by a number of river beds, some of which were long ago abandoned (see Flam, this volume). Below Mithankot, the river has changed its course many times and has been depositing much sediment during floods. The deposits are of comparatively recent origin. The coastal district is marked by many inlets and mangrove swamps.

The Indus river is mainly responsible for the formation of the surface soil of its plain. It brings its largely glacially derived sediment load of sand, silt and clay from the upper reaches of the northern and northwestern mountain system (Beg, 1977). The sediment deposits downstream with the decreasing gradient. The sediment load increases locally at tributary junctions and, by the time it reaches the middle Indus plain, the coarse sediment has largely settled out. For example, at Sehwan, 300 km upstream of the Indus mouth, the loss of sediment load is of the order of 20–50 per cent (Milliman et al., 1984).

The surface soils of Pakistan comprise the consolidated soils in the northern and the northwestern mountain systems and the unconsolidated soils in the Indus alluvial plain. The mountain soils form a separate belt; they are highly indurated and have little porosity. The unconsolidated sediment from the Himalayan foothills to the Indus delta is, in general, separated into gravels which were first deposited, followed successively by finer grades of sand, fine sand, silt and clay. Each flood or recharge of the streams has deposited fresh material on either side of the Indus or its tributaries which were flooding 40–50 km of their floodplains on neighbouring flat land (Table 13.2). The deposits have, therefore, been changing their texture from coarse to fine with the intensity of the floods. The Indus alluvium extends from the Salt range southward to the Arabian sea.

## GENERAL NATURE OF THE SOILS

The crystalline and sedimentary rocks of the northern region have been extensively folded and faulted. High angles of dip and closely packed folds are common there among the consolidated rocks, resulting in rapid overland flow, stream flow and erosion. Within the mountains, valleys with alluvial deposits are common. Some of the main depositional valleys are Mardan, Peshawar, Bannu–Kohat, Dera Ismail Khan and the Potwar plateau (Yeats and Lawrence, 1984).

The Peshawar valley is characterized by folds and faults of Precambrian–Tertiary era. The Precambrian rocks are composed of slates, shale and hard sandstone. These rocks are of Jurassic, Cretaceous, Palaeocene and Eocene age and are represented by limestone, sandstone, shales and claystone. These rocks form the bedrock under the alluvium which consists of sand, clay and gravel, and which is > 200 m thick.

The Bannu–Kohat basin shows the normal deposition pattern of coarse alluvium or mottled sand on the plains and moderately fine deposits in the lowest part of the basin. Thick clay zones occur quite commonly here. The Dera Ismail Khan area is primarily an alluvial plain which slopes from the northern and western parts of the mountain system to the Indus river. The rocks are of sedimentary origin and their age ranges from Precambrian to Pleistocene and Holocene.

The Potwar plateau represents a trough between the Salt range and the mountain foothills. It extends northwest to Attock and is bounded by the Indus in the northeast and south. The Siwalik system of the Himalayas, located on its north and east sides, was subject to folding and faulting during the Pliocene and early Pleistocene. The period of uplift and deformation alternated with periods of erosion. The mantle of silt, locally mixed with water-laid sand and gravel, was deposited in irregular troughs during the Pleistocene epoch. Thus it has a complex geological history of mountain building, alluvial and loessic deposition, and erosional cycles. It is underlain by interbedded conglomerates, sandstones, shales and siltstones. The uppermost wind-laid silty loess is of uniform texture. The loess deposits can, by their structure, be placed in two age groups: the middle and the late Pleistocene. The lower one is denser and mottled, the upper one is porous and of uniform colour. The present-day soil seems to have developed exclusively in the upper loess. The mottled loess is replaced in places by thick, sandy or stony piedmont deposits while, in others, the two loesses are separated by piedmont material. The extensive deposits of gravel in the northeastern part of the area seem to be stream gravels and glacial outwash from the lower Himalaya (Johnson et al., 1979; Recon. Survey, Rawalpindi, 1967).

The Indus plain starts south of the Potwar region and includes the piedmont areas that form a transition between the mountainous system of the Salt range and Himalayan foothills, and the Punjab alluvial plain. The Indus plain extends south to the Kirthar range and the coast. The area immediately to the south of the Potwar region is Gujrat which consists of an alluvial plain, with the exception of the Pabbi hills which form part of the Himalayan foothills of the Siwalik group. The eastern part of this region is a piedmont of the Himalayan foothills whereas there are river deposits on the western part. Three geomorphological surfaces can easily be recognized in this region: the oldest and highest surface belongs to the late Pleistocene, the middle surface belongs to the early and mid-Holocene, and the lowest surface to the Holocene where the deposition of new material continues. The eastern piedmont plain seems to have been formed by sheet-flood deposits of hill torrents such as the Bhimber and Bhandar streams and can be divided into the old piedmont plain, the young piedmont plain and the piedmont basin. The old piedmont plain in the north of Jalalpur and Gujrat occupies a high position and is characterized by gully erosion; the young piedmont plain is characterized by being nearly level; and the piedmont basin located around Dinga at the southwestern end of the piedmont deposits is characterized by its clayey soils (Recon. Survey, Gujrat, 1968).

The western part of the region can be divided into the Kirana Bar and flood-plain. The Kirana Bar has a uniformly level surface in its scalloped interfluve and the channel–levee remnants, which are old infilled stream beds. The floodplain can be subdivided into the cover floodplain, the meander floodplain, the young channel–levee remnants, the active floodplain and the braided river bed. The cover floodplain belongs to the Jhelum floodplain while the meander floodplain forms the major part of the Chenab floodplain. The young channel–levee remnants are the major river bed deposits occurring within the meander flood-plain while the active floodplain occurs in a narrow strip, along rivers and streams, which is flooded every year. The braided river includes the present beds of the river and streams.

South of Gujrat is the Gujranwala region which again consists of several hundred m of late Pleistocene and Holocene deposits. More than 60 per cent of the area consists of floodplain deposits of the Chenab and Ravi while the remaining area in the northeast is again a piedmont plain of the Himalayas. Several old basins and channel infills are located toward the southwest of the latter plain. Three ages of the deposits occur: the late Pleistocene, the subrecent or older Holocene and the recent or younger Holocene. The late Pleistocene surface occupies a high position in the piedmont plain. The subrecent deposits lie on it at some places while in the floodplains they are not clearly marked but are the main channel and old basin infills (Recon. Survey, Gujranwala, 1965).

Sheet-flood deposits from Basantar, Bain and Deg nala formed the piedmont plain in this region. It can be subdivided into the dissected piedmont plain of Pleistocene in the north of Shakargarh, the subrecent in the areas around Sialkot and north of Narowal, and the Holocene along the Deg nala south of Sialkot. The floodplain comprises landforms such as the old river terrace, subrecent deposits on the old river terrace, subrecent floodplains, and the younger Holocene floodplains. Late Pleistocene deposits in the old river terrace are extensive in the central part of the region. The subrecent deposits on the old river terrace cover the strip extending southwest from Wazirabad, widening toward the Sheikhupura border. The subrecent floodplain flanks the terrace on both sides while the younger Holocene floodplain occurs along the rivers where deposition continues (Recon. Survey, Gujrat, 1968).

The Sahiwal region in the south of Lahore is part of the Punjab plain where extensive sedimentation and deposition of alluvium has occurred. The deposit is > 350 m thick at places (Kidwai, 1966). The age of deposition is late Pleistocene, subrecent, and Holocene. The late Pleistocene landform occurs in the central high strip of land known locally as Ganji Bar. The subrecent deposits occur in the floodplains of the Ravi, Sutlej and Beas rivers and extend over a wide area of the region, while the Holocene deposits occur along the present rivers and the old bed of Beas (Recon. Survey, Sahiwal, 1968).

The soils of the region in the middle and lower Indus plain are part of the vast fluvial system. The entire land is located in the tectonic trough which has been filled by alluvium derived from the Himalayas by the Indus and its tributaries. The parent material is mixed calcareous alluvium ranging in age from late Pleistocene to youngest Holocene. It has been deposited on a basement of Tertiary rocks which outcrop sporadically. The rocks consist mainly of hard massive limestone of early Eocene. The soils are moderately to highly alkaline with a pH range of 8.2–10.0 depending on whether they are located in less alkaline or strongly alkaline zones. They have organic matter < 75 per cent. Their texture ranges from sandy to clayey, with clay dominating toward the south (Recon. Survey, Multan North, 1969, Hyderabad, 1970, Thatta West, 1979).

Variations in the land surface and in elevation, parent material, and degree and kind of soil development suggest that different parts of the plain were deposited in different ways and at different times throughout the Holocene. The area can be divided into three main physiographic units: Holocene floodplains, including active floodplains where the deposition of fresh sediments had continued until recently or is still going on; subrecent or older Holocene floodplains where the river activity ceased long ago; and the subrecent dissected river terrace, which was possibly formed during the early mid-Holocene and dissected later by flood channels. These physiographic units can be further subdivided into different landforms, as detailed below.

## Active and Holocene floodplains

These occur as a narrow belt mainly between the present course of the Indus river and its flood protection embankment. They are subject to seasonal flooding for 2–3 months, with the lower parts flooded longer and deeper than the upper. Patches of soil along the river course or its creeks are subject to erosion or burial by fresh sediments. Small patches occur outside the protection bunds and are not flooded. The main features of the active floodplain are that it has nearly-level to gently-undulating meander bars and levees, level plains, basins and channel infills. The soils are mainly laminated silts and clays. In the basins of the Holocene floodplains, soils are homogenized to ~ 45 cm depth. Sandy soils are not extensive.

## Subrecent floodplains

These form the major part of the region and are slightly lower than the Active and Holocene floodplains. They have been protected from the seasonal floods by embankments. Small patches occur inside the bunds and are inundated by floods. This portion of the plain has been subjected to great river activity during the past. The surface alluvium belongs to the sub-Holocene period, ranging from 300 to a few thousand yr old. The distinctive features of this unit are level plains, meander bars, levees and basins. Meander bars and levees occupy higher positions. Soils on the meander bars and levees vary from loamy, very fine sand to silt loam, or very fine sandy loam, whereas in the level plains and basins silty clay and clay occur. The soils on the meander bars and levees are either stratified or homogenized to shallow depths only, but in the level plains, basins and channel infills they are deeply homogenized.

A considerable part of the Multan area, for example, is occupied by saline–alkali soils which are generally porous, but some are dense and unstable. A small patch in the southwestern part has gypsiferous soils concentrated in the upper 30 cm. The pH values are 8.6–9.0. Soils with brighter coloured subsoils occur as levees and scattered clumps. These soils are loamy sands and loamy very fine sands in texture. At places they are also saline–alkali.

## Subrecent dissected river terraces

These are characterized by a scattered distribution mainly in the eastern and southern parts of the plain. They occupy slightly higher positions than the surrounding subrecent floodplains. They seem to have been deposited during the early period of the mid-Holocene and dissected later by the subrecent flood channels. The soil is gypsiferous, which is attributed to intermittent salinization by side seepage into the surrounding lower areas by floods for long periods that have made them hygroscopic. Level plains and basins are the salient features of the terrace. The soils in the level plains are mostly laminated, whereas in the basins and channel infills they are homogenized to ~ 60 cm. Much soil on the old river terraces occurs as an isolated island of saline–alkali silt loam, silty clay loam, and silty clays with pH ranges of 9.0–10.0. Calcrete kankars (nodules) occur on the surface or in the profile at some depth.

## Tidal plains

In the southernmost area of the country, tidal plains have been the active part of the river delta in the past but when the river shifted its course some centuries ago the plain was encroached upon by ocean tides. The landform consists mainly of tidal flats, tidal basins, ridges, tidal lakes and creeks. Tidal flats and basins are subject to frequent salt-water flooding. The estuarine plain occupies much of the delta, and has resulted from recent deposition from high tides into salt marshes.

## SOIL FORMATION PROCESSES

Factors affecting soil formation are climate, particularly temperature and precipitation; the nature of the parent material, e.g. texture, structure, chemical and mineralogical composition; topography of the area; living soil organisms and vegetation; and the length of time during which the parent material has remained at the site for soil formation. Climactic influences dominate soil formation. Some chemical reactions double their rates for every 10 °C in temperature. Biochemical changes by soil organisms are temperature sensitive as well. Topography can hasten or delay climactic forces. Rolling topography encourages natural erosion of surface layers. If this is extensive, deep soil formation is inhibited.

Soils of the Potwar area either have developed in transported materials, such as loess, old alluvium, mountain outwash, and recent stream valley deposits, or have formed *in situ* from underlying sandstones and shales. Climate is subhumid in the northeast, and changes gradually to semi-arid in the southwest. The original plant cover probably consisted of subtropical, open thorn thicket with short and medium grasses. Relief is variably sloping, gently rolling or undulating, nearly level, or depressional. Soil ages vary from late Pleistocene to youngest Holocene. Humans have also influenced soil formation. Removal of natural vegetative cover and subsequent intensive agriculture have accelerated soil erosion and have depleted organic matter.

Farther south, the parent material of the soils of the Gujranwala are mixed calcareous alluvium derived from a wide variety of rocks from the Himalaya. The source rocks are calcareous sandstones, shales and clays from the lower and middle Himalaya, and igneous and metamorphic granites, schists, gneisses and slates from the high Himalaya (Bakr and Jackson, 1964). The material has been transported and deposited by the major and minor streams of the region.

The area has both piedmont and floodplain deposits. The piedmont material originates from the Himalaya foothills. The foothills consist of sedimentary rocks that were eroded from igneous, metamorphic and old sedimentary rocks without much pre-weathering. Therefore, the piedmont and floodplain materials have a more or less similar mineralogy, and differences in the texture of the materials are considered as differences in parent material. Sorting of the material into different size fractions took place during transportation and deposition with the result that sandy materials, for example, occur as stream levees, and clayey deposits occur in back-slope basins or as channel fills.

The significantly different materials that occur in the Gujranwala area are: sands and loamy sands, fine- and medium-sandy loams, silt loams, loams, and silty clay loams, including silty clays. Differences in parent material explain soil differences where other soil-forming factors are uniform. In old deposits, for example, in the subhumid area and under good drainage, the Wazirabad and Pindorian series have developed in loamy sand; the Khair in sandy loam, and the Shahdara in silt loam.

The soils of the Lahore area have developed in mixed calcareous alluvium derived from the calcareous sandstones, shales and clays of the lower and middle Himalaya and from the crystalline rocks of the high Himalaya. The parent materials again range in size from clay to sand (Beg *et al.*, 1989). The climate of the area is semi-arid, subtropical and continental, approaching subhumid in the northeast. The original plant cover over most of the area was probably a tropical thorn forest, but most has been cleared away for cultivation. Relief is nearly level, with local depressions, convexities and gentle slopes. The age of soils ranges from late Pleistocene to youngest Holocene. Groundwater is fairly deep, but wherever large canals cut through underlying sandy strata, a considerable rise in the water table has occurred.

The sediments deposited in the Holocene and on active floodplains around Lahore are least affected by soil-forming factors. These soils have formed in Holocene sediments, part of which are still subject to annual accretion. They are calcareous and lack profile development, although some homogenization has taken place in the top soil series owing to biotic activity.

The soils of the meander and cover floodplains have mainly developed in subrecent sediments where the soil-forming factors have been operative over longer periods. The original stratification has been largely destroyed and a weak structure has developed in the upper 60–90 cm owing to activity of roots and organisms, and changes in volume through alternate wetting and drying of expandable soils. The substrate, however, is still stratified and the profiles are almost uniformly calcareous throughout.

The soil-forming factors have proceeded to greater depths and to higher degrees in late Pleistocene deposits occupying the old river terraces. Rain and groundwater wetting, followed by drying, have caused solution of part of the lime in the soil and reprecipitation as lime nodules or kankar, usually in definite zones. Kankar zones are well developed and fairly shallow in the margins of channels and depressions where carbonate-bearing groundwater evaporates from a shallow front. Part of the primary minerals in the subsoil have weathered to clay. Sesquioxides thus liberated have strongly coloured the subsoils. Humus incorporated by roots and soil organisms have made it darker to a considerable depth than the younger soils. The soils are completely homogenized to 150 cm or more, and a stable and porous structure has formed by the combination of calcium saturation and long-continued biotic activity. Part of the structure in the fine-textured soils results from swelling and shrinking of 2:1 lattice clays. These processes and characteristics are common to the calcareous and non-calcareous soils of the old river terraces.

In the northeastern part of Lahore, the fairly high rainfall has decalcified the late Pleistocene soils to 90–120 cm depth on level and concave sites and has caused some clay illuviation. Such soils are absent in the drier southwestern parts, except in depressions that collect rainfall. These processes have not been active on convex sites around the Bhalwal, Jaranwala, Hafizabad, Resalpur and Sindwan soil series where runoff has predominated over infiltration with the result that the soils are still calcareous throughout. The calcareous soils are more extensive in the central part of the area with relatively lower rainfall, where they also occupy level to nearly level sites, and in the southwestern part where they occur in the shallow depressions as well.

In some regions, mainly in depressions, groundwater has saturated the soils seasonally, causing mobilization and reduction of iron. Upon drying, the iron oxides tend to concentrate around pores and voids where air comes in and water evaporates. This makes the soil mass dark grey with brown mottles, as opposed to the uniform colours of the most soils mentioned above. Mottled soils comprise the Lalian, Bahalike, Pacca, Kamoke and Sagar series.

In some of the depressions and well drained sites mentioned above, leaching of salts has dominated over concentration owing to evaporation as groundwater moved into upper soil horizons by capillary action. Furthermore, the soils have been under continual homogenizing action of roots and soil organisms, which has produced a porous structure stabilized by calcium and organic matter. On the margins of channels and depressions, however, where most of the precipitation is lost to the adjacent lower sites without infiltration but evaporation of shallow groundwater proceeds normally, salts have concentrated in the upper part of the profile. During occasional leaching, a part of the salts is moved down in the profile again, and another part is taken by the runoff. Owing to the predominance of sodium bicarbonate over calcium and magnesium salts in much of the groundwater in the plain, these processes of evaporation alternating with occasional leaching over time have given rise to soils with high exchangeable sodium and high pH. Such soils are unstable when subjected to mechanical disturbance. Vegetation on these sites also deteriorates, and then the organic matter content of the topsoil decreases, which further lowers the structural stability. When this stage is reached, every rainstorm disperses and brings into suspension topsoil material which runs down the pores both in the topsoil and subsoil and is deposited as dense, commonly laminated infillings. The denseness typically does not result from clay illuviation but from washing in of complete topsoil material. No natric horizons appear to occur in spite of the high proportion of exchangeable sodium that occurs in the Firoz, Khurrianwala and Sindhelianwali soil series.

Some soils have been subject to the above series of processes for a short period and have most of their subsoil porosity intact. The topsoil is generally dense already, but the organic matter content is higher and the exchangeable sodium percentage lower than in the extreme soils described above. These porous saline–alkali soils comprise, for example, Gijiana, Jhakkar, Kasur and Niazbeg series.

The soils of the Multan area have, as in the case of the areas along the Indus nearer the Himalaya, formed in mixed calcareous alluvium derived from the Himalaya but deposited by the Ravi and Chenab rivers. The Beas river has also formed narrow strips along its old course (see Shroder, this volume, for locations). Some wind resorting has also taken place. During deposition, sorting was dependent upon proximity to the river, with sandy materials deposited in levees adjacent to active channels, medium-textured materials on the backslopes, and fine textures in the basins and abandoned channels. The original stratification of the alluvial deposits has been destroyed and most soils have structural horizons (Recon. Survey, Multan North, 1969).

The soil-formation processes in the middle and lower Indus plain comprising Sukkur, Hyderabad and Thatta regions have mainly involved homogenization (Recon. Survey, Hyderabad, 1970, Thatta West, 1979). The arid climate of the southern area, with rainfall of 100–150 mm, could have inhibited homogenization, but major parts of the floodplains have been flooded extensively in the past. The flooding, accompanied by a large sediment load, provided the required moisture and bonding material for this process. Because the general depth of homogenization depends on moisture and relief, the medium-textured soils are 45–90 cm thick and occasionally > 90 cm (e.g. at Sultanpur, Nabipur, Awagat and Bagh). The soils of the active floodplains in the middle plain are laminated and no homogenization has taken place there. New sediments are deposited by floods almost every year and thus the original stratification has not had enough time to be destroyed except for a few cm of the surface, as in the Shahdara, Rustam and Sodhara series. In the case of clayey soils, which occupy the lower plain and have more effective moisture, the soils are homogenized comparatively rapidly and their depths are 37–75 cm (e.g. the Matli, Miani, Shujabad, Pacca and Sindhu series). Some deeply homogenized soils (e.g. the Shikarpur and Shahpur series) also occur. These soils have homogenized to the extent of being placed in the subrecent class. The soils of the old river terrace have developed from sediments deposited in the late Pleistocene and these have homogenized to a depth of > 150 cm (e.g. the Juranwala, Faisalabad and Hafizabad series in the middle plain).

The soils of the lower plain have been affected by soil-forming factors and processes resulting in a definite profile development with distinct horizons. The alluvium was originally stratified but stratification has largely disappeared as a result of biotic action upon the parent material, as conditioned by time and surface relief. The parent material was deposited in a typical riverine pattern; i.e.

sand bars, silt levees on meanders, and fine silts and clays in basins or channel infills.

Moisture and relief are the main factors affecting the degree and direction of soil formation in the lower plain. Most of the profile development may be attributed to moisture from seasonal river floods prior to construction of flood embankments, rather than to rainfall. The embankments can now contain even high floods as occurred in 1988. The floodwater and sediments are, therefore, not available for soil formation.

Even minute differences in relief have affected the degree of soil formation significantly in this region. Areas occupying relatively lower physiographic positions receive more moisture as compared to those occupying higher positions. Consequently, the low-lying areas contain soils that are fairly deeply developed, whereas those occupying higher physiographic positions are either stratified or developed to a shallower depth. The water in low-lying areas moves down through the soils, leaching the excess surface salts into the groundwater and giving rise to non-saline soils at such places as Matli and Pacca. In the higher, mainly uncultivated areas, on the other hand, the salts accumulate at the surface by evaporation of groundwater owing to continuous upward capillary movement from the saturated subsoils. As a consequence, strongly saline or saline–alkali Jhakkar and Garhi soils have formed. The process of alkalization in the middle and southern plain has reduced plant and animal activity and has thus reduced the porosity of the soil.

Soil in the coastal zone has also formed from the material transported by the Indus and was at first low in salt content. Soils occupying tidal flats are now loaded with salt. The estuary plains are also characterized by level silty flats and clayey basins with saline surfaces owing to hygroscopic salts. Collection of tidal water has turned the flats swampy (Recon. Survey, 1979). Salinity intrusion is now claiming more land owing to reduction in flow of fresh water.

## RECENT CHANGES

Human activities in the Indus catchment have considerably degraded the surface soil over the past 50 yr. Indiscriminate tree felling for fuel and agricultural land clearing have resulted in soil erosion and silting of dams. The construction of dams and barrages across rivers and construction of an intricate canal network also has done considerable damage to the environment. Canals have brought water into arid areas, but many soils have been damaged by changes in soil–water relationships. The saline–alakali soils have been reclaimed at places where the water is in excess. At other places where irrigation water is insufficient to leach salts, the water table has risen owing to seepages from the canals and the soils have become saline. Almost 40 per cent of 15.3 million hectares of irrigated land thus has been lost owing to salinization and alkalization, out of which 4 per cent is disastrously waterlogged to within ~ 1.5 m of the surface.

Construction of dams, barrages and canals has reduced the sediment load from $> 675$ mt yr$^{-1}$ in the 1930s to $< 100$ mt yr$^{-1}$ recently. Over 250 mt yr$^{-1}$ used to flow into the estuaries but now almost nothing comes in because of the upstream diversions (Milliman *et al.*, 1984). That the sediment is arrested by structures is evident from measurements of 120 mt yr$^{-1}$ caught at Mangala dam between 1967 and 1973 (Saeed, 1974). Reduced flow has resulted in an increased groundwater salinity and in salt water intrusion from the sea in winter at Ghorabari, $\sim 100$ km upstream. This has inhibited growth of mangroves and fisheries.

Irrigation has fundamentally changed the natural soil-forming properties of the lower Indus plain and has modified the hydrology of the plain to such an extent that the microclimate over the irrigated area has become subhumid. Drainage properties of the soil are poor, particularly in places where clay is dominant. A vast area of land is waterlogged and new marshes have formed. A major ecological disturbance has taken over the surface soil of the lower Indus plain.

# 14

# FLUVIAL GEOMORPHOLOGY OF THE LOWER INDUS BASIN (SINDH, PAKISTAN) AND THE INDUS CIVILIZATION

*Louis Flam*

## ABSTRACT

Aerial photocomposites and field research reveal numerous post-Pleistocene fluvial remnants associated with former courses of the Indus river in the lower Indus basin of Sindh province, Pakistan. Geomorphologic and historical data were used to connect fluvial remnants and to delineate former courses of the Indus river from northern Sindh to the Arabian Sea. Sindhu Nadi was a prior course of the Indus river during 4000–3000 BC. Also at this time a second exclusive perennial river, Nara Nadi, existed in the east part of the lower Indus basin. Sindu and Nara Nadi were confluent in the southeastern lower Indus basin, with the prehistoric coastline occurring north of the present location. Ancient river course locations affected prehistoric Indus civilization ( ~ 2600–2000 BC) through flood hazards and access to water for irrigation.

## INTRODUCTION

The study of fluvial systems in prehistory is an integral component of geo-archaeology, with the objective of linking synchronic and diachronic cultural adaptations with fluvial landforms and their evolution. Archaeological research in the Indus river valley of Pakistan has documented the development of pre-historic cultures during the fourth and third millenia BC (4000–2000 BC), yet no detailed research has specifically addressed issues concerned with landforms and dynamics of the Indus river that had significance to human behaviour in pre-history. Descriptions of the physical setting of archaeological sites and the linkages between sites as living settlements of the past and their palaeoenvironments have been based on present conditions. The location of the main channel of the Indus river has been assumed by many to have remained unchanged since the Pleistocene. Until recently (Flam, 1981a, 1986a), Fairservis (1967) was the only archaeologist to recognize that the Indus fluvial palaeoenvironment may have been different from that of today, while Holmes (1968) explicitly delineated more recent, historical changes in the course of the Indus river and showed them to have been radically different from modern conditions. Furthermore, archaeological and historical studies have documented the effects of Indus river course changes on cultural developments in the lower Indus basin during historical times (Lambrick, 1975; Flam, 1986b), and this must have certainly been the case during the prehistoric period. The Indus river in both present and past has been a fluvial system of tremendous magnitude and dynamics. A more explicit approach to the behaviour of the Indus is needed to reconstruct its palaeoenvironment and the human adaptations to it. This paper uses scientific evidence to delineate the course(s) of the Indus river during 4000–2000 BC and suggests salient implications of river course locations for prehistoric cultures of the region.

## LOWER INDUS BASIN

The primary structural form of the lower Indus basin is a subsurface geosyncline or synclinorium of the interplate marginal convergence boundary that was forced down as a foredeep alongside the colliding edges of the Indian and Eurasian plates, while the Himalaya were forced up at the leading edge of the plates (Kazmi and Rana, 1982; Shroder, 1984; Figure 14.1). Formation of this trough began during the late Mesozoic Era as a marine bay formed between the converging plates. The bay continued into the Tertiary Period as it received sediments eroded from the rising Himalaya to the north and northwest (Farah and Dejong, 1979; Shroder, 1989c). By ~ 20 myr ago, the marine delta of an ancestral Indus was at least 300 km north of its present location (Kazmi, 1984). Further erosion in the Himalaya and Punjab throughout the late Tertiary and Quaternary filled the lower Indus basin with alluvium. Brinkman and Rafiq (1971) noted that particularly rapid aggradation of alluvial sediments in the lower Indus basin began during the early Holocene, with major sedimentation patterns commanded by the ancient river courses of the region. In contrast, Kazmi (1984) felt that many of the former courses of the Indus, and the possibly associated deltaic deposits, were instead Pleistocene in age, with a progressive westward shift of deltas from mid-Pleistocene to present. Whatever the timing, the alluvial plain of Sindh has aggraded as much as an estimated 600 m (Lower Indus Project, 1966).

In spite of a possible or apparent similarity of present or past environments in the lower Indus region, the prehistoric configurations of soil, relief and hydrology, as well as potential land use would have had considerable effect on prehistoric settlement patterns. Evidence is presented herein for an initial reconstruction of the palaeogeography and the ancient river courses of the lower Indus basin, primarily during 4000–2000 yr BC. This is accomplished from study of diachronic fluvial processes and landforms of the region during the Holocene.

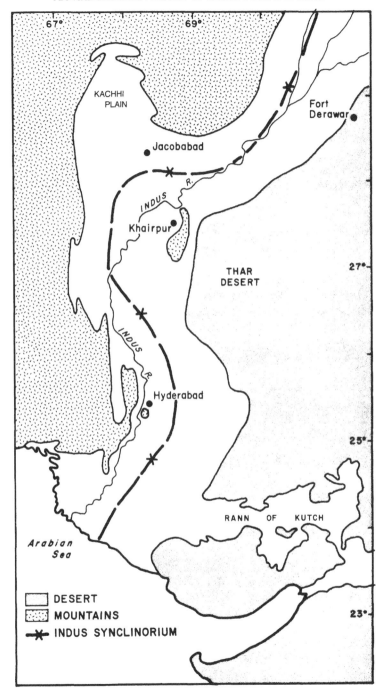

*Figure 14.1* The modern environs of the Indus river and the Indus synclinorium.

# GEOLOGY, LANDFORMS, AND SOILS

No physical barriers to flow of rivers in the lower Indus basin occur except for the Rohri hills near Khairpur and others at Hyderabad (Figure 14.2). Neither group of hills rises more than 61 m above the surrounding plain. The Rhori hills extend ~ 55 km south from Sukkur in northern Sindh province. Ganjo Takar hill extends ~ 22 km south from Hyderabad in southern Sindh. Much of the Rohri hills is hard, fissured and cracked limestone with flint. This flint commonly weathers out and covers the surface over a large area. It was a major resource for the manufacture of prehistoric stone tools (Biagi and Cremaschi, 1988). Other beds are nummulitic limestones, ironstones near Khairpur, as well as red or green clays with large quantities of gypsum. Ganjo Takar hill upon which Hyderabad sits has an escarpment all around it, with higher elevations to the south. It is an unfossiliferous, white chalky limestone.

The soils of the lower Indus basin are one of its most important geological features. Ecologically they play a vital role in the biotic community including, of course, potential human land-use patterns. The development of distinct soil profiles on stratified alluvial deposits, and the approximate dating of these soils, are vitally important for the analyses of ancient fluvial environments. Because soil development is a consequence of regional climate and vegetation, as well as of edaphic conditions including type of parent material, local hydrology, and topography, the soils may be a significant clue to past environmental conditions. The parent material of soils, the transportation of this parent material to various locales, and the transformation or development of the material into distinct local and regional horizons and profiles, coupled with the approximate timing of their development, can be a significant factor toward reconstructing the palaeogeography of the primarily alluvial lower Indus basin.

The major classes of soil parent material are those formed in place through weathering of consolidated or soft and unconsolidated bedrock, those that have been transported and redeposited, and organic deposits. Soil parent materials that eventually find their way into the lower Indus basin as transported alluvium primarily have their origin in the Himalaya where, in declining order of surface exposure, the bedrock is diverse metamorphic, igneous and sedimentary. Mid- to late Cenozoic and present-day climatic and weathering intensity, coupled with strong uplift (Hewitt, 1968a, b; Whalley et al., 1984; Goudie et al., 1984; Lawrence and Shroder, 1985) have contributed to extensive comminution of bedrock, soil development, and erosion that have resulted in high stream loads over a long period (R.I. Ferguson, 1984; Milliman et al., 1984; Shroder, 1989a, c). These processes in the Pleistocene created wide alluvial plains in the Punjab and to a lesser extent in portions of the lower Indus basin.

Subsequently during the early Holocene, wide plains were cut by the rivers of the Punjab, downcutting and removing the older sediments along their courses, so that steep-edged, old river terraces, or the so-called 'bar uplands' now remain between each pair of rivers (Brinkman and Rafiq, 1971; Figure 14.2). This

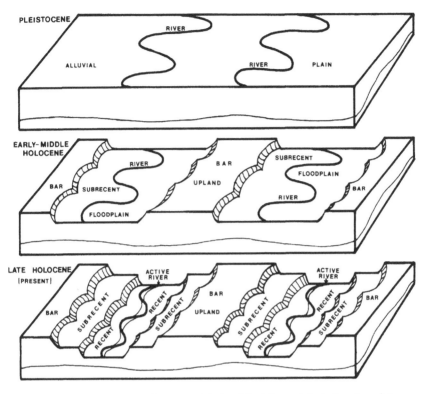

*Figure 14.2* Geomorphic development of the Punjab plains and the Indus river plains.

process of dissection or fluvial regradation of Pleistocene land surfaces continued through the mid-Holocene and into more recent times. Together with the continuous supply of freshly eroded rock waste from the geomorphically active Himalaya, the reworked alluvium of the Punjab and upper Sindh filled the lower Indus basin with plentiful alluvium distributed through the various ancient river courses extant at different times throughout the mid- and late Holocene. Once deposited, the alluvium was subject to aspects of the regional climate and the influence of the local biotic community. Most of the soils of the region show definite profile development with distinct horizons (Lower Indus Project, 1966; Baig, 1969; Rafiq, 1971).

The most important soil-forming process in the lower Indus basin is homogenization through bioturbation from plant and animal growth and decay in the soil (Brinkman, 1971). Although the degree of homogenization depends on many factors, when soils have the same sedimentary and hydrological history it is possible to estimate their relative ages from the degree and depth of homogenization. This approach was used to define landforms and their relative ages in the lower Indus basin and the Punjab (Lower Indus Project, 1965a, 1965b, 1966;

Ali, 1967, 1969; Alim, 1968; Baig, 1969; Akhtar, 1970; Akram, 1971; Ansari, 1971, 1972, 1973; Food and Agricultural Organization, 1971).

Five landforms have been defined in the lower Indus basin: Pleistocene remnants, the subrecent river plain and river terraces, the recent floodplain, and the active floodplain. During the early and mid-Holocene, the subrecent river plain was built up over the late Pleistocene levels, owing to sea-level rise and the sediment added to rivers by the dissection upstream (Brinkman and Rafiq, 1971). In the western part of the lower Indus basin, the subrecent river plain is a major landform unit comprising level plains, basins, infilled channels, and covered bars characteristic of abandoned river channels (Ansari, 1973). The early and mid-Holocene subrecent river terrace remnants rise ~0.3–1 m above the adjoining river plain. They are further recognized in the extreme western and eastern parts of the lower Indus basin by thicker and more homogenized soils than the soils of the main river plain. The subrecent river terrace is presumably a cover floodplain deposit from when the Indus river followed a considerably longer, more westerly course, probably near Jacobabad. This was shallowly dissected by subsidiary Indus channels when the river found a shorter, nearer and more easterly course (Brinkman and Rafiq, 1971). Late Holocene landforms are marked by recently abandoned river channels and levees, the recent floodplain, and by the active floodplain and course of the present-day Indus river. The significance of these landforms for the palaeogeography of the lower Indus basin is discussed below.

## RIVERS OF THE PLAIN

The Indus river is more or less centrally located in the upper part of the lower Indus basin, and is in the extreme western part of the region from past Hyderabad to the Arabian Sea (Figure 14.1). Throughout its long history the Indus has been a notably vagrant river of enormous magnitude, exhibiting extraordinary wanderings and mutations of its course in response to natural and human-induced environmental changes. One major occurrence in the geomorphological history of the Indus was the passage of the river's course through the gap in limestone outcrops near Sukkur (Figure 14.3). Formation of this 'Sukkur gap' was dated at AD 900–1300 from Arab texts (Lower Indus Project, 1965a; Lambrick, 1967; Holmes, 1968), and has ever since remained the fixed course of the Indus river. The full flow of the river probably passed through the Sukkur gap by AD 1252, and certainly by 1333 when it was flowing on both sides of Bukkur island. The repercussions of this event were large. In relation to previous courses, the new one exhibited greater sinuosity, a wider meander plain, enlarged channel dimensions, and patterns of increased alluvial deposition. Implicit in these changes are characteristic increases of regimen including sediment load, rates of discharge, and gradient of the river. The reasons for these changes are not well understood and require further detailed research (Schumm, 1985, 1986; Jorgensen et al., this volume).

The course of the Indus river in Sindh is held in its present-day location by

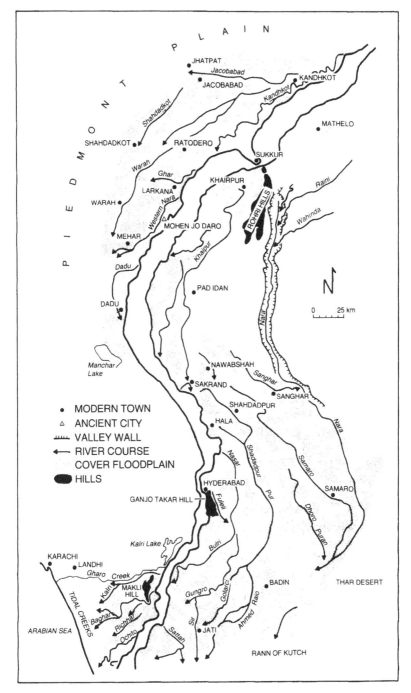

*Figure 14.3* Prior and ancestral courses of the Indus river in the Holocene.

flood protection embankments (*bund*) on either side of the river along its entire length. These embankments exaggerate the build-up of alluvial sediments along clearly and narrowly delineated zones, raising the bed of the river unnaturally above its floodplain, as well as usually but not always preventing any major changes of the river's course. The natural morphology of the river has been further modified by the attempted control of the Indus river's perennial and ephemeral tributaries in the Punjab and western piedmont regions of the Northwest Frontier province. In addition to the protective embankments in Sindh, a protective *bund* separates the alluvial plain of the river from the seasonal torrent streams of the western mountains and piedmont region. The hydrological regulation of the Indus drainage net by barrages and irrigation canals has reduced peak discharges, prolonged periods of water flow, and trapped alluvial sediments, which has largely put an end to natural floods. Canals have replaced the natural flood channels and field irrigation has replaced the natural covering activity of the river. Consequently, the human-induced alterations of sediment load, regime, and runoff reception have greatly altered Indus river morphology and have created an artificial and very recent landform (Memon, 1969). Thus, prior assumptions are erroneous in assuming that the ancient courses of the Indus, and the factors influencing the fluvial environment of the lower Indus basin in prehistory, were similar to those of today. Although recent reports (Fraser, 1958; Revelle, 1964) are significant contributions to study of the Indus, its landforms, and modern land-use patterns, care must be exercised in their uncritical application to the prehistoric situation.

Previous attempts to delineate ancient river courses in the lower Indus basin have used Greek, Roman and Arab geographies, histories or travel accounts, sometimes coupled with field observations to describe fluvial sequences. In addition, evidence and inferences from these studies, considered in light of geomorphic processes, have added a new dimension to the study of lower Indus basin fluvial history (Lambrick, 1975). In most previous studies, linear patterns of soil types and distributions of microrelief landforms were attributed to variations in the modern course of the Indus river. While attempting to trace ancient river courses, researchers commonly mistook small spillways of recent date for historical remnants of major channels. The regional significance of soil and landform patterns was commonly misunderstood and misinterpreted.

In order to trace the course of the Indus during 4000–2000 yr BC, it is necessary to approach the fluvial environment of the lower Indus basin in historical perspective by establishing a chronological sequence of fluvial events and considering evolution of the environment to its present configuration. Methods and theory for analysis of ancient sedimentary environments are sophisticated and precise (Pettijohn, 1962; Middleton, 1965; Schumm, 1968, 1972, 1977; Rigby and Hamblin, 1972; Selley, 1978). The criteria for the reconstruction of palaeochannel shape and dimensions, however, as well as the precise causal relationships of fluvial changes are lacking for the fluvial deposits of the lower Indus basin. Therefore, results of any effort to study the chronological sequence of former river courses in Sindh remain largely descriptive. The accuracy of description of these sequences is greatly enhanced by the use of historical, geomorphological and field checked, aerial photographic evidence to trace and date former river courses in the lower Indus basin.

Indistinct soil and landform patterns discerned from ground survey of aerial photographs are fairly easily delineated. The arid climate of Sindh and the lateral shifting of the Indus courses over the wide plain have preserved enough remnants of old river channels and their landforms to permit delineation of entire river courses. In light of this, Lambrick's (1975) work exhibits the greatest accuracy and laid the foundation for this study. The Lower Indus Project (1965a) and Holmes (1968) used the comprehensive methodology to delineate six major post-Pleistocene river course remnants in the northwest and northcentral plain of the lower Indus basin (Figure 14.3): (1) Jacobobad course; (2) Shahdadkot course; (3) Warah course; (4) Kandhkot course; (5) Khairpur course; and (6) present-day course of the Indus river. In the southern plain of the lower Indus basin, four major post-Pleistocene river course remnants have been delineated; (7) Sanghar course; (8) Samaro–Dhoro Badahri course; (9) Shahdadpur; and (10) present-day course of the Indus. In addition, in the extreme eastern (northern, central, and southern) plain of the lower Indus basin, another series of exclusive river course remnants have been delineated; (11) Wahinda course; (12) Raini course; and (13) Nara course.

Of the course remnants listed above, numbers 1–10 belong to former or recent courses of the Indus river, whereas those numbered 11–13 belong to the distinct and separate Nara river system. Following Butler (1950), Pels (1964) and Schumm (1968), the Jacobabad, Shahdadkot, Warah, Sanghar and Samaro–Dhoro Badahri course remnants can be considered the older or 'prior stream' courses of the present Indus river, and the Kandhkot, Khairpur and Shahdadpur course remnants can be considered younger or 'ancestral' courses of the Indus.

## JACOBABAD AND SHAHDAKHOT COURSES

The Jacobabad course (Figure 14.3), marked by subrecent river terraces, is traceable as the most northwesterly and the oldest of the former river courses of the lower Indus basin. Traces of this course emerge from the present Indus river near Kandhkot, and run in a westerly direction toward Jacobabad where the course's remnant deposits become mixed with Kachhi plain piedmont sediments and the course becomes untraceable any further west.

The Shahdadkot course remnant is traceable as a zone of bars and channels from southwest of Jacobabad down to Shahdadkot where it disappears (Holmes, 1968; Figure 14.3). This course is perhaps the remnant of a distinct and independent river course; i.e. separate from the Jacobabad course. Aerial photographs and maps of the prior river courses of the lower Indus basin permit speculation that the Jacobabad and Shahdadkot channels could have been two reaches of the same river. No other traceable remnants occur north or west of the

Shahdadkot course, except the Jacobabad course, and the two appear to align at their respective ends southwest of Jacobabad town. Farther south of the Jacobabad–Shahdadkot course, northwest of Warah town, Holmes (1968) discerned an additional river course remnant now occupied by Hamal lake. He concluded that this unnamed channel had the characteristic crescentic channel and bar patterns of Indus meander floodplains, and believed it to be a continuation of the Jacobabad, or possibly the Shahdadkot course, obscured in its intervening portions by piedmont alluvium of later date. Holmes (1968) thought that the connected pieces of the two river courses and the lake constituted evidence of a distinct river system running around the northwestern flank of the Indus alluvial plain, south to Manchar lake, now partly obscured by the advance of the piedmont alluvial plain.

## WARAH COURSE

The Warah course (Figure 14.3) occurs as a distinctive broad zone of extensive high bar deposits and clearly defined channels indicative of a major river (Holmes, 1968). Similar to the Jacobabad course, the Warah course had its origin near Kandhkot. Unlike the Jacobabad course, which ran due west from Khandkot, the Warah course turned toward the southwest a short distance west of Kandhkot. Running southwest of Kandhkot, the Warah course is clearly traceable as it passes west of Sukkur and Ratodero, through Warah, and west of Mehar towns. West of Mehar the alluvial deposits of the Warah course are intermixed with and obscured by piedmont plain deposits of the Kirthar mountains along the entire length of the trough extending from Jacobabad to Manchar lake (Bull, 1972; Ansari, 1973).

Previously the Warah trough had been thought to have been a major independent river course of the westernmost part of the lower Indus basin, but this is not precisely the case. The trough was named the Western Narah channel by Pithawala (1936, 1959), Fraser (1958) and Memon (1969), but should not be confused with the seasonal spillway of a later time called the Western Nara (Holmes, 1968). The trough was a major river course in that it was an extension and part of the Jacobabad and Warah courses. Raverty (1895) called it the Sind Hollow, and noted that it was also referred to locally as the *Ran* or marsh, the *Pat* or desert, or the *Dasht-i-Bedari*. Sind Hollow was the designation used by Panhwar (1969) and is used herein.

## KANDHKOT, KHAIRPUR AND SHAHDADPUR COURSES

The Kandhkot course is traceable in the northwestern part of the lower Indus basin (Figure 14.3). It is for much of its length a single channel, narrow, deep and winding, running parallel to the modern Indus river from Kashmor, through Kandhkot which is located on its bank, to south of Sukkur, where it is cut by the modern Indus (Holmes, 1968). The channel morphology suggests that it was a large river, probably the previous or ancestral course of the modern Indus before it flowed through the Sukkur gap (Lower Indus Project, 1965a). Two spill channels are traceable on the aerial photographs as offshoots of the Kandhkot course and the more recent Indus river: the Western Nara course and the Dadu course. Neither rather young channel has ever been a major branch of the Indus as previously supposed by some (Holmes, 1968).

The Kandhkot course, on the right bank of the modern Indus, is complemented on the left bank by the older Khairpur course (Figure 14.3). This course can be traced from west of Khairpur city southward to Sakrand. A short distance south of Sakrand begins the Shahdadpur course of the southern plain of the lower Indus basin. Holmes (1968) showed that the Kandhkot, Khairpur and Shahdadpur courses were reaches of the same river, basing his work on their similarity of channel form with narrow, deep channels, meandering with a tighter radius than the modern river, between high sandy bar deposits.

## SANGHAR AND SAMARO–DHORO BADAHRI COURSES

These two courses (Figure 14.3) are of special interest in that neither channel was part of the Kandhkot–Khairpur–Shahdadpur river, nor were they part of the (Eastern) Nara course (see below); but they may be a clue to the seaward path of the Jacobabad and Warah courses, respectively. The Sanghar course is discernible on the aerial photographs ~ 20 km northeast of Nawabshah, where it follows a southeasterly course to Sanghar town. From Sanghar the course runs eastward and joins the channel of the Nara course. Holmes (1968) noted that the Sanghar channel is narrow and appears to have been a stable river which seems to have carried the heavy floodwaters of a major river branch, or possibly a major distributary. The Samaro course remnants originate ~ 20 km southeast of Nawabshah and continue southeast through Samaro town to join the Nara course via the Dhoro Badahri channel.

The remaining courses of the southern and southwestern parts of the lower Indus basin which have been described by Holmes and the Lower Indus Project are pertinent to the later Arab and Muslim periods, and are beyond the scope of this paper (Flam, 1986b). These courses were distributaries of the ancestral and present courses of the Indus river.

## A CHRONOLOGY OF RIVERS

Thus far the discussion of the former courses of the Indus river in Sindh has focused on the use of aerial photographic interpretation and ground observations to recognize various fluvial remnants. To reconstruct the course of the Indus river in 4000–2000 BC, it is necessary to connect the various course remnants, trace entire lengths of former rivers from northern Sindh to the Arabian Sea, and suggest dates for the activity of the various rivers.

The most recent course of the Indus river is the easiest to define and date. As noted above, the course of the river through the Sukkur gap, and its associated landforms, is < 1000 yr old. Therefore it is unlikely that the prehistoric river can be equated with that of the present day.

Looking back in time, the river marked by the Kandhkot, Khairpur and Shahdadpur courses (hereafter just Kandhkot course) is the second river that can be documented. Lambrick (1975) used the writings of Greek and Roman geographers and chroniclers (Arrian, Megasthenes, Strabo and Pliny) to locate and date this early course of the Indus river in Sindh. He argued the following points that aid in locating the Greek period river. First, the reporters of Alexander's river journey through Sindh make no mention of the Indus course through the limestone hills near Sukkur and Rohri, an unusual occurrence of which they certainly would have taken note. Lambrick (1975) suggested that the river must have flowed either west or east around the Rohri hills, if it did not pass through them in the Sukkur gap as the present river does. Aerial photographs show that a spill channel of the Indus flowed to the east of Rohri at some time, but its dimensions are not large enough to indicate that the main river *ever* flowed in this channel (Lower Indus Project, 1965a; Holmes, 1968). Lambrick inferred that the Greek period course of the Indus did not flow too far west from Sukkur. This flow can now be confirmed by the location of the Kandhkot course. Second, Lambrick (1975) maintains that during the Greek period the Indus flowed east of the present-day city of Hyderabad and the Ganjo Takar hill. Aerial photographic and channel remnant evidence confirm this (Lower Indus Project, 1966; Holmes, 1968), as the channel is marked by the Shahdadpur course. Third, without the aid of aerial photographs, Lambrick (1975) produced a map, 'Sind in 326–325 BC', with an hypothesized path of the Indus river during the Greek period that matches almost exactly the Kandhkot course mapped from aerial photographs.

With the Kandhkot course defined as the Greek period location of the Indus, question arises concerning configurations of the islands Prasiane in northern Sindh and Patale in southern Sindh. Magasthenes (Fragment LVI in Pliny, *Hist. Nat.* VI. 23) stated that the Indus divided Sindh into these two islands. Alexander the Great travelled through Sindh in late summer 325 BC (Lambrick, 1975), when the Kandhkot course would have been at flood stage. Just as in the present day, perhaps the Warah course functioned as an overspill channel after it was abandoned as the main channel of the river. Prasiane would thus have been formed of the western Indus floodplain and, in the south, Patale would have formed between the main river and flood flow in another major distributary channel, perhaps the Samaro–Dharo Badahri course.

While historical evidence suggests an approximate date for adoption of the course of the present Indus river and the period of activity for the Kandhkot course, no historical dates are available for adoption of the Kandhkot course or of activity on the Jacobabad and Warah courses. Geomorphologic and sedimentologic concepts must be used to assess this. In this case the Indus is considered in light of its entire drainage system from the Himalaya to the

Arabian Sea. The river is also considered in dynamic equilibrium as a graded stream in which it will adjust appropriate sections of its length in response to changes of hydrologic regimen (discharge, sediment load, channel dimensions, gradient), tectonism, sea level, or sediment load change.

During the late Mesozoic and Cenozoic Eras, the Punjab plains and the lower Indus basin were a bay of the Arabian Sea. Uplift of the ancestral Himalaya–Hindu Kush increased gradients and resulted in deposition of plentiful sediment into the bay from northwest to southeast. By the late Pleistocene, extremes of uplift, weathering and erosion in the northern uplands resulted in rapid deposition along the shallow, strongly seasonal river systems. In the Punjab, deposition of mainly sandy sediments and windblown loess occurred in late Pleistocene, followed by deposition of silty alluvium and extensive soil formation in Holocene. Environmental adjustments that produced the main surface landforms in the Indus drainage system likely were a response to the warmer and more humid hypsithermal interval of the early and mid-Holocene. This presumably affected vegetal cover in the catchments of the drainage basin, thereby influencing runoff and modifying sediment yields and concentrations (Schumm, 1965, 1968, 1969). For the Indus system the early Holocene climate probably produced less erosion in the headwaters and larger, more constant river flows with less sediment load. Increased discharge possibly resulted in increased sediment size carried by the rivers. The Pleistocene river plains of the Punjab were incised by channel erosion toward a new equilibrium grade, which produced river terraces and raised interfluves or bar uplands (Figure 14.2). A rise in sea level to present levels probably also occurred, with probable movement of the coastline north to a location between Hyderabad and Sukkur. Because of the upstream dissection and rise in sea level, rapid aggradation of sediments occurred in the lower Indus basin. Downcutting was greatest in the north and decreased to about the Multan area; delta growth and coastal accretion was greatest in the south and decreased north to about Multan (Figure 14.4).

During mid-Holocene the relative magnitude of climate change further altered the hydrologic regimen of the Indus drainage system (Brinkman and Rafiq, 1971). The climate was similar to that of today, with a reduced rainfall from that of the early Holocene. The increased dryness reduced vegetation and increased erosion in the northeastern uplands. Climatic adjustment following the hypsithermal resulted in greater seasonal adjustments with higher flood peaks and shorter durations of flood flow: the prevailing conditions of the present Indus river. Increased erosion lessened river gradients. Sediment loads probably shifted from sands to silty-clays, and may have varied seasonally from low to high suspended loads. Later erosion and terracing continued in the river plains of Punjab, but more locally limited than in the previous phase. Sea level remained constant but the estuary plain and coastline of the lower Indus basin moved south rapidly, owing to delta building, so that the river plain encroached over the former estuary plain. Aggradation in the lower Indus basin continued but was more dependent on seasonal cycles and was less rapid than in the previous phase.

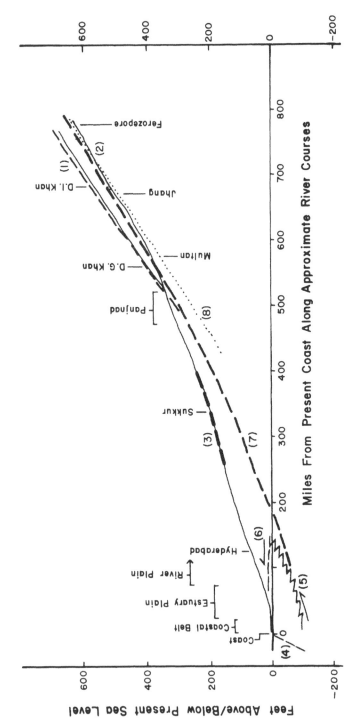

*Figure 14.4* Schematic longitudinal section of the Indus river and associated landforms.

Soil formation was interrupted where local flooding and accretion were occurring.

Alluvial infilling of the lower Indus basin has been the main landform development process since the Pleistocene. For the most part, Pleistocene landforms have been buried by later geomorphic processes; interestingly the prehistoric site of Mohen jo Daro is located on a Pleistocene high. Where these landforms occur they do not generally form patterns that can be traced as river courses, with the result that only Holocene history of landform development is clear.

In the northwestern subregion of the lower Indus basin the Jacobabad course is the oldest fluvial remnant and is also in the most northwestern location (Lower Indus Project, 1965a; Holmes, 1968). The Sanghar and Samaro–Dhoro Badahri courses are the two oldest remnants in the southern region; a pre-Greek period date for their activity has been inferred (Holmes, 1968). The Sanghar course is the oldest of the two and exhibits all the characteristics of an infilled deltaic distributary as encountered in Lower Sindh (Holmes, 1968).

Inasmuch as the early Holocene coastline probably was located north of Hyderabad, the Samaro–Dhoro Badahri course would therefore have been under the sea at this time. In the Jacobabad course, the broad zone of bar and channel deposits, the coarse-textured (sand?) sediment load, and the adjustment to grade of the course correlate well with expected early Holocene phases of the Indus drainage system. This fact and the relative dating of soil formation (pers. comm., M. Alim Mian and M. Akram, Soil Survey of Pakistan, 1976) allows rough estimation of activity of the Jacobabad course at ~ 10,000–6000 yr BP.

The Warah course in the northwestern subregion of the lower Indus basin is more distinct than the Shahdadkot course and is recognized above Ratodero as a broad zone of extensive high bar deposits and clearly defined channels. Below Ratodero relief is more subdued and the evidence is restricted to a few oxbow swamps and the deep channel of Drigh Lake west of Qambar. Below Warah the course is temporarily well defined again, before disappearing into the trough sediments of Sind Hollow (Holmes, 1968).

The Warah course is larger, more prominent and more recent than the Jacobabad course (Lower Indus Project, 1965a,b). In the southeastern subregion of the lower Indus basin, the Samaro–Dhoro Badahri course is compatible with the supposed mid-Holocene advancement of the riverine plain by delta building. As the delta encroached south it increased the length of the river and reduced the river gradient. To maintain its gradient *vis-à-vis* its amount and type of sediment load (silty-clay) and discharge (seasonal highs and lows), the river probably increased its sinuosity and channel dimensions. When compared to the Jacobabad course, these conditions are hypothesized for the Warah course. The river descriptions for the Warah course are in accord with the environmental conditions described for the Indus during mid-Holocene. In addition, if the Warah course is more recent than the Jacobabad course and yet older than the Kandhkot course, it seems temporally and spatially correct to place the Warah course 'between' the latter two. Thus a mid-Holocene date ( ~ 6000 yr BP or

~ 4000 BC) is assigned to the activity of the Warah, Sindh Hollow and Samaro-Dhoro Badahri courses. Furthermore, it is likely that these courses comprised a single river during the greater part of 4000–2000 BC that is referred to here as the Sindhu Nadi, thereby emphasizing its prehistoric regional integrity and distinction.

While river adjustments from the Jabobabad course to the Sindh Nadi can be attributed to relative changes in hydrologic regimen as a consequence of respective early and mid-Holocene environmental changes, the question of the transformation from the Sindhu Nadi to the Kandhkot course is more enigmatic. The mid- and late Holocene seem to have had similar environmental conditions, so that changes such as those noted from early to mid-Holocene cannot be the cause of these course changes. The Kandhkot course is a single, narrow and deep winding channel whereas the Sindhu Nadi is a broad zone of extensive deposits with a higher degree of sinuosity than the Jacobabad course but less than that of the Kandhkot course. The change of river course from Sindhu Nadi to the Kandhkot course seems to be of equal or greater magnitude than that between the Jacobabad and Sindhu rivers. With these latter two the southern reaches adapted to the temporal growth of the delta, and the upper reaches (Jacobabad and Warah remnants) adjusted their courses to an altered hydrologic regimen. The middle reaches of the two rivers remained the same: the Sind Hollow. In all sections of its length, the location of the Khandkot course differs from the path followed by the Sindhu Nadi. With the change in course, the delta shifted from an eastern to a central location in the lower Indus basin. Concurrently, the central and upper reaches shifted from a western to a central location. These transformations suggest a change in hydrologic regimen, which may be commensurate with the capture of the Sutlej river and its catchment by the Indus drainage system. If this was the reason for the river course change from the Sindhu Nadi to the Kandhkot course, it may well have occurred during the Mature Harappan cultural period of prehistoric Sindh. As a consequence of this river course change, some sites that had been a safe yet agriculturally viable distance from Sindhu Nadi were suddenly being inundated during the flood season. Their new proximity to the meander and cover floodplain of the Kandhkot course would have made their existence tenuous. This hypothesis requires more fieldwork but is a plausible cause for decline of the Indus civilization.

In summary, three former courses of the modern Indus river have been delineated in the lower Indus basin for the time period ~ 8000 BC–AD 1300. The earliest Holocene river, the Jacobabad course (Figure 14.3), was comprised of the Jacobabad, Shahdadkot, Sind Hollow and Sanghar course remnants of the lower Indus basin. The second river, of mid-Holocene age, is the Sindhu Nadi (Figure 14.3), and was comprised of the Warah, Sind Hollow and Samaro–Dhoro Badahri course remnants of the lower Indus basin. A still later river of mid- to late Holocene age is the Kandhkot course, which consisted of the Kandhkot, Khairpur and Shahdadpur courses. The most recent course is the modern Indus (Figure 14.3), which is post-AD 1300 to present.

These channels represent only main channels and not spill channels or distributaries which may have been important in flood years. Flow divergence to such secondary channels may have resulted in main channel avulsion in the past.

## NARA NADI

The Nara Nadi river system of the extreme eastern lower Indus basin is easily visible on satellite imagery and aerial photographs (Figure 14.3). In the northeast from Fort Abbas to Fort Derawar it is considered to be the Hakra river (Mughal, 1981, 1982). From Fort Derawar to southeast of the Rohri hills it is marked by the Raini and Wahinda remnants. From there to the Rann of Kutch it is the main Nara Nadi and has been excavated in modern times to serve as a main feeder canal of Sindh province.

The ancient Nara course has long been of interest to scholars of the evolution of the Indus and Gangetic river systems. Pascoe (1920a) and Pilgrim (1919) speculated about the possible existence of one great river system along the entire submontane region of the northern and northeastern Himalaya which drained into the Arabian Sea. This 'Siwalik' or 'Indo–Brahm' river was supposed to have carried the combined discharge of the Indus, Ganga and Brahmaputra rivers, and at first to have drained to the northwest along the rising Himalaya. Pleistocene tectonics and stream capture were presumed to have dismembered the single extensive river system and evolved into the Indus and Ganga master systems. Pascoe (1920a) noted that the last occurrence in this progression of geomorphic events was the capture of the Sutlej river by the Indus system, and the concomitant capture of the Yamuna river by the Gangetic system. This was supposed to have dried up the Ghaggar–Hakra–Nara course, although it should be noted that river capture and drying up of the captured stream are not single event phenomena. This version of the geomorphic history of the Indus has been challenged by a variety of workers (see Shroder, 1989a, 1989c, and this volume for a summary) but no consensus has yet emerged. In any case, the Hakra–Nara Nadi was a perennial river in 4000–2000 BC (Lower Indus Project, 1965a, 1966; Holmes, 1968; Wilhelmy, 1969; Akram, 1971; Ansari, 1971; Lambrick, 1975; Allchin et al., 1978).

The Hakra course from Fort Abbas to Fort Derawar is marked by a depression that has been partially infilled by sheet flood and local erosion since the Hakra lost its perennial flow. Akram (1971) defined a variety of landforms including old rolling sand plains, or undulating sandy ridges, of late Pleistocene age, and the subrecent floodplain that is elevated above his recent floodplain of the Hakra of early to mid-Holocene age. Southwest of Fort Derawar, east of Rahimyar Khan, the Hakra course becomes 'lost' beneath the sand dunes that have encroached upon the area. Remnants of the Hakra's course emerge where the dunes are less numerous. From Fort Derawar to the south, the Hakra can be aligned with the Raini and Wahinda remnants, which subsequently connect to and blend with the Nara channel.

Aerial photographs show the Raini and Wahinda area to have been one of great alluviation. Two distinct episodes of fluvial deposition occurred, one associated with the Indus and the other with the Hakra–Nara Nadi (Lower Indus Project, 1965a). The contours of this area fall from the northwest to the southeast, with the Nara course occupying the lowest contour line along the eastern edge of the lower Indus basin. This sharp fall in gradient probably accounts for the wide meander plain from the Hakra course near Fort Derawar, through the Raini and Wahinda area, to the northern reaches of the Nara course. It is likely that as the river's course turned toward the southeast its heavy seasonal floodwaters would have flowed over its banks on the river's left bank, flooding and depositing sediment over a large area to the east and southeast.

South of the Raini and Wahinda courses, the Nara Nadi flowed due south. At the present location of the Jamarao Head, the Nara course ran toward the southeast and followed a path which skirted the western edge of the Thar desert to the Rann of Kutch.

## COASTLINE OF THE LOWER INDUS BASIN

The present Indus river delta consists of two main parts; a coastal belt comprised of marine tidal deposits, and an extensive estuary plain deposited by Indus river distributaries (Jalal-ud-din *et al.*, 1970a, 1970b). The Indus delta has grown considerably since the end of the Pleistocene as alluvium has been deposited by the river's distributaries (Figure 14.5A). The present and ancestral Indus deltaic distributaries and spill channels are concentrated in the southcentral and southwestern portion of the lower Indus basin (Figure 14.3). The courses of these channels have been traced from aerial photographs and have been dated with corroborative historical evidence (Lower Indus Project, 1965a; Holmes, 1968).

Correlated with the prior river courses of the lower Indus basin, the deltaic areas of the Jacobabad course and the Sindhu Nadi were different from that of the ancestral and modern Indus. Following the Sanghar and the Samaro–Dhoro Badahri remnants, the ancient or prior deltas were most likely to have been located in the southeastern part of the lower Indus basin, instead of the southcentral and western location of the ancestral and modern delta. During the early Holocene the estuary plain and the coast of the lower Indus basin slowly moved south from probably some distance north of Hyderabad to their present positions, and river plain sediments gradually overran the northern fringe of the estuary plain (Pithawala, 1936, 1959; Brinkman and Rafiq, 1971). This early Holocene delta and coastline location seems congruent with the fluvial evidence of the Sanghar remnant, or distributary of the Jacobabad course (Figure 14.3). Rapid vertical and lateral accretion of sediments from the Jacobabad course and Nara Nadi in the eastcentral and southeastern lower Indus basin probably advanced the delta southeasterly during the early Holocene. During this delta building process, much of the southcentral and southwestern lower Indus basin was probably estuarine.

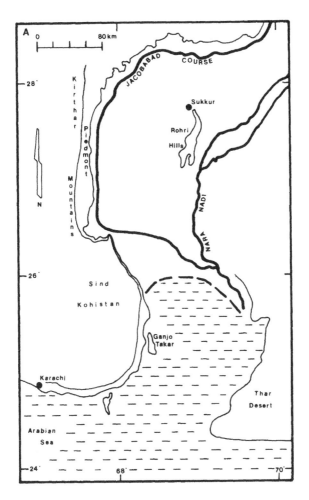

*Figure 14.5 A–C* A  The Jacobabad course and approximate coastline
~ 10,000–4000 yr BP.

The delta of the Sindhu Nadi in mid-Holocene was probably located in the southeast part of the lower Indus basin, following the Samaro–Dhoro Badahri course (Figure 14.5b). Raikes (1964) suggested that the coastline may have been near Amri in 4000–2000 BC. Lambrick (1967, 1975) rejected this notion and thought that the coastline was nearer the 9 m contour line of the lower Indus basin, or about 95 km inland from its present location. If Raike's estimate seems too far north, Lambrick's may be too far south. Lambrick based his estimate on existing contours and on estimated rates of alluvial aggradation for the Indus. He based his assumptions upon figures for the Egyptian Nile river and he also assumed constant aggradation, both of which are invalid procedures. Distri-

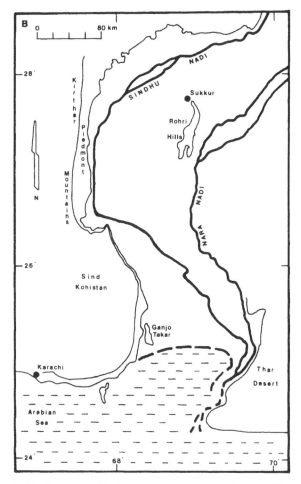

B The Sindhu Nadi and Nara Nadi and approximate coastline ~ 4000–2000 yr BC
( ~ 6000–4000 yr BP).

butary sediments continued to move the estuarine coastline south of the Hyderabad area throughout the late Holocene (Figure 14.5c).

Siveright (1907) argued that the Rann of Kutch was a delta of the Nara Nadi. A wide variety of workers have concluded that the Rann of Kutch and the Little Rann of Kutch were extensions of the Gulf of Kutch or inlets of the Arabian Sea (Blanford, 1876; Fedden, 1884; Siveright, 1970; Spate and Learmonth, 1967; Allchin *et al.*, 1978). Thus, terrestrial Kutch would have consisted of an island or islands completely surrounded by a tidal sea. Further field research is required for a proper assessment of the role of the Rann of Kutch in the palaeogeography of the lower Indus basin.

Because of the uncertainty discussed above, any delineation of the coastline of

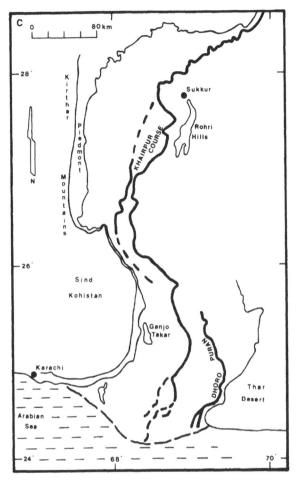

C  The Kandhkot course and approximate coastline ~ 2500–2000 yr BP.

the lower Indus basin in 4000–2000 BC is only provisional, awaiting further research (Figure 14.5b). The precise delineation of the Sindhu Nadi and Nara Nadi delta and the coastline of the Arabian Sea have been purposely left vague.

## ARCHAEOLOGICAL IMPLICATIONS

The location and fluvial behaviour of the Sindhu Nadi and the Nara Nadi had important consequences for human settlement and subsistence in the lower Indus basin in 4000–2000 yr BC. Prehistoric site distributions can be grouped according to their proximity to the ancient river courses and coastline of the region. Of the 25 presently known prehistoric sites located on the alluvial plain of the lower

Indus basin, 9 (including Mohen jo Daro) were close to Sindhu Nadi, 9 were close to Nara Nadi, 4 were close to the prehistoric coastline as proposed herein, and only 3 were close to neither major rivers nor the coastline but may have been located near a subsidiary channel of the Sindhu Nadi.

Cultural developments in the lower Indus basin can be divided into two broad phases (Mughal, 1973, 1988): Early Harappan ($\sim$ 3300–2600 BC) and Mature Harappan ($\sim$ 2600–2000 BC). In the lower Indus basin, the Early Harappan phase is represented by the single Kot Diji type site (Khan, 1965; but see Mughal, 1971, 1973, 1988 for explicit definition of this phase), and the Mature Harappan phase is best known from the excavations at the sites of Mohen jo Daro (Marshall, 1931; Mackay, 1938) and Chanhu Daro (Mackay, 1943). The single Kot Diji site of the Early Harappan phase is remarkable because adjacent regions (e.g. Kirthar mountains, Sindh Kohistan and Baluchistan) were densely occupied. During the Mature Harappan phase in the lower Indus basin, 24 settlements were occupied (Flam, 1981a, 1981b, 1986a). This dramatic change in settlement pattern has yet to be fully explained, but the Sindhu Nadi probably played an important role. On present evidence this change in settlement pattern seems not to have been related to a possible introduction of a major irrigation system during the Mature Harappan phase. To date, no evidence of such a system for the prehistoric period has been discovered, in spite of rigorous search. The traditional *sailabi* method of natural river inundation, which is dictated by the flooding cycle and patterns of water distribution inherent in the rivers and floodplain contours of the region, was probably the means of irrigating cultivated areas of the alluvial plain during the third millennium BC (Flam, 1986a). The reason for the settlement of the alluvial plain of the lower Indus basin may be related to changes in subsistence requirements, a shift from overland to riverine socioeconomic interaction routes, population growth, or other factors as yet unresearched (Flam, 1976, 1981b, 1986a; Shaffer, 1982).

Perhaps the most important archaeological implication associated with the delineation of the Sindhu Nadi and the Nara Nadi is relevant to the location of Mohen jo Daro. The supply of domestic and agricultural water needs explanation. It is probable that the innumerable wells in the settlement provided domestic water. If the Indus river was as close to the site as is the modern Indus, there would have been no necessity for the wells or for their long maintenance through the life of the settlement, as indicated by their considerable depth. Their numbers and locations reveal that they were not just a luxury item in élite houses but the source of water for all inhabitants.

The Sindhu Nadi was located $\sim$ 25 km northwest of Mohen jo Daro. In the northwestern lower Indus basin the regional contour gradients fall from northwest to southeast (Jorgensen *et al.*, this volume). During flood stage the greater amount of overbank floodwater would have flowed directly to the area of Mohen jo Daro, but would not have flooded the town because it was built on a raised Pleistocene landform. Deutsch and Ruggles (1978) have documented the magnitude of modern Indus floods, which are regulated with engineered controls on the river's discharge through barrages, canals, and embankments. They found that the floodwaters of the Indus covered an area of 10–20 km wide on both sides of the modern Indus river. Thus, with no flood controls in prehistory, the Sindhu Nadi would have sheet-flooded the Mohen jo Daro area, which, along with use of small, distributary irrigation channels, would have permitted *sailabi* cultivation.

Changing river courses may have been instrumental in causing abandonment of many sites in the lower Indus basin. If the river shifted its course from the Sindhu Nadi to the Kandhkot course during the prehistoric occupation of Mohen jo Daro, then the settlement would have been in a precarious and unviable location too close to the unrestrained flooding of the ancestral Indus river.

## SUMMARY

Natural transformations coupled with human attempts to regulate the fluvial environment of the lower Indus basin have brought great changes in the environmental configurations of the region during the Holocene. Equating modern configurations with those of 4000–2000 BC is erroneous and misleading when used to interpret the ecology of cultural dynamics and adaptation in the lower Indus basin.

A chronological sequence of fluvial events has been presented to reconstruct fluvial configurations in the lower Indus river basin in 4000–2000 BC. Two of the oldest, or prior, river courses of the Indus have been delineated: the Jacobabad (Figure 14.5a) and Sindhu Nadi (Figure 14.5b) courses. In addition, one younger, or ancestral, course has been delineated: the Kandhkot course (Figure 14.5c). The Sindhu Nadi is recognized herein as the former course of the Indus in 4000–3000 BC. In addition, the Nara Nadi is recognized as a perennial river that flowed in the eastern part of the lower Indus basin during the same time period as well. Traces of the Sindhu and Nara Nadi indicate that in the southeastern part of the basin the two rivers were confluent south of the present town of Naukot. Also during 4000–2000 BC the delta of the Sindhu–Nara Nadi confluence appears to have occurred in the extreme southeast of the lower Indus basin. The prehistoric coastline was probably located somewhere between the Rann of Kutch on the east side of the lower Indus basin and between Hyderabad and Thatta on the west.

Archaeological evidence reveals that the lower Indus basin was not extensively occupied until the middle of the third millennium BC. Evidence of a major canal or embankment system to control floodwaters of Sindhu and Nara Nadi for winter crop cultivation is lacking. Presumably, therefore, traditional inundation irrigation or *sailabi* was used at that time. Mohen jo Daro, the largest and most important prehistoric settlement in the lower Indus basin, was located ~ 25 km southeast of the Sindhu Nadi course. The city may have been abandoned after the river shifted its course closer to that of the present day, placing Mohen jo Daro in a hazardous position close to the brunt of the annual flood.

287

# 15

# MORPHOLOGY AND DYNAMICS OF THE INDUS RIVER: IMPLICATIONS FOR THE MOHEN JO DARO SITE

*David W. Jorgensen, Michael D. Harvey, S. A. Schumm and Louis Flam*

## ABSTRACT

Map and gauge records of the Indus river reveal a dynamic river which varies significantly through time and with distance downstream. Tectonic disturbance of Indus valley slope is reflected in varying channel planform, channel dimensions, and frequency of flooding. Specific morphologic evidence of tectonic deformation includes: an anastomosing pattern in a reach of backtilt or subsidence, the development of highly sinuous meanders in a region of forward tilting or steep valley slope, a low sinuosity, temporally variable pattern in a subsiding foredeep, and the presence of asymmetrical meanders in a reach of suspected cross-valley tilt. The morphologic variation of the Indus river, plus downstream loss of water and sediment to the alluvial plain, produce a range of flood depths and flooding variability, and temporal fluctuations of water surface elevations. The middle reach of the river near the ancient city of Mohen jo Daro is characterized by great variability of channel and valley morphology, both temporally and spatially. This dynamic behaviour is suggested as a cause for variable rates of sedimentation on the Indus plain. These conditions plus the likelihood of channel avulsion are preferred as reasons for the abandonment of Mohen jo Daro in contrast to those which involve more unusual and improbable tectonic and geormorphic conditions.

## INTRODUCTION

The Indus river (Figure 15.1) is responsible for the creation of the lower Indus alluvial plain. Its influence has had a major impact on the lives of past and present inhabitants. It is a river that demonstrates remarkable spatial variation in the 800 km distance of its lower course. Moreover, it is a dynamic river exhibiting considerable temporal variation of channel planform, flooding behaviour, and flow path. This dynamism has undoubtedly been a hallmark of the Indus river

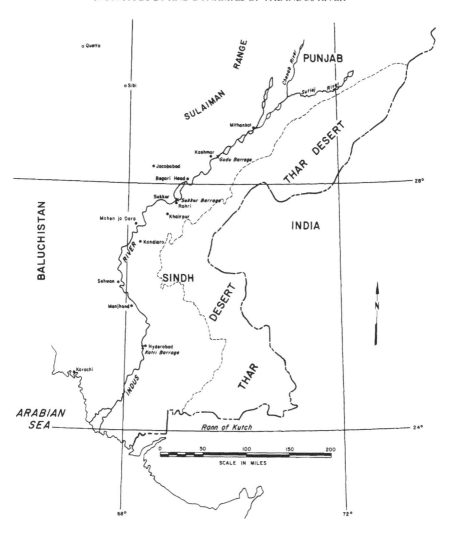

*Figure 15.1* Location map.

since the close of the last glacial phase, and it must be considered in any geomorphological and archaeological investigations of the Indus plain. The morphology and behaviour of the lower Indus river has not been related previously to the geologic and tectonic conditions that affect central and southern Pakistan. However, previous geomorphic and geologic research on other rivers has shown the following four points. First, big rivers are affected by major, continental-scale tectonic weakness and movements (Potter, 1978). Second, alluvial river channel patterns are sensitive to active movement of individual geologic structures (Adams, 1980; Burnett and Schumm, 1983;

Ouchi, 1985; Schumm, 1986). Third, river profiles reflect tectonic uplift and subsidence (Bendefy et al., 1967; Popp, 1971). Fourth, channel migration is affected by tectonically influenced erosion and sedimentation (Jorgensen, 1989). There are several goals in the following analyses and discussions. The first is to evaluate the tectonic and geologic controls on the morphology of the Indus river. This geomorphic description will be used to address a second goal, which is to explain the Indus river hydrologic conditions that are changing with time and distance downstream. This smaller scale hydrologic analysis and the improved understanding of the dynamics of the Indus river will be used to address the final goal, which is a discussion of the abandonment of the Harappan city of Mohen jo Daro.

## Geology

Present behaviour and past migration of the Indus river and its tributaries are inextricably linked to the collisional tectonics of the Himalayan orogen. The uplifted terrain serves as the headwaters and sediment supply for a river which drains 950,000 km$^2$ (Tamburi, 1974) and annually deposits 1 million m$^3$ of sediment on the alluvial plain and 175 million m$^3$ in the Arabian Sea (Gibbs, 1981). Leaving the Himalayan orogen near the Salt range fold belt, the Indus river flows entirely in the tectonic region referred to as the Indo–Gangetic foredeep (Seeber and Armbruster, 1979). The foredeep consists of sediments of Tertiary–Holocene age, which were derived from uplifted northern and western mountain belts and deposited on the tectonically loaded and subsiding Indo–Pakistan cratonic margin (Hunting Survey Corp., 1960; Kazmi, 1984). The part of the foredeep and river covered in this report, which extends from below the five rivers confluence to the top of the modern delta (Figure 15.2), is bordered to the west by two fold-mountain belts, the Sulaiman Arc to the north and the Karachi Arc to the south, and by the Kirthar range fault zone (Sarwar and DeJong, 1979). These regions remain seismically active, as northward convergence of India into the Eurasian continent continues to produce left-lateral offset and compression in Baluchistan. Modern seismicity in the fold belts is concentrated along the eastern edge (Quittmeyer et al., 1979), indicating continued eastward thrusting, loading of the crust, and foredeep subsidence.

Stability of the Indus plain is affected by uplift and subsidence of segments of the Indo-Gangetic foredeep. Principally, differential movement consists of subsidence in the Kirthar and Sulaiman foredeeps relative to adjacent zones of upwarp (Figure 15.2). The modern Indus river flows in the axis of the Sulaiman foredeep north of Mithankot but it currently crosses the Kachhi foredeep only between Kandiaro and Sehwan (Kazmi and Rana, 1982). Rapid subsidence and sedimentation in formerly occupied sections of the Kachhi foredeep have obliterated former channels of the Indus river near Jacobabad (Holmes, 1968). South of Sehwan, the Indus river now flows along the Thatta–Hyderabad high, west of former courses and a zone of tectonic downwarp. The Ganjo Takar hills

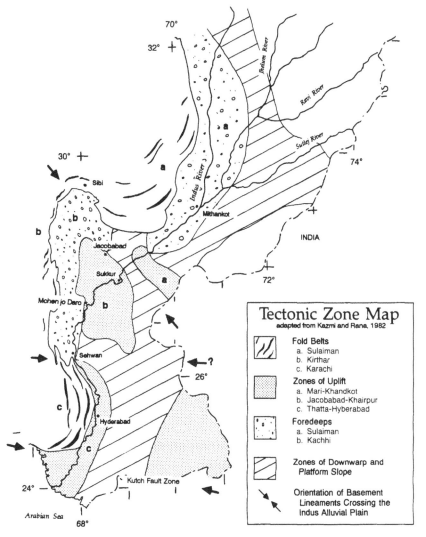

*Figure 15.2* Tectonic zone map (adapted from Kazmi and Rana, 1982; Haghipour *et al.* 1984; Biswas, 1987).

(Figure 15.3A), which consist of uplifted limestone bedrock east of the river, may be the result of basinward propagation of the fold belt. Lastly, active movement is indicated by low-level seismicity in the Indus plain which is related to development of new frontal thrusts, bending of the lithosphere in response to loading, and/or active basement faults transverse to the fold and thrust belts (Quittmeyer *et al.*, 1979).

In addition to epeirogenic subsidence and upwarp, east–west oriented

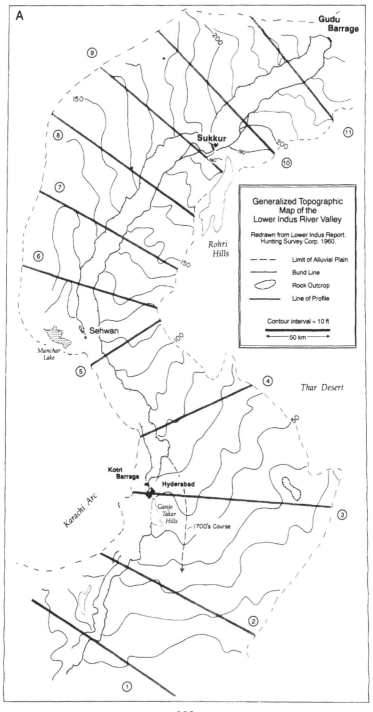

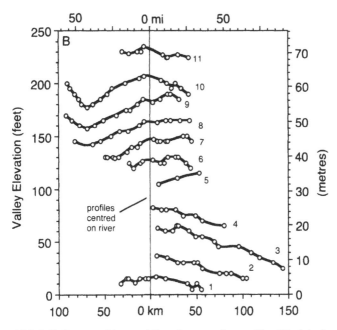

*Figure 15.3 A, B* Topographic map (A) and cross-valley profiles (B) of the lower Indus river alluvial plain.

structural lineaments have importance for the development and behaviour of the Indus river (Figure 15.2). The northern and southern margins of the fold and thrust belts are terminated by zones of active, transform or strike-slip faulting (Sarwar and DeJong, 1979). These transform zones tend to align with basement structures that continue east, under the Indus plain (Seeber and Armbruster, 1979). The transform zone at the southern margin of the Karachi Arc may connect with the Rann of Kutch fault zone (Kazmi, 1979), which was the site of a 6 m fault scarp in the 1800s (Oldham, 1926). The transform zone at the northern margin of the Karachi Arc apparently has no recognizable eastern continuation, but it marks the point where the modern Indus river crosses from the Kachhi foredeep on to the Thatta–Hyderabad high, where the Indus plain shifts abruptly eastward, and where Manchar Lake is formed at the southern end of the foredeep between the modern river and the piedmont (Figure 15.3A). Deformation of Pleistocene-age gravels in this region also points to recent uplift and tectonism.

The east–west structural orientation of the southwest margin of the Sulaiman Arc coincides with structures which continue across the Indus plain and have an effect on the Indus river. The Quetta or Sibi re-entrant, northwest of Jacobabad, has been a zone of thick sediment accumulation (Kazmi and Rana, 1982), and it may have formed in a zone of weakness, which extends southeast across the Indo–Pakistan continent (Haghipour *et al.*, 1984). In the vicinity of the Indus river near Sukkur, this structural zone is expressed as two elongate domes, the

Jacobabad–Khairpur to the south and the Mari–Khandkot to the north (Kazmi and Rana, 1982), which are separated by a 500 m deep trough (Voskresenkiy *et al.*, 1968). The domes may represent shoulder uplift on the margins of a faulted graben (Haghipour *et al.*, 1984), which may relate to the Cambay graben to the southeast (Biswas, 1987). The uplifts are identifiable on gravity surveys (Glennie, 1956), they are associated with low-level seismicity (Snelgrove, 1967; Quitt-meyer *et al.*, 1979), and they have been areas of stratigraphic thinning (Tanish *et al.*, 1959; Kazmi and Rana, 1982). Eocene-age limestone crops out in the southern uplift as low hills (Rohri Hills, Figure 15.3A) which have been breached by the Indus river.

To summarize, the Indus river enters the lower alluvial plain in the axis of the subsiding Sulaiman foredeep and then crosses a region of relative upwarp. This region consists of two uplifts (Mari–Khandkot and Jacobabad–Khairpur uplifts) separated by a graben-like trough. The river crosses the smaller, northern uplift near Begari Head (Figure 15.1) and then encounters the larger, southern uplift near Sukkur. Leaving the gap in the limestone hills near Sukkur, the Indus flows down the southwest flank of the uplift before entering the Kachhi foredeep near Kandiaro. Below Sehwan the river crosses the Thatta–Hyderabad upwarp until it reaches the Arabian Sea.

## Hydrology and geomorphology

The lower Indus river is fed by snowmelt streams which drain the Himalaya mountains and by piedmont streams which flood during late summer monsoons. The mean annual flood at Sukkur for the period 1902–83 is 18,100 m$^3$ s$^{-1}$. Estimates of the annual sediment load prior to construction of barrages on the lower Indus and dams on the upper Indus range from 270 to 600 million tonnes (Milliman *et al.*, 1984). The median grain size of the floodplain sediment ranges from silt and fine sand at the delta to fine sand at Sukkur (Kazmi, 1984). Coarse sand and gravel deposits of the late Pleistocene Tandojam formation are buried > 30 m below the floodplain (Kazmi, 1984).

Water and sediment discharge have been dramatically reduced after comple-tion of several water projects in the mid-1950s. In fact, Kazmi (1984) attributes shrinking of the modern delta to an 80 per cent reduction in water and sediment load. Prior to development of flood-control structures, discharge fluctuations led to frequent overbank flooding and occasional channel avulsions into lower elevations on the floodplain (Snelgrove, 1967). Construction of dams on upstream reaches has reduced flood peaks, and construction of barrages (Table 15.1) has increased water transfer to the alluvial plain through an extensive network of irrigation canals (Irrigation and Power Dept, 1978). Construction of flood-control bunds or levees has both constrained the channel to its present course by preventing the avulsion of the channel into natural flood channels and increased the proportion of flood discharge confined to the presently active meander belt.

The history of the lower Indus river is largely buried beneath the Indus plain. During the last glacial advance, the lower sea level and increased precipitation caused the Indus river to cut a 120 m to > 200 m deep trench into the alluvial plain (Kazmi, 1984) as far upstream as Mithankot. During this time the alluvial plain closest to the Himalayan uplift was aggrading (Brinkman and Rafiq, 1971). In the Holocene, the sea rose and caused the delta to be located above Hyderabad (Pithawalla, 1959). The Indus river has since filled the trench, extended the delta, and incised the alluvial plain in Punjab. The disappearance of terraces near Kashmor (Figure 15.1) marks the downstream change from recent incision to deposition (Holmes, 1968).

Several former courses of the Indus river have been identified on aerial photographs, maps, and through historical accounts (e.g. Hunting Survey Corp. 1960; Holmes, 1968). Flam (1981b) and Kazmi (1984) consolidated them into three or four main courses that have covered nearly the whole of the alluvial plain (see Flam, this volume). In the region near Hyderabad, the river has moved west four or five times by avulsion; the last taking place in 1758 (Figure 15.3A), when the Indus abandoned a course east of Hyderabad for the lower, present course to the west (Holmes, 1968). To the north, between Sehwan and Kashmor, the Indus has had four major courses, three east and one west of the present course. Aerial photographic mapping by the Hunting Survey Corporation (1960) defined floodplain, bar, and abandoned channel deposits, which cover the entire alluvial plain. During the mid-Holocene, the Indus river captured what is now the drainage area of the Sutlej river, causing abandonment and drying up of ancient channels east of the Rohri hills near the margin of the Thar desert (Wilhelmy, 1969).

There are only two areas of incision with identifiable terraces in the lower Indus plain (Holmes, 1968). One extends along the left bank (viewed looking downstream) from south of the limestone hills at Rohri to near Kandiaro, partially including the last ancient course of the Indus. The upstream margin of the terrace is shown by the elevation increase through the Sukkur gap (see the repeated 190 ft contour of Figure 15.3A). The other terrace is located on the far-left valley margin downstream of Gudu barrage in the Ghotki area but it ends north of the Rohri hills. In all other areas, aggradation and subsidence have resulted in burial of older deposits.

At Sukkur, the Indus river passes through the westernmost of two gaps in the Eocene limestone outcrop (Holmes, 1968). The gap existed prior to occupation by the present Indus river and so it must have been cut during a previous period of incision which was caused by either the tectonic rise of the limestone beneath a previous Indus course or the lowering of a previous Indus course on to the limestone during a time of overall degradation such as the beginning of a glacial phase.

The Indus river currently flows on an elevated alluvial ridge, suggesting that Holocene aggradation is continuing (Figure 15.3A). This alluvial ridge varies in relief from zero near Gudu barrage up to 6 m between Sukkur and Sehwan. From Sukkur to Kandiaro, the alluvial meander-belt ridge is up to 6 m high between

*Table 15.1* Location of gauges and barrages in km from the sea.

| Gauging station | Valley distance (km) | River distance (km) Av. 1972–83 |
|---|---|---|
| Mithankot | 728 | 1019 |
| Rojhan | 679 | 962 |
| Kotla | 650 | 926 |
| Gudu | 642 | 920 |
| Gudu barrage | 640 | 918 |
| Machka | 630 | 907 |
| Sarhad | 593 | 858 |
| Unhar Head | 575 | 835 |
| Begari Head | 564 | 819 |
| Sattabani | 546 | 797 |
| Rajib | 539 | 785 |
| Raza jo Goth | 539 | 785 |
| Panwari | 532 | 774 |
| Bukkor | 518 | 750 |
| Sukkur barrage | 514 | 746 |
| Bachal Shah | 511 | 743 |
| Begarji | 502 | 734 |
| Ruk | 497 | 722 |
| Akil | 462 | 658 |
| Mohen jo Daro | 428 | 600 |
| Kamal Dero | 407 | 563 |
| Pat Machi | 380 | 525 |
| Lalia | 355 | 493 |
| Daulatpur | 331 | 455 |
| Sehwan | 316 | 441 |
| Bhagotoro | 303 | 422 |
| Khuda Bux | 283 | 393 |
| Sann | 266 | 372 |
| Manjhand | 250 | 355 |
| Hala | 236 | 334 |
| Khanote | 224 | 320 |
| Salero | 219 | 307 |
| Jakhri | 213 | 299 |
| Manjhu | 211 | 292 |
| Ghallian | 202 | 282 |
| Bada | 193 | 274 |
| Kotri barrage | 188 | 269 |
| Detho | 187 | 267 |
| Kotri | 180 | 260 |
| Mehrani | 169 | 247 |
| Naro | 162 | 239 |

| Hajipur | 156 | 230 |
| Jherruck | 143 | 214 |
| Aghimani | 90 | 140 |

the right bund and Jacobabad, but it is less pronounced along the left bank. Former courses of the Indus river, especially the 1700s course east of Hyderabad, can be observed both on the topographic map and in profile as alluvial ridges (Figure 15.3A). From Hyderabad to the sea, the river crosses a smooth, fan-shaped delta plain.

## Aggradation and degradation

Both geomorphic and hydrologic evidence suggest that the Indus river has aggraded and continues to aggrade some reaches below Mithankot. Using a minimum value of 200 m for the depth of burial of a late Pleistocene (10,000–15,000 yr BP) channel, the estimated rate of vertical aggradation averaged over the Indus alluvial plain during the Holocene is 13–20 mm $yr^{-1}$. However, measurements made at the Mohen jo Daro excavations require an average aggradation rate of 5 m since ~ 2370 or 1750 BC or 1.2–1.3 mm $yr^{-1}$ (Allchin, 1976). Pithawalla (1959) suggests an aggradation rate of 1 ft per century (3 mm $yr^{-1}$) for the entire Indo–Gangetic plain. The average rate of aggradation has clearly decreased with time since the last glacial.

Change in suspended sediment load between Sukkur and Kotri during 1931–84 also provides a minimum estimate of present-day aggradation. Despite an increase in average annual sediment concentration from 0.102 per cent by volume at Sukkur to 0.113 per cent at Kotri, sediment discharge decreases downstream owing to a loss of discharge (Figure 15.4A). Average annual sediment loss amounts to over 31 million $m^3$ or ~ 24 per cent of the sediment passing Sukkur (Figure 15.4B). Sediment loss prior to construction of Kotri barrage (1931–55) was 25 million $m^3$ annually. Assuming that this material was dispersed over a 75 km wide area between Kotri and Sukkur through canals and by natural overbank flow, aggradation is 1.5–1.8 mm $yr^{-1}$. Even if most of the sediment is currently trapped between the flood-control bunds (15–20 km wide), aggradation is still less than 5 mm $yr^{-1}$. Therefore, the modern distribution of Indus river suspended sediment on the floodplain approximates mid-Holocene aggradation but it is less than early Holocene aggradation.

## Regional tectonic deformation and river migration

Significant deformation of the Indus alluvial plain has accompanied subsidence in response to the tectonic loading from the west. Ongoing seismicity in Baluchistan suggests that loading and subsidence are continuing yet the Indus river currently flows in the axis of foredeeps only north of Kashmor and north of Sehwan (Figure 15.2). The Indus river flows down the axis of the Sulaiman

297

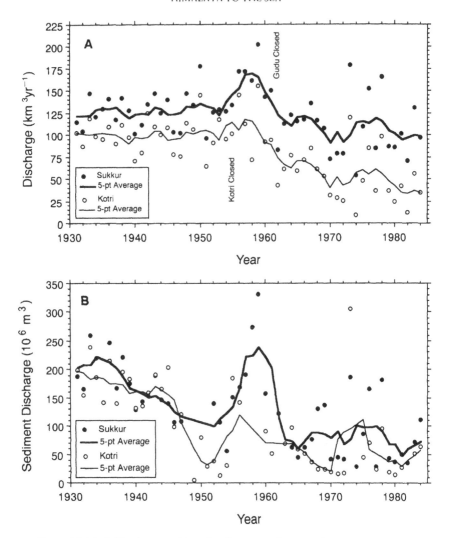

*Figure 15.4 A, B* Yearly water (A) and sediment (B) discharge for Sukkur and Kotri stations.

foredeep, which may have aided capture of the drainage of tributary river streams east of the foredeep. However, near Sukkur and along the eastern margin of the Karachi Arc, the Indus river has migrated to positions of relative tectonic upwarp away from zones of subsidence. This condition probably results from a sedimentation rate that is greater than the rate of tectonic or epeirogenic movement. A foreland basin subsidence rate of 0.3–0.5 mm yr$^{-1}$ obtained from Plio–Pleistocene sediment thickness in northeast Pakistan (Burbank *et al.*, 1988) is nearly an order of magnitude smaller than late Holocene rates of aggradation.

Therefore, rapid aggradation allows the river to shift to topographically low areas regardless of their tectonic position. In addition, reoccupation of previously cut gaps in the Rohri hills requires that the rate of aggradation exceed the rate of relative uplift near Sukkur.

Two foci of avulsions occur for the four major Indus courses, one near Gudu barrage and the other near Sehwan; both seem to be near major tectonic boundaries. The northern focus is on the upstream margin of the tectonic saddle which separates the Sulaiman and Kachhi foredeeps. The northern focus also marks the change from post-Pleistocene incision to aggradation (Holmes, 1968), and thus it may also be a function of the depositional hinge-line of the lower Indus alluvial plain. The location of this focus may further relate to the additional sediment contributed by the eastern tributaries or to deposition upstream of the Mari–Khandkot uplift. The downstream focus is located near Sehwan where the river crosses from the Kachhi foredeep to the Thatta–Hyderabad high. The cross-valley profiles suggest a change from aggradation (convex cross-valley profiles 6–9 of Figure 15.3B) to a short reach of degradation or stability (concave or flat cross-valley profile 5).

Lateral migration and avulsion of the Indus river have produced significant changes in the relative slope of the valley and channel. The trend of the late Holocene Indus river has been toward shorter and steeper courses. The topography of the alluvial plain east of the Karachi Arc suggests that the river is now at a higher elevation than during previous courses. An east–west profile drawn by Kazmi (1984) shows that the easternmost ancient course is 9–12 m below the present course (see profile 3, Figure 15.3b). However, the later courses became the shortest path to the Arabian Sea as delta growth extended the former shoreline. Similarly, the modern course and one ancient course of the river between Gudu barrage and Sehwan follow the shortest path south rather than to the west of Sukkur. Thus, the modern Indus plain can be thought of as two inland fans, one focused north of Sukkur and the other near Sehwan. Changes in water and sediment discharge as a result of river capture, Holocene hydrologic change, and gradual extension of the shoreline have accompanied shift of the Indus to the shortest, steepest path to the sea.

To summarize, geomorphic and hydrologic data show that most of the lower Indus river has been aggrading throughout the Holocene, but the rate has decreased in the late Holocene. Avulsion and channel migration near Sukkur and Hyderabad have positioned the modern river over areas currently disturbed by relative tectonic upwarp. Continued aggradation has built an alluvial ridge which promotes avulsion and facilitates sedimentation away from the river. Avulsion has been artificially controlled by the construction of levees, but water and sediment are distributed across parts of the alluvial valley through an extensive network of canals.

# TECTONIC CONTROL OF INDUS RIVER MORPHOLOGY

The Indus river has a wide, braided channel pattern near the confluence with the five rivers of the Punjab near Mithankot, but it becomes narrower and meandering downstream until it reaches the upstream limit of the Indus delta below Hyderabad. Departures from this general trend and irregularities of the longitudinal profile are related to past depositional patterns, past and present tectonic movements, and human activities. Geomorphic evaluation of the Indus river is hampered by size of the river and by the lack of detailed measurements. Unfortunately, cross-sections and longitudinal profiles of the river channel are not available. Therefore, geomorphic analyses are based on water levels taken at 49 gauges along the lower Indus river. All profiles and slope measurements are based on gauge heights, which in turn are tied to local benchmarks. Channel maps surveyed during low-water months record the position of the main channel(s) or thalweg(s), but they cannot be used to infer channel dimensions or shape.

## Channel and valley profiles

Alluvial rivers, those flowing in and on material which they transport, tend toward smooth equilibrium longitudinal profiles which decrease in slope downstream. However, water surface profiles of the Indus river decrease to about river kilometre ('rkm') 700 and then slope remains relatively constant (Figure 15.5A,B). Slope does not continue to decrease because there are no significant tributary contributions of water and sediment below Mithankot and because there are increasingly greater losses of water to evaporation, canal withdrawal, and storage. The profiles are irregular near barrages, which disrupt the flow of water and sediment, and they steepen again below Hyderabad (Figure 15.5A,B). One might expect that a river as large as the Indus, where instantaneous peak discharge has been as high as 34,000 m$^3$ s$^{-1}$ and which carries up to 600 million tonnes of sediment annually, could maintain an equilibrium profile with ease, yet two areas of significantly anomalous slope disrupt the profile. Low slope occurs upstream of Sukkur between rkm 750 and 850 (Figure 15.5B), and slope increases rather than decreases below rkm 400 (Bhagotoro gauge). The relatively high river slope between Sukkur and Mohen jo Daro (rkm 745 and 600, respectively) is in part a result of the shorter, steeper modern river course at this latitude (Flam, this volume). The lowest channel slope occurs where the river crosses the Kachhi foredeep upstream of Sehwan, between the axes of the Karachi Arc and Jacobabad–Khairpur uplift (rkm 440–570, Figure 15.5B). The irregular water surface profile persists despite the ability of an alluvial river to vary its slope by aggradation, degradation, and pattern change.

Construction of the Kotri barrage in 1955 and Gudu barrage in 1962 to provide water for irrigation canals has caused aggradation of the channel bed

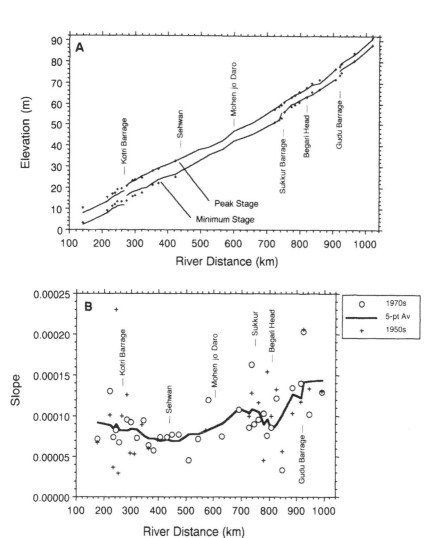

*Figure 15.5 A, B* Water surface profiles of the Indus river.

A river profile prepared using the average river distance from the period 1973–85 showing the average maximum stage for 3 years that have a discharge of ~ 17,000 m³s⁻¹ and the average minimum stage for the same years. The crosses mark the similar features for 1951–3 and are included to denote the change that has occurred since closure of the Kotri and Gudu barrages. Note the draw down of the recent water level near Kotri barrage and the build-up of silt 30 km upstream of the barrage.

B Slope of the Indus river water surface profile. Sections including barrages are omitted.

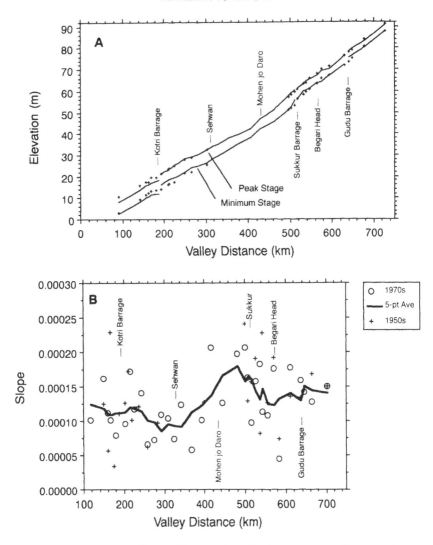

*Figure 15.6 A, B* Water surface profile (A) and profile slope (B) using valley centre line distance in the place of channel distance (see Figure 5 for explanation).

20–50 km upstream since the 1950 channel survey (Figure 15.5B). Some channel scour has also taken place below Kotri barrage during the same period (rkm 230–260). The Sukkur barrage, which is just downstream of the gap in the limestone outcrop, was closed in 1931. The channel bed above Sukkur is anomalously high, owing to either the sediment storage upstream of the barrage or bedrock outcrop.

The relatively smooth peak-stage water surface profile is similar to the valley or projected profile (Figure 15.6A) except between valley kilometre ('vkm') 350 and 500. The valley profile, constructed by projecting channel elevations to the valley centre line, is more variable, and generally steeper (Figure 15.6b). In part, comparison of the profiles demonstrates the degree to which the channel pattern of the Indus river compensates for steeper reaches of the Indus alluvial plain by increasing sinuosity, which increases channel distance (see below).

The influence of the uplifted Eocene limestone near the Rohri hills (vkm 510) produces a convexity in the valley profile with a flatter slope above and steeper slope below (Figure 15.6A). Valley slope nearly doubles from above Sukkur to below Sukkur (vkm 540–480, Figure 15.6B). The low slope reach near Sehwan is accentuated on the valley profile (vkm 250 and 400) relative to the river profile; between vkm 420 and 300, valley slope decreases from $\sim 0.00015$ to $< 0.0001$. Like river slope, valley slope again increases below Hyderabad ($\sim$ vkm 220) to the head of the modern delta (vkm 90). Other small profile anomalies occur near the Mohen jo Daro and Begari Head gauges (vkm 428 and 564 respectively) but elevations based on a single gauge are subject to considerable error. The Begari Head gauge is located at the downstream end of a small convexity which produces low slope followed by high slope. Similarly, the Mohen jo Daro gauge is the centre of a small convexity which is shown on both the valley and river profiles.

## Channel pattern

The Indus river patterns seen on Landsat images, topographic maps, and low-water channel maps can be grouped in three or four categories. On the largest scale, the channel tends to be multi-thread above Sukkur and single-thread below Sukkur (Figure 15.7). From Mithankot (just below the confluence of the five Punjab rivers) to Machka the multi-thread channel is clearly braided. It is characterized by large bars or islands (3–8 km long) and frequently shifting channels. Between Unhar Head and Sukkur, the islands or bars become less frequent and larger. This pattern is perhaps best described as anastomosing, because it consists of a larger main channel and smaller, anabranching flood channels. The channels in this reach tend to be more permanent than the braided channels of the adjacent, upstream reach. Anabranching flood channels have been found to receive gradually greater proportions of flood discharge and eventually become the main channel as the former main channel becomes more sinuous and fills with sediment (Schumann, 1989; Schumm and Erskine, 1989).

The single-thread section of the river, from Sukkur to the sea, can be subdivided into reaches displaying a variety of meander styles and rates of change as follows.

1　From the confining bedrock at Sukkur to Kamal Dero, the channel is highly sinuous with large, mobile meanders. Meander development in this reach varies in time and location. It was highest in the reach between Ruk and Akil during the 1970s but, during later surveys, it was highest near Mohen jo Daro. The amplitude of the large, relatively regular meanders ranges from 5 to 13 km and the wavelength from 8 to 16 km.

2    Between Kamal Dero and Sehwan, the channel pattern is generally straighter than it is upstream but it is highly irregular, exhibiting variability in the location of clusters of meanders or multiple channels through time. The meander clusters appear to be of short duration and they are about half of the size of upstream meanders. For example, a series of well defined meanders just upstream of Dualatpur, which appear on a 1979 Landsat image, had just begun to develop on the 1973–4 map but they had been already partially cut off by 1985–6 (Figure 15.7), leaving a single, large irregular bend.

3    From Sehwan to Hala, where the course of the Indus follows the northwest limit of low hills related to the Karachi Arc, the channel is comprised of strongly asymmetrical meanders. Individual meanders migrate toward the left (east) bank and downstream. This pattern causes the sharpest channel bends at the western margin of the floodplain, against the piedmont slopes. The movement suggests that deposition occurs preferentially on the eastern side of the active floodplain, whereas erosion and cutoffs occur along the western valley margin. Preferential cutoff of meanders on the western bank is consistent with the tendency of historic and prehistoric channels to avulse to courses farther to the west at this latitude.

4    Below Hala the Indus does not appear to have migrated significantly during 1959–85. Where the channel is confined between consolidated rock near Hyderabad, it is relatively straight and inactive, except for a very active bend at Naro. Above and below the confined reach the channel is more sinuous but temporally stable relative to upstream reaches. The stability of the pattern may reflect the effect of more consolidated materials, the decrease in steam power by continual loss of flow upstream, or the increase in cohesiveness of bank materials caused by a decrease in sediment size.

At the largest scale, the downstream change from a braided to meandering pattern partially reflects the loss of water as a result of modern floodplain irrigation and former, natural overbank flooding. Decreased flow can move the channel closer to or over a braiding–meandering threshold (Figure 15.8), which is defined by discharge and slope (Leopold and Wolman, 1957). R.T. Ferguson (1984) improved the discriminate function by including sediment size, albeit for coarser grained streams ( > 2 mm). By extrapolation of the grain-size relationships, the channel near Gudu should be at or near the pattern threshold (Figure 15.8). On the other hand, smaller mean annual floods, a lower channel slope, and a small increase in grain size near Kotri barrage all contribute to the stability of the meandering pattern.

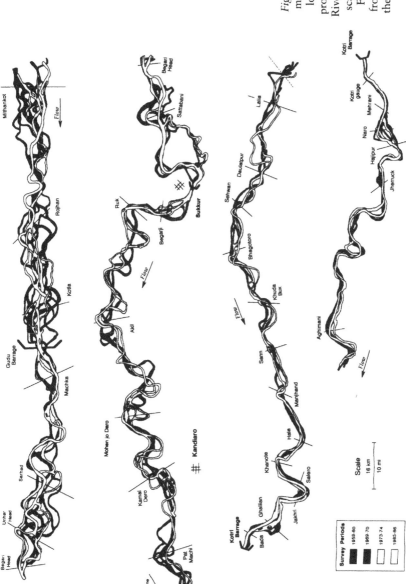

*Figure 15.7* Channel maps based on four low-water surveys provided by the Indus River Commission at a scale of 1 inch mi⁻¹. Flow is by section from the top right to the bottom left of the figure.

## Tectonic control

Observations that the Indus river profile contains areas of high and low slope and that channel pattern is variable suggest that active or recent deformation of the alluvial plains affects the river. Tectonic deformation affects river behaviour by controlling the valley slope down which the river flows once it avulses to a new position. In addition, deformation requires that the river distribute or redistribute its sediment load unequally, which in turn requires or results in different channel patterns because of the dependence of pattern on sediment transport and deposition. Relationships between valley slope and channel pattern have been studied in laboratory channels (Schumm and Kahn, 1972) and on major rivers (Schumm *et al.*, 1972; Adams, 1980). Channel sinuosity is the ratio of channel distance to valley centre line distance and expresses the degree of meandering. Sinuosity increases with increasing slope until it decreases to one as the channel becomes braided above a certain threshold slope (inset, Figure 15.9). Sinuosity of the Indus river was computed from gauge-to-gauge, main-channel distance which was supplied by the Indus River Commission for the period 1962–86. The sinuosity–slope characteristics of some reaches of the Indus river follow the previously defined relationship in that the four meandering reaches discussed show a positive correlation between slope and sinuosity (reaches 4–7, Figure 15.9). However, the braided reaches (1 and 2, Figure 15.9) are not steeper than the meandering reach below Sukkur (reach 4, Figure 15.9). The reasons for this anomalous relationship and for the presence of an anastomosing reach above Sukkur probably stem from the effect of relative subsidence and uplift on sedimentation. They are addressed below in downstream order as they relate to specific tectonic elements.

Channel braiding in the upstream reaches can be an indication of high sediment load in addition to high valley slope and discharge. High sediment transport rates lead to frequent channel shoaling and bar development (Carson, 1984), but the presence of terraces in these braided reaches suggests that they are transporting sediment rather than depositing it. Yet gauge records (see below) suggest that aggradation was occurring prior to barrage construction at the Machka gauge and therefore this may be a cause of shoaling and braiding. The reach just above Sukkur is more convincingly a site of sediment deposition. Aggradation is indicated both by the anastomosing pattern and by the appearance of the alluvial ridge just north of the uplift at Sukkur. An anastomosing channel pattern has been related to rapid aggradation (Smith, 1983, 1986), *especially in regions of low regional slope and subsidence. The anastomosing pattern is well developed in the trough between the Jacobabad–Khairpur and Mari–Khandkot uplifts, where the valley slope is less than either upstream or downstream (vkm 520–560, Figure 15.6B), as a result of backtilting above Sukkur or subsidence of the trough.

The Mari–Khandkot uplift, which has no topographic expression, is not particularly apparent in terms of river characteristics; the channel is slightly

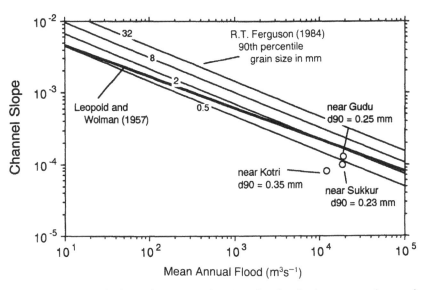

*Figure 15.8* Slope–discharge discriminant functions showing the downstream change of the Indus river pattern sensitivity.

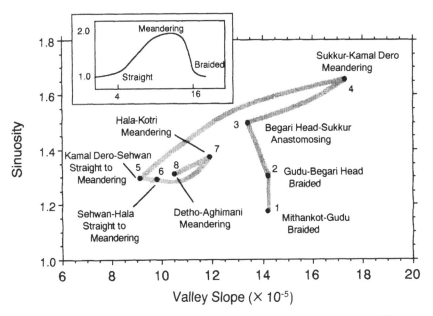

*Figure 15.9* Slope–sinuosity relationships of the Indus river. The shaded line follows adjacent reaches in downstream order. Value of sinuosity is the mean for the period 1962–86. The inset shows the pattern–slope relationship for the Mississippi river (Schumm *et al.*, 1972).

Table 15.2 Coefficients of nonparametric statistics[a] of specific gauge records for discharge of 5660 m³ s⁻¹.

| Gauge | Period of record | Sen's slope estimate[b] (mm/yr⁻¹) | Upper 90%-ile slope[c] | Lower 90%-ile slope[c] | Mann–Kendall 'S' statistic[d] | Number of data points | Probability of slope ≠ 0 |
|---|---|---|---|---|---|---|---|
| Machka | 1932–85 | 14 | 19 | 10 | 1886 | 108 | 1.000 |
| Machka, pre-barrage | 1932–61 | 10 | 21 | 3 | 334 | 60 | 0.967 |
| Machka, post-barrage | 1962–85 | 46 | 61 | 30 | 502 | 48 | 1.000 |
| Bukkor | 1901–83 | 7 | 12 | 1 | 1252 | 170 | 0.906 |
| Bukkor, pre-barrage | 1901–31 | −8 | 3 | −18 | −188 | 62 | 0.745 |
| Bukkor, post-barrage | 1932–83 | 23 | 26 | 20 | 3431 | 108 | 1.000 |
| Bachal Shah | 1932–85 | 17 | 20 | 13 | 2070 | 108 | 1.000 |
| Korri | 1932–85 | −15 | −12 | −19 | −2116 | 102 | 1.000 |
| Korri, pre-barrage | 1932–54 | −4 | 10 | −9 | 28 | 46 | 0.202 |
| Korri, post-barrage | 1955–84 | 4 | 10 | −1 | 166 | 56 | 0.758 |
| Rojhan (RB) | 1950–85 | −9 | 4 | −20 | −208 | 70 | 0.707 |
| Rojhan (LB) | 1950–85 | −18 | −9 | −28 | −580 | 70 | 0.997 |
| Sarhad (RB) | 1950–85 | −16 | 0 | −30 | −348 | 71 | 0.915 |
| Sarhad (LB) | 1950–85 | −6 | 10 | −21 | 70 | 70 | 0.440 |
| Unhar Head | 1950–85 | 1 | 9 | −7 | 72 | 70 | 0.281 |
| Begari Head | 1950–85 | 2 | −12 | −6 | 105 | 71 | 0.395 |
| Rajib | 1950–85 | 51 | 61 | 40 | 1211 | 71 | 1.000 |
| Raza jo Goth | 1950–85 | 32 | 45 | 20 | 792 | 69 | 0.995 |
| Mohen jo Daro | 1968–85 | −46 | 12 | −82 | −71 | 31 | 0.767 |
| Manjhand | 1950–85 | 37 | 46 | 28 | 1002 | 66 | 1.000 |

[a] from Gilbert (1987, p. 217).
[b] Slope of relationship between time (x) and gauge height (y).
[c] Upper and lower 90 percentile confidence interval of relationship of gauge height to time (column 3).
[d] Sum of positive (+1) and negative (−1) changes of gauge height between all possible pairs of data points.

straighter (Figure 15.7 near Begari Head), perhaps from incision, and there is the slight water surface profile convexity discussed above. The most significant local deformation of the Indus plain takes place near Sukkur. Continued relative rise of the Jacobabad–Khairpur high is indicated by the isolated terrace remnants between Khairpur and Sukkur and downstream of Gudu barrage (Holmes, 1968), by seismicity (Snelgrove, 1967; Quittmeyer et al., 1979), and by dramatic change in the pattern and behaviour of the Indus at Sukkur (Figure 15.7).

In the oversteepened reach below Sukkur (vkm 420–500, Figure 15.6B), the exceptionally high sinuosity reflects both the lower sediment load and the high valley slope caused by the Jacobabad–Khairpur high. Deposition of sediment above Sukkur reduces bar development and increases the tendency for a meandering channel to develop at the higher slope below Sukkur. Adams (1980) related increasing sinuosity of the Mississippi river to active forward-tilting of the valley. Active forward-tilting of the Indus plain on the downstream flank of the Jacobabad–Khairpur uplift may be caused by continued subsidence of the Kachhi foredeep relative to the Sukkur region (Figure 15.2).

The high sinuosity and steeply sloped reach below Sukkur contrasts sharply with the next two reaches, which extend from Kamal Dero to Hala (Figure 15.7). Here, where the Indus crosses the Kachhi foredeep and then the upstream flank of the Thatta-Hyderabad upwarp, channel and valley slopes are low, sinuosity is low but highly variable, and there is aggradation (Manjhand gauge, Table 15.2). A straight or slightly sinuous planform maintains the steepest possible channel slope across the region of low valley slope, especially between Kamal Dero and Sehwan.

The section of the river which crosses the Kachhi foredeep (Kamal Dero to Sehwan, Figure 15.7) also demonstrates the greatest temporal and spatial instability of channel pattern. The straight pattern is episodically interrupted by isolated reaches of high sinuosity or braiding (Figure 15.7). Periods of meander growth may represent the episodic delivery of high sediment loads and subsequent initiation of deposition and channel migration (Jorgensen, 1989). The sediment can be supplied episodically from upstream by channel pattern change and local scour or by periods of high sediment load which are related to hydrologic change. Regardless of the source of sediment, the transport capacity of the steep reach below Sukkur is greater than across the foredeep and so the channel will aggrade across the foredeep. Aggradation is greatest at the boundary between tectonic elements, which is just downstream of Mohen jo Daro.

Lateral river migration toward areas of relative uplift may represent a hydraulic response to lateral or oblique tilting of the alluvial plain. Nanson (1980) demonstrated that the Red river of Canada is migrating toward the side of the valley being raised by isostatic rebound following glacial retreat. The asymmetrical meanders south of Sehwan at the northeast edge of the Karachi Arc suggest that the Indus river is actively migrating or eroding to the west, which is the uptilt side of the valley. Conversely, Alexander and Leeder (1987) proposed that rivers migrate down tectonic slope, leaving cutoff and abandoned meanders

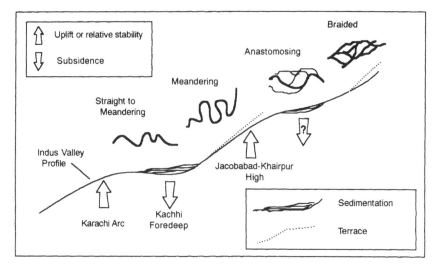

*Figure 15.10* Cartoon showing the relationship of the Indus valley profile and channel pattern to tectonic elements. Note that the slope and pattern changes are exaggerated for illustration.

on the upslope side. If they are correct, the Indus is migrating west into a regional low, which is adjacent to local, structurally produced topography. While this interpretation is possible for the reach from Sehwan to Hyderabad, it is untenable for the reach southwest of Sukkur where the channel migrated or avulsed from the foredeep to the Jacobabad–Khairpur high. Unfortunately, the key to structural or tectonic control of lateral migration by the Indus river lies in as yet unavailable geodetic data and not in geomorphic interpretation.

It is clear that the tectonic elements of the Indus plain still have significant correlation to the geomorphology of the Indus river despite the great changes brought about by human activity during the last century (Figure 15.10). All of the significant river pattern changes take place at tectonic boundaries, including the different types of meandering found between Sukkur and Kotri. In addition, anomalously steep and gently sloping reaches of the river conform to suspected reaches of relative uplift and subsidence, and avulsions may mark the points of relative upwarp in contrast to basin subsidence. Therefore, downstream changes of channel hydrologic characteristics, which will be addressed in the next section, can be linked in part to the effects of tectonically affected slope and sedimentation patterns.

## HYDROLOGIC VARIATION OF THE INDUS RIVER

As a result of erosion and deposition, channel migration, and pattern change, the elevation of the Indus river water surface demonstrates *several scales of vari-*

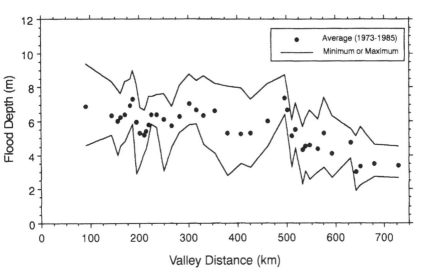

*Figure 15.11* Average, minimum and maximum flood depths for the period 1973–85.
Flood depth is computed from the maximum gauge reading minus the minimum gauge
reading for each water year.

ability. The largest scale involves advance and retreat of the delta position
accompanying incision and filling of the trench that was excavated during the
last glacial advance (Kazmi, 1984). These major erosional and depositional
events far overshadow any changes in the Indus plain during the late Holocene.
Nevertheless, pattern change and episodic sediment transport have significantly
influenced the water surface elevation of the Indus river during the short period
for which data are available. The following section describes the influence of
channel morphology and hydrology on variability of the Indus river stage at
different time scales. Patterns in this variability will then be documented in space
and time. It is the downstream and temporal variability of the Indus river that has
the greatest bearing on questions of the habitability of Mohen jo Daro.

## Flood depth

Downstream variability of the Indus river planform, channel cross-section,
sediment size, and discharge causes differences in water depth during annual
flood peaks (Figure 15.11). In general, flood depths increase downstream, despite
continual loss of water to the floodplain, evaporation and storage. The wider,
braided channels upstream have a lower flood peak relative to the minimum
stage than the narrower, meandering channels. Pattern change at the tectonic
boundary near Sukkur is evident in the change in the flood depth at this point.
Anomalously high flood depth occurs just below Sukkur (vkm 500) and near
Bhagotoro (vkm 300). The former represents incision of the channel into the

Jacobabad–Khairpur uplift, but the latter is probably a function of the low channel slope, and hence velocity, on the upstream side of the Karachi Arc. Based on a short period of record (12 years), the highest variability in flood depth occurs between Pat Machi and Mohen jo Daro (vkm 380–428) and near Unhar Head (vkm 575). The former contains the planform type which displays rapid temporal and spatial change in sinuosity. However, the sinuosity of the latter reach is relatively constant (Figure 15.7), and the variation must represent changes in the cross-section shape or aggradation and degradation.

## Water surface fluctuations

Two scales of change in water surface elevation are important for the study of the Indus river; long-term raising or lowering of the water surface as a result of channel aggradation or degradation and short-term oscillations about these long-term trends. Steady erosional or depositional trends may be related to tectonic uplift or subsidence, whereas the short-term variability relates to channel and sediment transport dynamics. These topics are addressed in the following section using specific gauge analyses, in which the stage or water surface elevation for a given discharge is determined through time (Inglis, 1949). The implicit assumption is that without change in velocity, slope, roughness or width, an increase in stage with time implies aggradation and a decrease implies erosion or scour. Needless to say, a specific gauge analysis can have different interpretations but it is a good tool for determining mean bed elevation changes of an alluvial river. For example, a spatial and temporal pattern of aggradation and degradation has been successfully documented on the Red river (Harvey *et al.*, 1988), and long-term aggradation or degradation has been detected in association with tectonic subsidence or uplift, respectively (Volkov *et al.*, 1967).

### Long-term aggradation and degradation

The Indus river gauges for which long-term records are available (1900 or 1930 to present) are unfortunately located near barrages, which themselves are located in areas where bedrock control is likely. Specific gauge analyses at these gauges indicate anomalously high rates of aggradation upstream of the barrages, where water and sediment are held in the barrage pool. Specific gauge heights as reported by Inglis (1949) also rise downstream of barrages, where the reduced duration of high flows when the barrage gates are fully open is insufficient to remove all of the temporarily deposited sand and silt. Not surprisingly all of these gauges indicate aggradation for the period after barrage closure (Table 15.2 and Figure 15.12), although the Kotri record is not statistically significant. Only the Machka gauge below Gudu barrage showed significant aggradation prior to closure, which has continued at a greater rate since that time. This pre-barrage aggradation suggests that braiding above Begari Head may be a function of aggradation. Inglis (1949) suggests that 'retrogression' or erosion preceded the

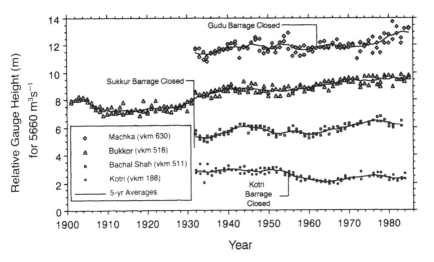

*Figure 15.12* Specific gauge height for 200,000 cfs ( ~ 5660 m³s⁻¹) discharge showing the long-term trends of gauges near barrages. Each year contains two points, one for the rising stage and one for the falling stage. The 5-year line is the average for the 10 points in the surrounding 5-year period. The calculated slopes for pre-and post-barrage, and entire records are listed in Table 15.2.

period of post-barrage aggradation for the Sukkur barrage. Erosion also characterized the Kotri site prior to closure, which degraded rapidly after 1955 followed by aggradation. The Kotri barrage is located near the axis of the Thatta–Hyderabad upwarp. Obviously, no trend for the Indus river as a whole should be inferred from these records.

Specific records for gauges located away from barrages are available at only a few sites and for shorter time periods (Table 15.2). Gauges in the braided reach above Sukkur (Rojhan, Sarhad) all show net degradation or no significant change during a period of aggradation near the barrages. Scour in this upstream reach may indicate downstream progression of the incision, which has characterized the Indus plain above the confluence of the five rivers during the Holocene. The Begari Head and Unhar Head gauges are located on the Mari–Khandkot uplift, but they do not suggest either aggradation or degradation. Change of water surface elevation may have also been caused by upstream dam construction and sediment traps that have caused the river to erode sediment from channel bed and banks (Milliman *et al.*, 1984). Analyses by Inglis (1949) indicate that the river bed upstream of Sukkur began to degrade soon after the construction of flood-control bunds that raised the stage or erosional potential of floods of a given size. Completion of flood-control bunds in the braided, degrading reaches during this period may have produced a similar response. Lastly, the sediment that is so obviously being stored near the barrages may simply have come from these intervening reaches.

Water surface elevations in the anastomosing reach just upstream of Sukkur and the Rohri hills (Rajib and Raza jo Goth gauges, Table 15.2) are rising at the extraordinary rate of 51 and 32 mm which is even greater than the rise at Bukkor gauge just upstream of Sukkur barrage. These gauges may be influenced by the backwater from the barrage just 40 km downstream and therefore the deposition represents the effect of the barrage rather than tectonic subsidence. Unfortunately, the record at Mohen jo Daro (vkm 428) is short and does not have a significant trend. Lastly, aggradation is indicated at Manjhand (Table 15.2), which is located on the upstream flank of the Karachi Arc (Figure 15.2).

Altogether, the trends of aggradation and degradation at the scale of ~ 50 yr present a confusing picture at best. Rates of aggradation have been altered by barrages and bunds, and they can be substantially higher than all Holocene estimates of aggradation. None the less, the data suggest erosion is or was occurring at gauges on tectonic upwarps (pre-barrage data Kotri and Sukkur, Table 15.2) and aggradation over the Sulaiman foredeep (Machka gauge at the Gudu barrage) and upstream of the axis of the Thatta–Hyderabad upwarp. However, shorter records from gauges in the braided reach indicate erosion (Table 15.2). Short-term records that approach estimates of late Holocene aggradation ( $< 5$ mm yr$^{-1}$) are not statistically significant.

## Short-term water surface fluctuations

Two controls are important in periodic or episodic rise and fall of stage for a specific discharge: (a) scour and fill of bed material caused by variation in sediment transport into and through a reach; and (b) change of channel sinuosity causing rise or fall in water surface elevation owing to increased or decreased slope and velocity. These two mechanisms often work in concert: lengthening channel distance by meander extension causes a decrease in sediment transport capacity and initiates upstream sediment deposition (Harvey, 1989). Water surface fluctuations at this scale are demonstrated by specific gauge records. Temporal variation of stage is computed at each station as a smoothed deviation from the trend of the record.

The longer records demonstrate that rises and falls in stage are partially synchronous for a considerable distance along the river (Figure 15.13). Two periods of stage rise occur during the 53-yr span. There is a lag between the upstream (Bukkor, vkm 518 and Bachal Shah, vkm 511) and downstream (Kotri, vkm 180) gauges on the order of 3–5 yr, which suggests that the change in water surface is related to downstream propagation of aggradation and degradation. However, the same two general periods of stage rise at the upstream Machka gauge (vkm 630) do not precede the rise at the other downstream gauges, as would be predicted. Thus, these more or less synchronous rises and falls in stage suggest the influence of discharge and sediment delivery on the whole of the Indus river (Figure 15.4). For example, periods of rising specific gauge heights (1960–70, Figure 5.13) tend to occur during periods of decreasing water and

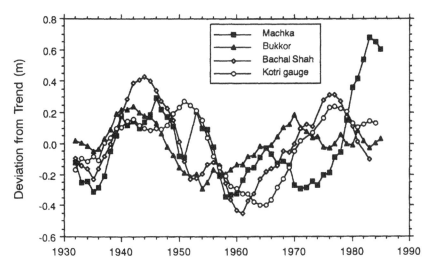

*Figure 15.13* Oscillation of the 200,000 cfs ( ~ 5660 m³ s⁻¹) specific gauge height about the trend reported in Table 15.2 for the stations with the longest period of record. Each line is the 5-year moving average of the deviation from the trend.

sediment discharge at Sukkur and Kotri (e.g. 1960–70, Figure 15.4). Inversely, falling specific gauge height (e.g. 1977–83 or 1950–60) occurs during increasing water and sediment discharge. These trends might be explained by an exchange of sediment between the flow and channel margins. Bed and bank materials are a significant source of sediment on the lower Indus river (Milliman *et al.*, 1984) and, therefore, an increasing sediment delivery would correlate to decreasing specific gauge height. Conversely, deposition and increasing gauge height would occur when flows become less able to carry the volume of sediment brought to the lower Indus in Sindh. Clearly, the change in sediment delivery on the decadal scale has had an influence on the patterns of erosion and deposition.

Similarly, shorter specific gauge records collected away from barrages (Figure 15.14) tend to rise during the later period of deposition (1970–75), but they also emphasize a slightly smaller time scale, during which the effect of different channel patterns becomes evident. Gauges in the braided reach (Figure 15.14A) show a rather consistent rise and fall, whereas stage levels in the meandering and anastomosing reaches (Figures 15.14B) demonstrate a greater variability and are less synchronous. This difference may relate to the fact that some meandering reaches undergo changes of sinuosity that do not occur in the braided channels (Figure 15.15). Sinuosity is especially variable between Kamal Dero and Sukkur (vkm 407–512), where the Indus descends the flank of the Jaobabad–Khairpur uplift, and near Lalia (vkm 355), where the channel crosses the Kachhi foredeep. Although sinuosity is higher near vkm 650, variability of sinuosity is moderate or low. A lower variation in sinuosity occurs in the

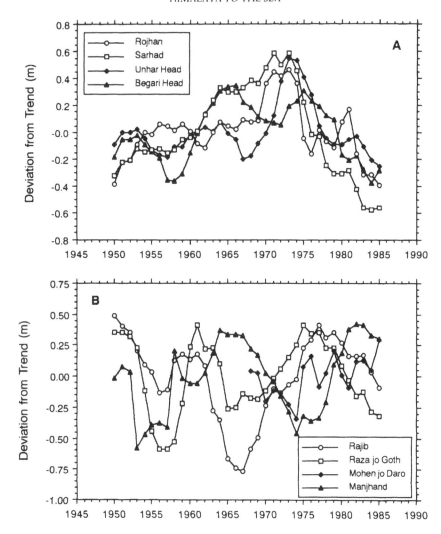

*Figure 15.14 A, B* Oscillation of the 200,000 cfs ( ~ 5660 m³s⁻¹) specific gauge height about the trend reported in Table 15.2 for the stations with shorter gauge records. Each line is the 5-year moving average of the deviation from the trend.

A  Braided reaches.

B  Anastomosing and meandering reaches.

downstream reaches (vkm 100–300), which are characterized by less active meandering (Figure 15.7).

The downstream differences in meander development and cutoff can have an affect on the variability of water surface elevation by changing channel slope and roughness (Harvey, 1989). Assuming constant channel roughness and channel

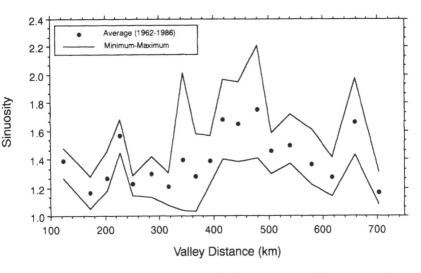

*Figure 15.15* Sinuosity (channel distance/valley distance) for 10–20 km sections of the Indus river for the period 1962–86, as calculated from gauge-to-gauge river distances supplied by the Indus River Commission.

widths (at a moderately low stage of $\sim$ 5660 m$^3$ s$^{-1}$) of 1000–1500 m, the change of stage from sinuosity-related slope change alone is up to 0.55 m for the meandering reach but < 0.2 m for the wider and straighter, multi-thread reaches. In addition, increase of sinuosity also adds to the form roughness of the channel, thereby causing even greater increases in stage. This latter effect cannot be estimated from the available records.

The remaining topic regarding the hydrologic variation of the Indus river concerns both the link between sedimentation and channel pattern, and how change in one reach affects reaches farther downstream. An analysis of specific gauge height and sinuosity in both time and space (Figure 15.16) suggests a relationship exists between pattern change and erosional or depositional episodes. The variability of the 5660 m$^3$ s$^{-1}$ gauge height, relative to the mean of the period 1972–85, is considerable in both time and space (Figure 15.16A), but two patterns are apparent. First, the rise and fall of the water surface elevation tend to be opposite to that of neighbouring upstream and downstream reaches during a 3–5 year period. For example, the rising stage in 1983–5 near Mohen jo Daro (vkm 428) corresponds to falling stage near Manjhand (vkm 250), and the falling stage in 1977–80 near Mohen jo Daro corresponds to rising stage near Manjhand. This pattern is repeated at many locations during the 13-yr record. Second, there is a general trend of specific gauge rise, which shifts downstream during the 13-yr period. Whereas the gauges in the reaches above Sukkur (vkm 514) tend to be at their highest during the early part of the record, those below Sehwan rise near the end of the record. The pattern does not show downstream

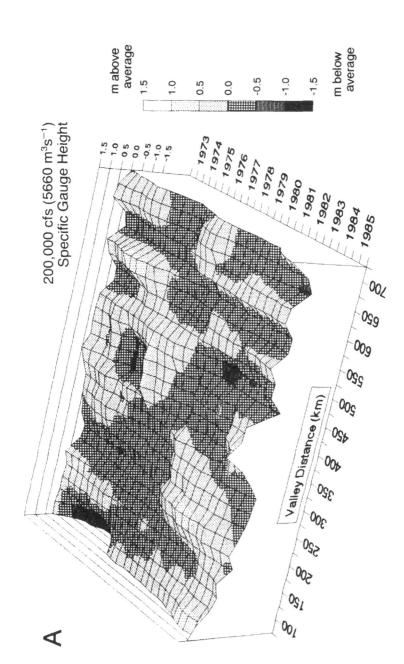

*Figure 15.16 A, B* Change with time and distance downstream of the 200,000 cfs ($\sim$ 5660 m$^3$s$^{-1}$) specific gauge height (A) and sinuosity (B). Plotted data points are the deviation from the mean of the period plotted, smoothed (5-pt) over time and subsequently smoothed to valley distance grid using 'inverse distance' weighting (Davis, 1986) raised to the power 1. Specific gauge records consists of two readings per year for rising and falling stage respectively, whereas sinuosity consists of one point per year.

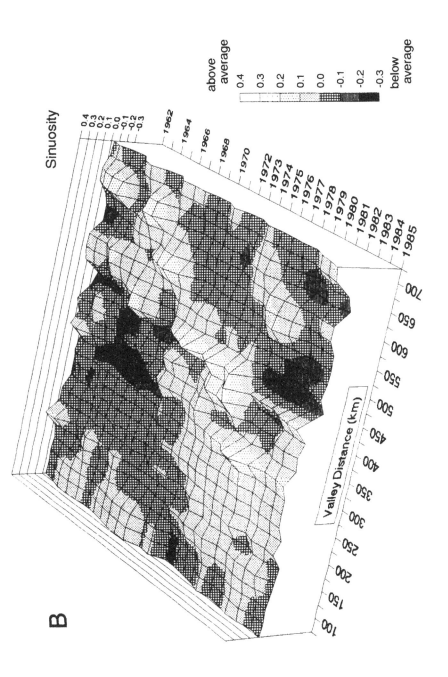

B

Sinuosity

Valley Distance (km)

above average

0.4
0.3
0.2
0.1
0.0
-0.1
-0.2
-0.3

below average

progression of the high specific gauge readings but rather their rise and fall at one location, followed by rise and fall at the next location farther downstream.

Change of sinuosity (Figure 15.16B), on the other hand, appears to have a downstream progression (note that data are available for a longer period). Values of sinuosity for the braided reaches were computed for the principal thalweg rather than the pattern as a whole, which approaches a value of unity. The highest sinuosity migrated from between Sukkur and Mithankot (vkm 500–700) in the early 1960s to below Sehwan (vkm 316) by the early 1980s. The high sinuosity correlates reasonably well with specific gauge rise in the meandering reaches below Sukkur (vkm 514). Here, change in hydraulic roughness, backwater effects, and possible sediment deposition all could contribute to the synchronous increase of specific gauge height and sinuosity. On the other hand, high sinuosity of the main channel in the braided and anastomosing reaches above Sukkur (above vkm 514) corresponds to low specific gauge height, especially during 1973–6. The increase of sinuosity in braided reaches correlates to increase in bar or island size. The size of braid bars has been observed to increase during degradation (Germanoski, 1989), which perhaps explains the reversal of the stage–sinuosity relation found in meandering reaches.

Whether the water surface changes are transmitted downstream in waves or episodic jumps is unclear. The longer gauge records suggest that channels along the length of the river aggrade and degrade with changes in hydrology (Figure 15.13). Alternatively, downstream propagation of stage changes caused by pattern change is suggested by the pattern of sinuosity (Figure 15.16B). Ashmore (1987) found that sediments accumulated in aggradational zones, which caused channel migration and increased braiding at regularly spaced intervals. Rather than migrate downstream, these waves of aggradation tend to erode then reform at the next location downstream. Similarly, Macklin and Lewin (1989) did not find obvious downstream translocation of an aggradational wave in a valley disturbed by mining-induced sediment loads. Indus river specific gauge height arguably is related to aggradation and degradation, which reflects the episodic, downstream movement of sediment (Figure 15.16A). The modern Indus river system is characterized certainly by large changes in sediment transport (Figure 15.4B). This variability may be initiated by upstream supply, which then triggers the downstream pattern changes that in turn control channel dynamics farther downstream. This downstream dependence of pattern and stage is clearly shown in the spatial and temporal behaviour of the Indus river, which in addition demonstrates variability at a number of scales.

The foregoing geomorphic analyses have established the relationship of the Indus river pattern and profile to the tectonic and geomorphic controls that have shaped the lower Indus alluvial plain. In turn, this adaptation of channel pattern and behaviour to tectonics has a significant effect on the hydrology of the channel. Rates of aggradation or erosion, flood depths, and water surface levels are quite variable but they are closely related to the tectonic segmentation of the Indus valley. Some sites undergo constant sedimentation while others are characterized by episodic erosion, deposition and avulsion. The modern spatial and temporal variability, although greatly affected by human influence, has probably characterized the Indus river for much of the late Holocene.

## THE ABANDONMENT OF MOHEN JO DARO

The analysis of Indus river morphology, sedimentation trends, and hydrology provides a basis for the evaluation of hypotheses advanced for the abandonment of the ancient city of Mohen jo Daro. The causes that have been postulated for the demise of the Indus (Harappan) civilization can be grouped broadly into three general hypotheses. First, the Harappans were conquered and destroyed by Aryan invaders (Marshall, 1931; Mackay, 1938; Piggott, 1953; Wheeler, 1968). Second, the Harappans abandoned Mohen jo Daro and the civilization collapsed after a series of floods caused by the downstream damming of the Indus river near Sehwan and the creation of subsequent lakes that engulfed the site in water and silt (Raikes, 1964, 1965; Dales, 1964, 1965, 1972; Dales and Raikes, 1977; Raikes and Dales, 1977; Possehl, 1967). Third, the Harappans were forced to abandon their settlements because the Indus river avulsed, thereby eliminating the local source of water and annual overbank flooding (soil moisture recharge) and its attendant sedimentation that replenished the soil fertility (Lambrick, 1967). The latter two hypotheses were postulated because of dissatisfaction with the first, while the first is beyond the scope of this paper.

Two of the three hypotheses have a common thread; that the demise of Mohen jo Daro was related to the dynamics of the Indus river in the period ~ 2350 BC to ~ 1800 BC. However, the inferred behaviour of the river and its location during the period of Harappan occupation has never been investigated. All of the archaeological reports based on excavations at Mohen jo Daro have a common element: the presence of flood deposits below the present ground elevation at the site. Archaeological excavations and drill hole data at the site indicate that there is ~ 10–20 m of sediment beneath the current floodplain surface that contains evidence of occupation. This sediment overlies sterile floodplain sediments.

In addition, there are alleged flood deposits 8–12 m above the present floodplain (Raikes, 1965; Raikes and Dales, 1977). The mechanism for their emplacement provides the basis for the often heated controversy that has been generated within the archaeological community. Raikes (1964, 1965), Dales (1964, 1965, 1972) and Raikes and Dales (1977) have argued that these deposits could have been emplaced only as a result of slackwater deposition in a low-energy paludal environment. Possehl (1967) postulated an oxbow environment but his hypothesis is untenable because, in order for an oxbow to exist at the site of the city, the city would have been destroyed by lateral migration of the river, since by definition the oxbow is a cutoff channel segment. Lambrick (1967) suggested that the alleged flood deposits are, in fact, aeolian deposits and of neither fluvial nor paludal origin. The presence of aeolian deposits along the margins of the present course of the Indus river and its former courses provides a reasonable basis for this interpretation. The Indus river floodplain alluvium is composed of fine sand, silt and clay. Fine sands and silts are optimal grain sizes for subsequent wind transport (Bagnold, 1941). The hydraulic roughness provided by structures at the site of Mohen jo Daro would have provided the necessary conditions for aeolian deposition.

Raikes and Dales (1977) disagreed with Lambrick's interpretation and appear to base their arguments primarily on the slackwater nature of the sediments. Dales and Raikes (1977), in a rejoinder to Possehl's (1967) oxbow hypothesis, used size analyses and Atterburg limits to determine the origin of the deposits. As pointed out by Lambrick (1971), these criteria are inherently incapable of uniquely determining the origin of the deposits. In the absence of any data on the internal stratification of the deposits, grain size analyses alone cannot be used to differentiate between fluvial, paludal or aeolian deposits especially because the range of grain sizes that occur along the lower Indus river floodplain is restricted to fine sand, silt and clay (Memon, 1969). However, the size analyses provided by Dales and Raikes (1977) indicate that the deposits are composed mainly of silt with some fine sand and clay ( ~ 15 per cent), which is a distribution that is characteristic of wind blown (loess) deposits (Ruhe, 1983). Much of the existing controversy could be resolved if sedimentological investigations had been conducted in the course of the various excavations.

The most controversial of the hypotheses for the demise of Mohen jo Daro is that proposed originally by Raikes (1964, 1965) and its subsequent modifications by Raikes and Dales (1977). Based on an idea by Sahni (1956) that tectonic events may have been responsible for the emplacement of some high elevation deposits in the Ganjo Takar hills near Hyderabad, Raikes postulated that the demise of Mohen jo Daro may have been the result of a tectonically emplaced 'dam' that backed water and sediment upstream as far as Khairpur. The location of the 'dam' was hypothesized to be near Sehwan and it would have been ~ 45 m high and 80 km long. A fault scarp (Allah bund) formed in 1819 along the Kutch Fault Zone was initially used as an analogue for the 'dam'. However, the dimensions of this feature (height ~ 5 m, width ~ 24 km, length ~ 80 km) and the fact that it was breached by overbank flows that collected in the dry course of the eastern Nara river in 1826 make it a poor analogue. Raikes (1965) then suggested that the mud volcanoes and ridges along the Makran coast (Snead, 1964) may be analogues for the postulated 'dam' at Sehwan. The use of these mud diapirs as an analogue suffers from the fact that they persist in their high relief forms in an arid climate, but erode rapidly in a humid climate (M.A. Stevens, pers. comm., 1987, based on observations of similar structures in Bangladesh). Raikes (1965) suggested that the 'dam' was formed and breached at least five times during the period of occupation of Mohen jo Daro, which implies that there were at least five diapiric events in the vicinity of Sehwan within the period of Harappan occupation. Regardless of whether the analogue is appropriate, the validity of the hypothesis has been questioned on the basis of the absence of evidence for the 'dam' (Lambrick, 1967; Possehl, 1967). Raikes (1965) speculated that there may be remnants of the diapirs at Sehwan, but our observations indicate that the evidence is suspect. Although east-dipping (N.5E, 36E) interbedded sandstones and shales were observed at Sehwan, no evidence for diapirs was observed.

Raikes (1965) suggested that the 'dam' may have been formed from upwardly displaced alluvial sediments which overlay the mud diapirs. Although this scenario may be possible, it appears unlikely because known diapirs have a surface expression. The alluvial 'dam' hypothesis suffers also from the fact that Raikes' calculations require that it be permeable (Raikes, 1965). The initial ponding of the waters of the Indus river behind the 'dam' would have generated a significant hydrostatic head at the 'dam' and the potential for piping and subsequent failure of the 'dam' would have been high. Levees (bunds) constructed with alluvial sediments along the modern Indus river fail primarily as a result of piping (Irrigation and Power Dept, 1978). Raikes and Dales (1977) suggested that, if the entire modern flow of the river were being trapped, the maintenance of shallow water depths behind the 'dam' could have resulted from the presence of single or multiple spill channels across the 'dam'. The assertion that a doubling of the slope of the river on the downstream face of the 'dam' would not cause breaching of the 'dam' is hardly credible. If the sediment were being deposited upstream then the flow through the spill channels would have been relatively sediment free and downstream erosion would have been significant. Therefore, the contention that the steeper slope is within the range of equilibrium slopes for rivers similar to the Indus is a mischaracterization of the relationship between channel slope and sediment transport characteristics (Lane, 1957; Schumm, 1977).

The hypotheses that have been used to explain the demise of the city of Mohen jo Daro, with the exception of Lambrick's, have failed to recognize that the Indus river is dynamic and that the dynamics of a meandering river significantly affect the patterns of deposition on the floodplain in both time and space. The presence of floodplain deposits below the ground surface at the city is not in dispute (Raikes and Dales, 1977). Rates of deposition used to support or dispute the effects of overbank flooding at the city average 2–5 mm yr$^{-1}$. However, overbank deposition for single events of up to 1 m have been reported for the Mississippi river (Kesel et al., 1974) and exceptionally high rates of deposition on a floodplain can also occur as a result of breaching of natural levees by flood-flows. Resulting splay deposits can extend over large areas (Coleman, 1969) and may be as thick as 3 m (Farrell, 1987). The reach near Mohen jo Daro (vkm 428) is characterized by an anomalously high range of flood depths (Figure 15.11), which suggest that flood events and deposition rates are highly variable at this location.

The use of average rates of deposition also masks the fact that the rate of floodplain sedimentation in a meander belt is ultimately controlled by the rate of lateral migration of the river (Ritter, 1978). The rate of deposition exponentially

323

decreases away from the river (Bridge and Leeder, 1979; Allen, 1985), and as a river migrates laterally toward a location on the floodplain, the rate of sediment deposition increases.

Meander growth and the resulting increase in the amplitude of the bend cause energy losses in the flow through the bend and this is expressed as backwater upstream of the bend. The backwater causes an increase in the water surface elevation upstream of the bend, which in turn increases the frequency of over-bank flows (Harvey, 1989). Specific gauge analyses suggest the reach near Mohen jo Daro (vkm 350–450, Figure 15.16A) has had greater water surface fluctuations than most other reaches of the river, in part owing to the high and very variable sinuosity (Figure 15.16B). There is no a priori reason to suspect that the historical river was not as dynamic. In sum, deposition rates at Mohen jo Daro during the period of Harappan occupation would have been highly variable and dependent on the location of the river and its planform. Periods of high floods and rapid sediment deposition on the Indus plain would have been a serious problem for an agricultural society, contributing to abandonment of the city.

Channel adjustments in a meandering river take place at two scales: lateral migration of individual bends and large-scale avulsions. The Indus river has occupied a number of different courses during the Holocene as is evidenced by the presence of former channels (Lambrick, 1967; Holmes, 1968; Memon, 1969; Flam, 1981b; Flam, this volume; and others). Each of the former courses is characterized by a meander-belt ridge that was formed as a result of vertical accretion (Galloway and Hobday, 1983). Elevation of the meander-belt ridge above the surrounding floodplain is ultimately responsible for the avulsion of the river (Reading, 1978). Numerous remnants of split-flow channels are also present on the modern Indus river floodplain and each of these channels is likely to have produced a natural levee of its own (Fisk, 1947). Holmes (1968) reports that relatively minor spillway channels can have levees that are up to 3 m high and 3 km wide. The presence of interconnected topographic ridges on the floodplain has formed shallow flood basins which are capable of damming floodwaters (Hunting Survey Corp., 1960) and which may have influenced flooding and ponding in the vicinity of Mohen jo Daro during the period of Harappan occupation. The presence of thick clay-dominated deposits within the stratigraphic section at the city has been cited as evidence for the 'dam' hypothesis (Raikes and Dales, 1977). Observations of the active channel during a period of very low discharge at Mohen jo Daro indicated that 1-m thick mud drapes were present where the waters had been ponded locally during flow recession. It is therefore, quite reasonable to infer that thick clay deposits (Raikes and Dales, 1977) could have been deposited in local floodbasins that were formed by intersecting natural levees on the floodplain.

The particular tectonic setting of the river near Mohen jo Daro is one that can be subject to high rates of sedimentation. If the former river occupied a position similar to that of the modern river, several tectonic/geomorphic elements that affect the river today would probably have produced similar results in the past. At the largest scale, river and valley slope are high down the southwestern flank of the Jacobabad–Khairpur high and decrease across the Kachhi foredeep. The tectonically induced slope change causes high rates of deposition and development of a wedge of alluvial material, as the deposition alters the slope and allows transport of material further into the foredeep (Figure 15.10). Mohen jo Daro lies just upstream of the tectonic transition from uplift to foredeep, and might have experienced accelerating rates of deposition as the foredeep–uplift transition aggraded. Disappearance of a terrace beneath alluvium of the active meander belt just east of Mohen jo Daro near Kandiaro (Holmes, 1968) further supports increased deposition in this reach. Second, the temporal variability of sinuosity, which characterizes the river as it crosses the Kachhi foredeep, reflects episodic sediment delivery or accelerated meandering. Thus, Mohen jo Daro lies just upstream of the northern margin of the foredeep and, therefore, it lies in a zone which has the potential for exceptionally high rates of deposition that are spatially and temporally variable. Last, the channel and valley profiles of the Indus river (Figures 15.5A and 15.6A) show that Mohen jo Daro may be located on a small-scale convexity on the Indus plain. If the first inhabitants of Mohen jo Daro were searching for a relatively flood-free location, they would have selected a high point on the Indus plain. However, through the centuries, deposition on the flatter slope above Mohen jo Daro caused a rise of the river bed and increased flooding, which made the site less desirable as time passed and eventually may have led to abandonment during a period of high floods.

The controversy surrounding the demise of the city of Mohen jo Daro appears to be based primarily on two factors. First, there are no sedimentological data to differentiate the environment of deposition of the deposits. Grain-size analyses and Atterburg limits are not unique descriptors of depositional environment when the available range of materials is narrow. Given the available data, the sediments could have been deposited in either fluvial, paludal or aeolian environments. Second, the controversy is essentially dependent on the interpretation of the origin of the upper 8–12 m of the deposits. If, in fact, these sediments were deposited in a paludal environment (Raikes, 1964, 1965; Raikes and Dales, 1977), that environment could have been produced by local deposition. This would obviate the need to postulate a tectonically induced 'dam' as the cause for the deposition.

In conclusion, Lambrick's (1967) hypothesis is the only one which has been postulated to date that agrees with the processes that characterize the modern Indus river setting. Deposits below the current floodplain elevation are of fluvial origin, and the variability in the composition of the deposits and their thicknesses can be explained in terms of floodplain deposition as controlled by the meander dynamics of the river. Aeolian deposition (Lambrick, 1967) is a reasonable explanation for the above floodplain sedimentation based on the presence of significant aeolian deposits along the course of the present and former courses of the river. Channel avulsion, as the primary cause of abandonment of the city

(Lambrick, 1967), fits the known behaviour of the river, as recorded by the presence of numerous former channels and the time scale of major avulsions of many meandering rivers (1000 years; Leeder, 1978). The location of the present course of the Indus river at Mohen jo Daro is the fundamental cause of much of the controversy. It is unknown whether the river has reoccupied a former course or if the present location of the river with respect to the location of Mohen jo Daro is the result of lateral migration within a new meander belt. If the river had remained in a course away from Mohen jo Daro there would be no controversy over the reasons for the abandonment of the city.

## CONCLUSION

The Indus river continues to exhibit the dynamic temporal and spatial variability that has been characteristic of its history. The behaviour of the Indus river and morphology of the Indus plain is related to distinct geomorphic segmentation of the valley by tectonic elements. Although the Holocene and modern rates of sediment accumulation are probably greater than the rate of tectonic deformation, the Indus river responds to changes of valley slope by clear changes of channel pattern. As a result, the Indus river distributes its sediment load unevenly across the alluvial plain. This varying pattern of floodplain sedimentation is associated with variable rates of lateral channel migration, variable flood depths and flood frequency, and scour and fill of the channel bed. The observed changes of channel pattern will also be reflected in the distribution, grain size and sedimentology of alluvial deposits.

The Indus river has the greatest variability in the middle reaches between Sukkur and Sehwan, where avulsion and active meandering affect the river course and level at several scales. The morphology and dynamic behavior of the present river in this reach support the conclusion that the ancient city of Mohen jo Daro was abandoned as a result of river behaviour. Avulsion was a function of tectonically controlled subsidence and aggradation. In addition, channel pattern change and lateral meander migration influence the short-term water surface levels and rates of erosion or deposition. A period of increased flooding and sediment deposition followed by avulsion of the river away from the city is a reasonable explanation for its abandonment.

# 16

# UPLIFTED MARINE TERRACES ALONG THE MAKRAN COAST OF PAKISTAN AND IRAN

*Rodman J. Snead*

## ABSTRACT

The Makran coast of Pakistan and Iran is dominated by spectacular marine terraces of greater size and extent than most other coasts of the world. This terraced coast extends > 1000 km from near Karachi, Pakistan, to Jask, Iran, near the entrance to the Strait of Hormuz. Although wave-cut cliffs, nips, tombolos, beach ridges, marshes, lagoons, sand dune colonies and stream wadis also occur along this coast, the most striking features are marine terraces which rise as isolated, flat-topped hills with wave-cut levels bevelled on their seaward side. Submarine, structural and stream terraces also occur but marine terraces are by far the most extensive and comprise about half of the Iranian Makran coast and about a third of the Pakistan coast. These terraces rise at least 245 m above the flat coastal plain of Iran and up to 460 m along the Pakistan coast. Most of the terraces represent small horsts along faults parallel to the coast. Because of the diverse elevations the terraces cannot be correlated precisely between the different areas, except for the closely spaced Konarak and Gurdim terraces. Terraces sampled were formed during two time periods, at ~ 30,000 $^{14}$C yr BP and 23,000 $^{14}$C yr BP.

The history of the Makran marine terraces involves both sea-level and tectonic movements. About 30,000 yr BP, uplift along parallel coastal faults combined with rising sea levels to produce a series of wave-cut levels. A possible still-stand at ~ 23,000 yr BP produced another wave-cut terrace level which was subsequently uplifted by renewed faulting. After the lower sea levels of the late Pleistocene, when the now submerged terraces, at existing depths of 55–91 m below sea level, were cut, a marine transgression again flooded the terraces and erosion was renewed. Coastal progradation connected most of the terraces with the coastal plain. The history of the Makran coast involves complex interrelationships of tectonic unrest, sea-level change, and coastal sedimentation and erosion. Erosion, especially near the terrace areas, dominates the coast.

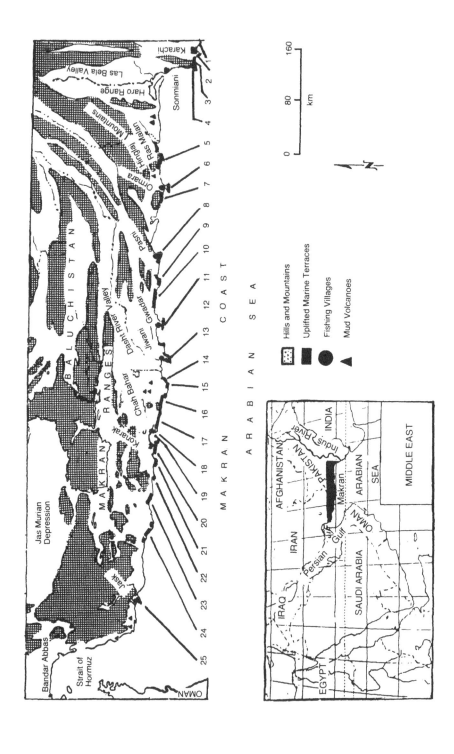

*Figure 16.1* Location map of the Makran coastal region showing 25 terrace and platform areas. See Table 16.1 for a description of the terraces.

*Terraces of Pakistan*

1 Clifton Hills
2 Manora Rocky Headland
3 Oyster Rocks
4 Hawks Bay to Cape Monze (Ras Mauri)
5 Hinglaj Mtns and Ras Malan
6 Ormara Terrace
7 Ras Sakanni/Kamagar Hills
8 Ras Zarain/Ras Jaddi Headlands
9 Ras Shamal/Bandar Point
10 Jabal-i-Mehdi Platform
11 Gwadar Terrace
12 Ras Pishukan Platforms
13 Jiwani Terrace

*Terraces of Iran*

14 Gavater Terraces
15 Ras Fasteh Terrace
16 Beris Terrace
17 Nishar area
18 Chah Bahar Terraces
19 Tis Terraces
20 Konarak Terrace
21 Gurdim Terrace
22 Ras Tang Terraces
23 Meydani Terraces
24 Gohart Terraces
25 Jask Terrace

Table 16.1 Structural and marine terraces and platforms of the Makran coast of Pakistan (nos 1–13) and Iran (nos 14–25).

| Name of area (nos: see Figure 16.1) | Highest elevation | Strandline height of wave-cut cliffs relative to sea level (m) | Flat or dipping | Rock type | Age (approx.) |
|---|---|---|---|---|---|
| 1. Clifton Hills | 29 m | 0.8 km inland of present shoreline 3–6 m above present sea level | Irregular hills: not good platforms | cgl/ss | Pliocene |

CHARACTERISTICS
Series of small bedrock outcrops: marine ss/cgl. Wave-cut scarps and sea caves.

| | | | | | |
|---|---|---|---|---|---|
| 2. Manora Rocky Headland | 6 m | 9.14 m directly on present beach | Flat | cgl/ss | Pliocene |

CHARACTERISTICS
A wave-cut bench (platform), 3.2 km long. Upper 3 m capped by Pliocene cgl. Storm waves have cut out small caves and a sea arch.

| | | | | | |
|---|---|---|---|---|---|
| 3. Oyster Rocks | 2 m | 9.1 m | Jagged rocky points | cgl/ss | Pliocene |

CHARACTERISTICS
Six small rocky islands ~ 1.6 km offshore at entrance to Karachi harbour.

| | | | | | |
|---|---|---|---|---|---|
| 4. Hawks Bay to Cape Monze | 12–15 m | Become lower to E: 6–9 m at Hawks Bay | Dip 3° E (seaward) | ss/ls/sh | Pliocene/ Pleistocene |

CHARACTERISTICS
Found on the east flanks of the Kirthar hills as low dipping bedrock beds. Appear as small cuestas on aerial photographs. Recent oyster beds found 2.4–3.0 m above present sea level. Faint trace of an uplifted marine terrace at ~ 15–30 m and even up to 46 m above sea level. Dirt track to lighthouse at tip of Cape Monze runs along these poorly recognized terraces. Terraces disappear at tip of Cape Monze. On west side, cliffs come directly to the sea; no clear evidence of a terrace level. Rocky spurs extend into the sea. Pediment surfaces extend from the rocky ridge.

| | | | | | |
|---|---|---|---|---|---|
| 5. Flanks of the Hinglaj Mtns and Ras Malan | 611 m | 91–152 m on E; 183–244 m to W | Perched platforms on rock face, gentle dip to W | ss/ms/sh: pink & white | Pliocene/ Pleistocene |

CHARACTERISTICS

Hinglaj mountains the highest mountain chain to come to the Makran coast. Inland mountains reach 1189 m. Structurally synclinal ranges with a high, steep edge on the coast. Highly dissected region, appearing to be a series of uplifted marine platforms or large benches on the east flanks of Hinlaj and Jabal Haro ranges. Cliffs at Ras Malan reach 305 m. Cliffs to the west of Ras Malan are 244 m near outermost point and 183 m farther west. A very difficult area to traverse; platform nature of this stretch of coastline discernible only on aerial photographs.

| | | | | | |
|---|---|---|---|---|---|
| 6. Ormara Terrace | 474 m | Inner cliffs 427 m; seaward side 183–305 m | Dips 3–5°S (seaward) | ms/ss/cgl/ shelly ls | Pliocene/ Pleistocene |

CHARACTERISTICS

Structure is uplifted fault-block (horst) 12.9 km long. Tapers from 0.9 km wide on the west end to a point on the east end. Along with Gwadar platform, large tombolos connect these former islands to mainland. Ormara is higher, reaching 474 m and more massive than Gwadar. Surface of Ormara dissected with steep, deeply cut canyons.

| | | | | | |
|---|---|---|---|---|---|
| 7. Ras Sakanni/Kamagar Hills | 318 m | 12–15 m | Small rocky, largely flat platform | ms/ss | Pliocene/ Pleistocene |

CHARACTERISTICS

On east side of Ras Sakanni a low rocky platform faces West Bay. Although little known about this stretch of coast, aerial photographs indicate platform is probably of marine origin.

| | | | | | |
|---|---|---|---|---|---|
| 8. Ras Zarain/Ras Jaddi Headlands | 127 m | 91 m | Rocks dip 20° N | ms/cgl/ shelly ls | Miocene – Pleistocene |

CHARACTERISTICS

Both are uplifted bedrock platforms. A distinct 4.6–6.0 m platform extends in front of Ras Jaddi. Sea currently incising this platform. Some uplift may have occurred during 1945 earthquake (Sondhi, 1947; Snead, 1967). Wave-cut unconformity exists between the ss with shell fragments and the fine conglomerate at 3 m.

Table 16.1 continued.

| Name of area (nos: see Figure 16.1) | Highest elevation | Strandline height of wave-cut cliffs relative to sea level (m) | Flat or dipping | Rock type | Age (approx.) |
|---|---|---|---|---|---|
| 9. Ras Shamal/Bandar Point | 195 m | 140–183 m | Platform dips to SW | ms/ss | Miocene |

CHARACTERISTICS
About 56 km of coast fringed by rugged, deeply eroded mountains. On aerial photographs there appears to be a marine platform at the east end with cliffs rising to 183 m.

| 10. Jabal-i-Mehdi Platform | 410 m | 152–183 m | Has dips of 5–15° SW | ms/cgl/ shelly ls | Pliocene/ Pleistocene |
|---|---|---|---|---|---|

CHARACTERISTICS
A deeply dissected 410 m platform which may be of marine origin. Wave-cut cliffs of 152–183 m form an imposing landform when seen from the sea. When it rains, streams flow off cliffs in a series of hanging valleys which may represent successive stages of uplift or varying rock hardness.

| 11. Gwadar Terrace | 145 m | Inner cliffs 107–22 m. Outer cliffs 85 m to E; 21 m to W | Dips 3° SW | ms/cg/ shelly ls | Pliocene/ Pleistocene |
|---|---|---|---|---|---|

CHARACTERISTICS
Flat-topped rocky platform connected to mainland by large tombolo. Similar to Ormara but not as large. Some rocks appear to have been recently uplifted above the sea. Only a few small streams have cut back into flat-topped surface. Extensive marine cg and shelly ls.

| 12. Ras Pishukan Platforms | 15–18 m | 7.6 m | Several rocky outcrops | cg/shelly ls | Pleistocene |
|---|---|---|---|---|---|

CHARACTERISTICS
Two small uplifted platforms form rocky headlands on west side of Gwadar West Bay. Similar to platforms at Jiwani and Gwadar but much smaller.

| 13. Jiwani Terrace | 135 m | 26–30 m | Dips 5° W–NW | ms/ss/cgl/shelly ls | Pliocene/Pleistocene |
|---|---|---|---|---|---|

CHARACTERISTICS
Uplifted tilted platform forming a prominent rocky headland. Has nearly straight cliffs on eastern side. Upland has been eroded into jagged badland topography on eastern side. Where the sandstones have formed a capping, the softer mudstones weather faster and leave a tilted, pillar-like mushroom-capped topography. The large number of non-fossilized marine shells point to this region as being recently uplifted. Platform is so recent it has not been deeply dissected where the marine conglomerates and shelly limestones occur. Except in the east, where the mudstones and sandstones do not exist, the surface has not been deeply dissected like the Ormara platform. This is the westernmost platform in Pakistan.

| 14. Gavater Terraces | 93 m | 24 m and 44 m | Slight dip 1° NW | ms/shelly ls | Pliocene/Late Pleistocene |
|---|---|---|---|---|---|

CHARACTERISTICS
Group of very prominent platforms which appears to have been much more extensive at one time. Heavy stream erosion has cut headward into terraces, producing finger-like protrusions. Platforms are mainly concentrated on east side with a 1° dip. Lowest 15.2 m terrace level is covered by recent flood silt from Dashtiari river which is less consolidated than material making up higher terraces (Falcon, 1947). The lighter mudstones are prominent in aerial photographs. Carbon dating of highest level, 5780 ± 115 yr BP, may be wrong; that of lowest level, 23,600 ± 650 yr BP, may be correct. Late Pleistocene shelly limestone forms a 3 m cap over the underlying mudstones. Three wave-cut levels occur: 91 m; 46 m; and 15.2 m.

| 15. Ras Fasteh Terrace | 44 m | 44 m | Almost level | cgl/shelly ls/ms/ss | Miocene/Pliocene/Late Pleistocene |
|---|---|---|---|---|---|

CHARACTERISTICS
This terrace is the easternmost seaward promontory of the Iran coast. Extends east as a round, flat-topped headland for 1.4 km. Terrace is highly dissected on the south, seaward side. Spectacular step-like slumping occurs along one stretch.

| 16. Beris Terrace | 52 m | 37–50 m Low, narrow bench at 15.2 m | Slopes 1°S | cg/ms/shelly ls | Miocene/Pliocene/Late Pleistocene |
|---|---|---|---|---|---|

*Table 16.1 continued.*

| Name of area (nos: see Figure 16.1) | Highest elevation | Strandline height of wave-cut cliffs relative to sea level (m) | Flat or dipping | Rock type | Age (approx.) |
|---|---|---|---|---|---|
| CHARACTERISTICS<br>Terrace flat and featureless. Terrace complex may have been much more extensive because outliers occur inland at 61 m and 91 m and slope at 1° toward terraces at the coast. Mountain ridge of Mazar Kuh, 8 km inland of Beris, may also be connected, although not 230 m above sea level. Large parallel, transcurrent strike-slip fault, running east–west, can be found at Beris. This is the most spectacular fault along the entire Makran coast. | | | | | |
| 17. Nishar area | 15 m raised | Inner cliff 122 m | Extremely dissected | ss/ms/ shelly ls | Pleistocene/ Holocene deposits |
| CHARACTERISTICS<br>The 15 m terrace, really a raised beach, is covered along the immediate coast by beach ridges and longitudinal dunes. | | | | | |
| 18. Chah Bahar Terraces | 50 m | 3.0–5.4 m | Nearly flat | cgl/shelly ls | Pliocene/ Pleistocene |
| CHARACTERISTICS<br>One of the most complex of all the terrace areas. Appears to have had alternating history through both tectonic movements and sea-level changes. Faulting of platforms has produced prominent structural terraces, also found on the Tis terraces. Many rocks appear to be uplifted portions of the coastal mudstone formation, locally capped by terrace deposits. A series of at least four narrow terraces dip west and disappear into the sea. At other locations, dune sand occurs throughout cross-sections. Sand deposits well laminated with cross-stratification. Main low coastal platform, 32 km long, is 3–5 m above mean sea level with a series of higher, uplifted terraces up to 50 m. Marine terrace farthest seaward 400–800 m wide. | | | | | |
| 19. Tis Terraces (non-marine) | 227 m | 15–40 m | Highly dissected. Number of different levels | ss/ms/sh | Pliocene/ Pleistocene |

Imposing geomorphic features consisting of flat-topped hills higher than Chah Bahar terraces. Also, T's terraces are more dissected, possess more levels, and appear to be of different origin. Structural terraces lacking shell material; therefore, not wave-cut marine terraces. Main terrace levels at 30 m and 70 m.

| 20. Konarak Terrace | 86 m | 12 m W side; 24 m E side; 40 m inner E cliff | Average dip 1° S (seaward) | Structural terrace resistant ss. Marine terrace cg/shelly ls/ss/ms | Tertiary structural terrace/ Pleistocene/ Holocene marine terrace |
|---|---|---|---|---|---|

CHARACTERISTICS
One of the best developed terraces of Iran. Studied most extensively by research team. Elongated block paralleling the coast: 18 km long, 300–900 m wide. Terrace surrounded by high cliffs making it difficult to reach the top. Two major sections: NW area an eroded structural terrace composed of late Tertiary sedimentary rock; E portion a marine terrace faulted at several locations. Structural terrace made up of several resistant ss beds which are synclinal, plunging slightly to SE. Entire complex an uplifted horst block.

| 21. Gurdim Terrace | 53 m | 49 m on inner E cliffs; outer seaward cliffs 21 m on W, 18 m on E | Flat terrace levels, eroded into irregular shapes | Brown ss overlying ms | Pleistocene/ Holocene |
|---|---|---|---|---|---|

CHARACTERISTICS
Terrace flat-topped, 16 km long, 310–320 m wide. Example of a simple, uncomplicated marine terrace. Flat-topped platform 19.3 km long, paralleling the coastline 450–320 m wide surrounded by vertical cliffs. A wide 12.8 km tombolo, displaying well-developed beach ridges, connects this terrace with the coastal plain. A 46 m massive mudstone layer is overlain by 6 m of brown ss. Contact is a wave-cut surface with coarser wave-deposited materials. Many small stream channels have indented the surface but no significant stream dissection apparent. Wave erosion has produced slumping of the seaward cliffs. This terrace is of tectonic origin, a horst uplifted to its present position with little deformation. Similar to Konarak terrace in age, materials, and slope. About 33 m lower than Konarak terrace.

*Table 16.1 continued.*

| Name of area (nos: see Figure 16.1) | Highest elevation | Strandline height of wave-cut cliffs relative to sea level (m) | Flat or dipping | Rock type | Age (approx.) |
|---|---|---|---|---|---|
| 22. Ras Tang Terraces (Figure 16.10) | 139 m | 9 m | Tilted hills | shelly ls/ ms/sh | Pliocene/ Pleistocene |

CHARACTERISTICS
Group of irregular, flat-topped and tilted hills extending 9.6 km inland and 6.4 m on either side of tombolo leading to the low 6–9 m Tang terrace. Similar to the Jask terrace farther west, this extends south into the sea and is connected to mainland with a small tombolo. Its marine origin is indicated by the shelly limestones and marine fossils in the mudstones. Higher and more distinctive marine terraces occur on three inland hill areas. These inner platforms rise to elevations of 104 m, 125 m and 139 m respectively. The higher terraces are more eroded with east–west trending elongations. A large number of faults can be depicted on aerial photographs, supporting the theory that these terraces have been uplifted by tectonic forces. Highest terraces appear to be ~ 30,000 yr BP, intermediate levels are 20,000 yr BP, and the lowest Tang terrace is Holocene. Unusual beach ridges extend right across the bedrock terraces from the non-bedrock coastal plain. This is one of the clearer indications of tectonic uplift.

| | | | | | |
|---|---|---|---|---|---|
| 23. Meydani Terraces | 76 m | 43–46 m | Slight tilting to E. Some erosion into irregular shapes | ms/shelly ls | Miocene– Holocene |

CHARACTERISTICS
Terrace area ~ 186 km² of flat-terrace levels with most eroded into irregular shapes. Two major terrace elevations occur. Erosion greatest on highest terrace in western portion. Resistant shelly limestone surface cap overlies the massive lighter-coloured mudstones. Three major terrace heights can be distinguished: 43–46 m at the coast going up to 76 m inland. Terraces extend inland ~ 16 km. Terraces faulted at several locations with prominent north–south lineation through central portion of bedrock area. Movement along fault appears to be right-lateral strike slip with vertical component of movement, since not only does the eastern block of the fault protrude at the coastline (it has been somewhat modified by wave erosion) but also the terraces are apparently tilted to the east, away from the fault trace. Inland, beach ridges occur, indicating wave action was once more extensive on the terraces, and marine action predominated.

| 24. Gohart Terraces | 195 m | 11 m W side, 5.5 m E side | Tilted surfaces eroded in several dip directions | ms/ss/sh | Pliocene/ Pleistocene |

CHARACTERISTICS
Extensive group of complex terraces forming several marine headlands at the coast. At least two major levels, a low level near the coast and a higher level inland reaching 183 m or more. Faulting can be seen on aerial photographs. These appear to be structural terraces, formed by erosion of resistant beds in the coastal mudstones. Bands of sandstone interbedded with the mudstones. The coast has prograded nearly 3.2 km from terraces.

| 25. Jask Terrace | 6 m | 4.2–5.4 m | Gentle dip N | ms/ss/ shelly ls | Pliocene/ Pleistocene |

CHARACTERISTICS
Westernmost of Makran terraces. Jask headland extends as a 3.2 km-long promontory into Gulf of Oman near entrance to Strait of Hormuz. Terrace partly the result of tectonic movements indicated by the terrace dip that does not conform to a normal sea floor (which would dip offshore to the south). A possible wave-cut unconformity exists between the shelly limestone and the underlying mudstone. Low cliffs extend along the present shoreline.

# INTRODUCTION

Marine terraces are among the most striking features along the Makran coast of Iran and Pakistan (Figure 16.1; Table 16.1). They rise as isolated, flat-topped hills with wave-cut levels bevelled on their seaward sides. These terraces rise at least 245 m above the flat coastal plain of Iran and up to 260 m above the Pakistan Makran coast (Hunting Survey Corp., 1960). Aside from being imposing geomorphic features, these terraces are fairly extensive, making up about half of the 90 km-long Makran coast of Iran and nearly a third of the 160 km-long coast of Pakistan.

Ancient writings about the journeys of Alexander and Nearchus mention the Makran coast but do not describe it (Arrian, 1879). During the British colonial era, a number of travellers described expeditions to the area (Blanford, 1868–73, 1872; Stiffe, 1873; Lees, 1928; Harrison, 1941, 1944). Asrar (1953) defined stratigraphy, described marine and fluvial terraces, and interpreted some structure. Siddiqi (1959) described terrace levels near Karachi. A more comprehensive geomorphological analysis of the Pakistan Makran coast was published by Snead (1964, 1966, 1967, 1968, 1969) and Snead and Frishman (1968). Snead (1970) also prepared a detailed analysis of the Makran coast of Iran. The only publication on the Iranian coastal terraces is a short description by Falcon (1947) who traversed the area by camel in 1931–2. He used an aneroid barometer to determine approximate elevations of the terraces, recognized their generally similar elevations, and briefly described their rock types.

# GEOLOGY OF THE MAKRAN REGION

The Makran region of Pakistan and Iran is an 800 km-long part of the Baluchistan basin comprising the elevated central part of the forearc of an active continental margin (Figure 16.2). It is separated on the west from the oil-producing Zagros region by the Oman Fracture Zone (Farhoudi and Karig, 1977), and on the east by the Ornach–Nal and Chaman transform fault system. The region was formed as the Indian plate to the west of the Makran moved northward in its collisional course with Asia along a transform fault expressed as the Owen fracture zone, the Murray ridge, and the transform fault system and axial fold belt of Pakistan which passes through the Las Bela plain. In the north the Makran is bordered by the Mashkhel depression, which is an upper slope or forearc basin south of the Chagai and Ras Koh arcs. The southern limit of the Makran is marked in the Arabian Sea by the filled trench at an active subduction zone (Jacob and Quittmeyer, 1979). The eastern and western boundaries separate the Makran from older terraces with distinct deformational styles and histories. In effect, the Makran region represents a gap along the southern border of the Eurasian plate along which the ancient Tethys Seaway did not completely close (Shearman, 1976).

Coastal Makran and the area to the north are an accretionary wedge of deformed sediments, ranging in age from late Cretaceous to Holocene, which

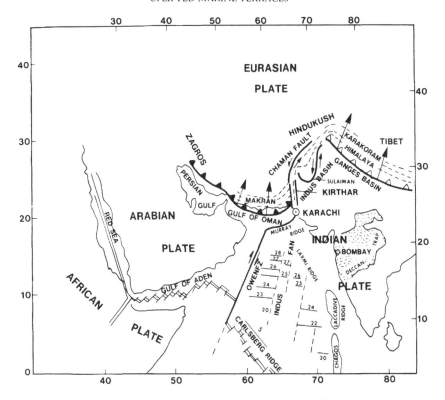

*Figure 16.2* Tectonic setting of the Makran region (modified from Jacob and Quittmeyer, 1979). The boundaries between the major plates are as follows.

Open tooth marks: thrusting associated with continental collison zone;
Solid tooth marks: thrusting associated with an oceanic subduction zone;
Double line: spreading on a mid-oceanic ridge or rifting on continents;
Single solid line: transform fault, Owen fracture zone, Murray ridge;
Lines numbered 20, 22, etc: seafloor magnetic anaomalies;
Dashed lines: fracture zones separating magnetic anomalies.

Arrows indicate relative sense of plate motions (Haq and Milliman, 1984).

have piled up at an oceanic subduction margin (Figure 16.3A, B). The structure and depositional setting have been compared to a typical near-shore arc model composed of upper-slope deposits followed by lower-slope and trench deposits, progressively deformed by continued subduction (Farhoudi and Karig, 1977). As an arc–trench system, however, the Makran is hardly typical; in fact it is perhaps largely anomalous (Jacob and Quittmeyer, 1979). The arc–trench gap is ~ 500 km, far wider than most systems. A possible dipping earthquake (Benioff) zone is weakly developed and extends to a depth of only 80 km. Focal mechanism solutions indicate points of tension in the descending oceanic slab. The subduction rate is ~ 5 cm yr$^{-1}$. Volcanic centres are absent along the arc

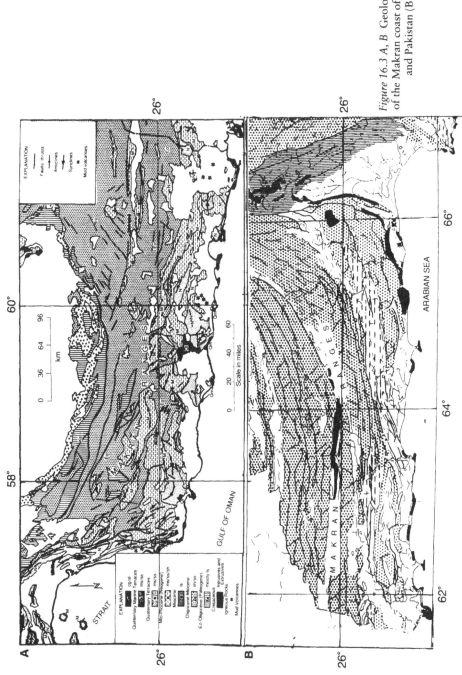

*Figure 16.3 A, B* Geologic maps of the Makran coast of Iran (A) and Pakistan (B).

and occur only far inland close to the border with Afghanistan. The bulk of the rocks of the folded ranges of central Baluchistan in both Iran and Pakistan are made up of mid-Eocene to late Oligocene rocks of the Hoshab and Siahan shale and Panjgur formation, which consist of abyssal muds and turbidities, possible marking trench sedimentation (Figure 16.3A, B). Additionally, a very large part of the accretionary prism is exposed and a significant volume of post-middle Miocene sediment is a shallow shelf deposit, not trench and slope deposits as thought by Farhoudi and Karig (1977). Coastal Makran provides a perhaps unique area of a well-exposed and young accretionary margin. The uplifted marine terraces are just part of this active tectonism. The depositional and deformational history is clear in at least the coastal area, as compared to others that are older or with more intensely deformed interiors.

Distinct differences in style and intensity of tectonism occur in various areas of the Makran. Imbricate thrust wedges and steep asymmetrical tight folds in the Oligocene sediments mark the surface expression of the accretionary prism. Reverse faults, steeply dipping to the north or northwest, are roughly parallel with the fold axes but less perfectly aligned. The folds generally strike east–west, parallel with the regional trend; they are narrow, tightly folded asymmetric anticlines and synclines, many of which are overturned and truncated against high angle reverse faults (Ahmed, 1969). In southern areas, near the coast, the density of reverse faults decreases noticeably (Hunting Survey Corp., 1960; Vita-Finzi and Ghorashi, 1978).

The principal outcrops of the Makran region are severely deformed turbidities of the mid- to late Oligocene Panjgur formation, overlying the abyssal Hoshab and Siahan shales of late Eocene to early Oligocene age. A few scattered outcrops of Mesozoic and Palaeogene sediments, including the sporadic exposure of Eocene reefoid limestones, also are exposed near the Iranian border (Hunting Survey Corp., 1960).

The southern area, the region of the terraces, is dominantly less deformed younger sediments of early Miocene–Pleistocene age, which are considered to be disharmonically deposited in an accretionary forearc basin. The exposed sequences of Tertiary clastics in the Makran are subdivided into five stratigraphic units of which the most southern and most recent are early–late Pleistocene marine shoreline deposits (Ormara and Jiwani formations) which make up the basic rocks of the coastal terraces (Figure 16.4). Associated with and underlying the Pleistocene marine formations are Plio–Pleistocene neritic, massive and monotonous sequences of calcareous mudstones (Chatti mudstone). This makes up the coastal terraces of both Pakistan and Iran. Inland of these formations, but at the coast west of Ormara in the Kalmat Khor area, are late Miocene–Pliocene deltaic and prodeltaic sediments (Talar/Hinglaj formation) which were deposited on the east and west flanks of the Pasni anticlinorium. This feature is made up of early to mid-Miocene sediments of the Parkini formation which were deposited in a shallow marine environment in an accretionary forearc basin and which are restricted to the southern Makran (Raza et al., 1981; Raza and Alam, 1983).

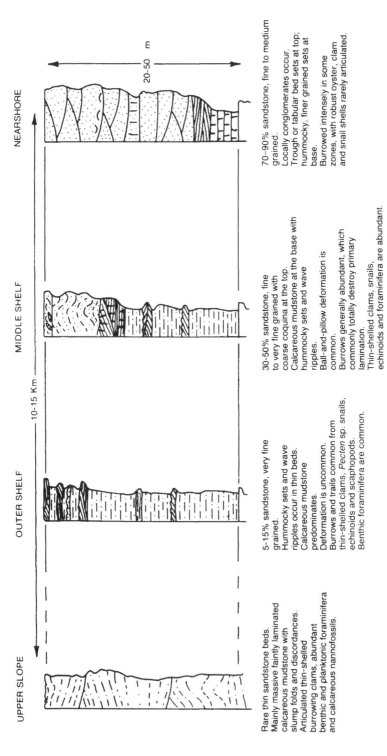

Figure 16.4 Shelf–slope facies transitions from near shore to upper slope within a single depositional cycle (Harms *et al*, 1984).

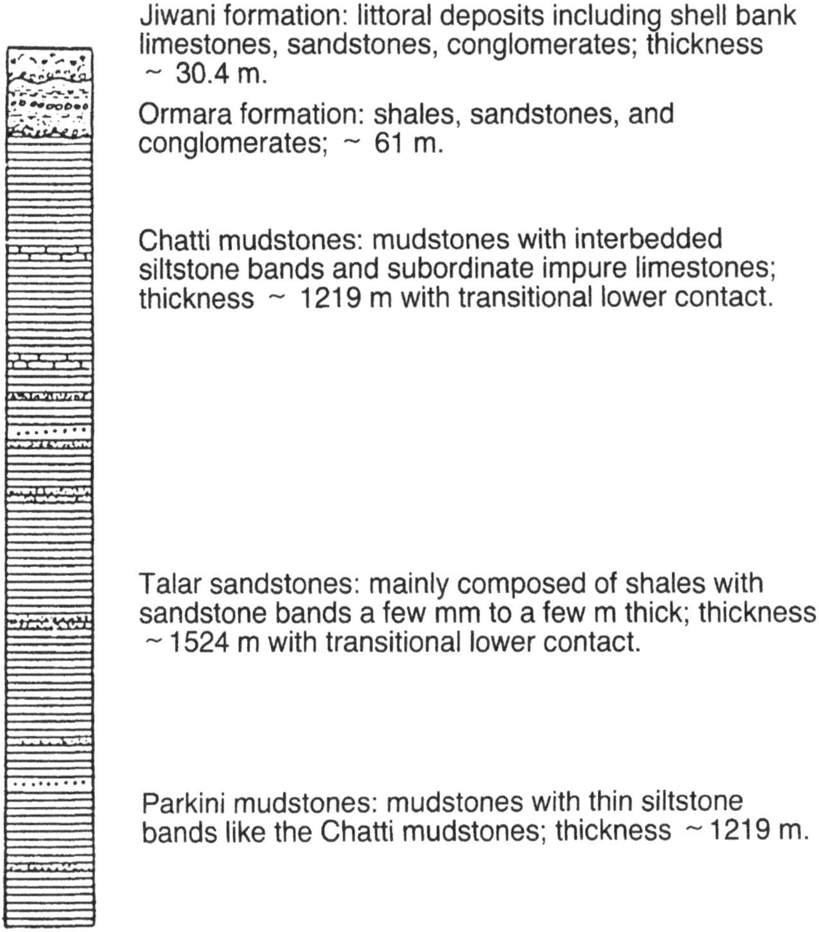

Jiwani formation: littoral deposits including shell bank limestones, sandstones, conglomerates; thickness ~ 30.4 m.

Ormara formation: shales, sandstones, and conglomerates; ~ 61 m.

Chatti mudstones: mudstones with interbedded siltstone bands and subordinate impure limestones; thickness ~ 1219 m with transitional lower contact.

Talar sandstones: mainly composed of shales with sandstone bands a few mm to a few m thick; thickness ~ 1524 m with transitional lower contact.

Parkini mudstones: mudstones with thin siltstone bands like the Chatti mudstones; thickness ~ 1219 m.

*Figure 16.5* Stratigraphic section of the Makran coast at Jiwani, Pakistan (after Little, 1972).

The Jiwani formation, of Pleistocene–Holocene age and named for outcrops just to the east of the Iranian border in the middle of the Makran coastal zone, is important because it represents the cap on the coastal terraces (Figure 16.5). It can usually be distinguished from the underlying formations because of its coarser sediment (shells, sand, cobbles) and prominent unconformity (wave-cut surface) with the underlying strata (Hunting Survey Corp., 1960; Little, 1972; Plate 16.1).

*Plate 16.1* Cross-section of the uplifted Tis terrace in Iran showing sharp unconformity between the Pleistocene conglomerates and shelly limestones that cap the tilted Talar sandstones of Oligocene–Miocene age. (Photograph: Snead, 1968)

## EVIDENCE OF REGIONAL UPLIFT

Epeirogenic movements have elevated the Makran coastal region as a whole since the time of the main Himalayan orogeny. An effect of such movement has been the constant change of base level and the local patterns of erosion and deposition. Partly as a result of this process, many areas of alluvium have reverted to erosion and stream dissection, leaving alluvial remnants that are generally known as the 'Subrecent deposits'. Yet factors other than epeirogenic movement complicate the relationships of erosion and deposition to such an extent that it is difficult to ascertain the time of uplift or to order the stages of the uplift (Hunting Survey Corp., 1960).

Along the coast of the Arabian Sea, the raised beach terraces (Jiwani formation) of the Jiwani, Gwadar and Ormara headlands, and Astola island of the Pakistan coast, and the Konarak, Chah Bahar and Gavater terraces of the Iranian coast indicate relatively recent uplift (Figure 16.1). Uplift of the Ormara headland amounts to ~ 460 m. Many intermediate stages between the highest littoral deposit and present sea level are marked by wave-cut benches on the faces of the headlands, as well as by emerged strand lines that are prevalent on beaches along the coast. Moreover, local inhabitants claim that the shore near Pasni per-

*Plate 16.2* Slickensides indicating faulting which is so common along the Makran coast. This exposure occurs at Gadani, north of Cape Monze, Pakistan. (Photograph: Snead, 1964–8)

manently rose ～4.5 m during the earthquake and tsunami of 1945, which destroyed many coastal settlements (Sondhi, 1947; Hunting Survey Corp., 1960).

Entrenched drainage, including incised meanders in the Hingol river and Ras Malan regions, indicates gradual coastal emergence. All indications of coastal epeirogenic movement indicate emergence; no evidence of subsidence or temporary resubmergence is known. The raised beaches and elevated shell beds were considered evidence in the older literature that the submarine cliff lying 16–32 km south of the coast was a fault scarp (Medlicott and Blanford, 1879; Wadia, 1939).

## FAULTING ALONG THE MAKRAN COAST

Strike-slip and normal faults with substantial displacement can be identified in several areas, whereas reverse faults are not conspicuous. Two major strike-slip faults occur in the area. One forms a large rift zone to the east of the Hingol river. This fault trends N10°E, has a left-lateral displacement, and an apparent offset of

5 km. It apparently represents a tear fault in the folded Plio–Pleistocene sheet, required as fold axes swing northward along the Las Bela axial fold belt (Figure 16.2). In trend and sense of offset, the Hingol river (Las Bela valley) fault is similar and related to the very large Ornach Nal and Chaman fault zones, which reflect the more rapid northward drift rate of the Indo–Pakistani plate relative to Baluchistan (Haq and Milliman, 1984). A second major strike-slip fault was observed near the Iran–Pakistan border by the writer while flying over the Makran coast.

Several large normal faults have been observed in surface exposures or seismic sections in the western Pakistan Makran. North of Gwadar, two such faults can be clearly discerned. The fault south of the Kulanch syncline has a sinuous course which dips as gently as 30°S and which can be traced east–west at least 100 km. The low dip is directly related to coastal uplift and northward tilting, especially on the north flank of the Koh Dimak Dome, at a time after fault displacement. Offset may be as much as 2500 m. Also it is possible that erosion has cut deeply enough into the fault to expose the lower, more gently dipping portion of a curved surface. An explained concentration of normal faults occurs in the central part of the Makran coast. Uplift of similar amounts has occurred along much of the coast and apparently at about the same rate (Plate 16.2). The largest of the recent earthquakes was centred along the Pasni anticlinorium, but the relationship between the seismicity and the normal faults is not clear. The existence of these large normal faults and the suggested tensional regime from the seismic data can be used as evidence that the very recent movements in the Makran are not from a single subduction motion (Haq and Milliman, 1984).

## TERRACE TERMINOLOGY AND MORPHOLOGY

Terraces on the Makran coast are wave-cut platforms covered with beach and near-shore sediment, which are now elevated above the level of present wave action through either elevation of the land or lowering of the sea. Features on the terraces similar to those on the present shore platform can be identified at most locations. Wave-cut unconformities have been observed at most locations in the field and have been reported in the literature (Falcon, 1947; Hunting Survey Corp., 1960; Little, 1972). Remains of marine animals are plentiful and include corals, large clams, oyster colonies, and many other robust wave-resistant forms. Many of these deposits are relatively thin, representing areas with temporarily low rates of detrital supply (Plate 16.3). Some beds are quite young but now cap terraces as much as 500 m above sea level, thus attesting to recent uplift (Haq and Milliman, 1984).

Much of the contemporary shore platform does not have a sediment layer because it is still in an early stage of formation. As the platform becomes larger, wave energy is reduced, allowing sediment to collect on the platform. Beaches are formed, and it is these lithified beach sands, pebbles and broken shells that occur above the wave-cut unconformities of the terraces. These terrace cappings have a

*Plate 16.3* Holocene marine oyster shells imbedded in the surface conglomerates of the Gwadar terrace in Pakistan, 90 m above mean sea level. (Photograph: Snead, 1964–8)

distinctive shelly sandstone and conglomerate which has a variable thickness of 1.5–6.1 m (Falcon, 1947), although ~ 30 m has been measured just east of the Iranian border at Jiwani, Pakistan (Hunting Survey Corp., 1960). Some of this sediment cap appears to be former wind-blown coastal dunes cemented by $CaCO_3$ to form eolianite. Sediment size and thickness on the wave-cut platform are greatly affected by relative movements between land and sea. If sea level is greater than uplift of the land, a new higher wave-cut level is formed, and sediments much finer than the littoral deposits will cover the former platform. If slow uplift is greater than sea-level rise, a new wave-cut platform is formed at a lower level on the wave-cut side, leaving the former platform to be covered by coastal dunes. Both situations occur along the Makran coast of Iran and Pakistan (Little, 1972).

The environment of the Makran coast is conducive to shore-platform development. Wave energy is strong enough to cut into the sandstones of the headlands where the best platforms occur. Where the weaker, finer grained sedimentary rocks outcrop, wave attack easily bevels a platform surface, but the rock breaks down to its constituent sediments and no hard surface is visible. The nip, however, still may be present indicating cliff retreat. The sandstones present a fairly uniform material for the waves to attack, so structural complications are not important. Joints perpendicular to the shore are loci of accelerated erosion and may cause interruptions in the longshore continuity of the platforms (Figure 16.4). As Edwards (1941) noted, the widths of wave-cut platforms are inversely related to the coastal cliff heights. As they are undercut by waves, higher cliffs will produce more talus than lower cliffs, and thus talus will temporarily protect the cliff from further erosion. Coastal cliffs backing the wave-cut platforms along the Makran coast are generally < 6 m high.

Tidal range is also important in width of platforms because it governs the depth of effective wave erosion. The higher the tidal range, the greater the possibility for platform development. The Makran coast has an average tidal range of 1.8–2.7 m. Because wave-action abrasion is effective only to ~ 12 m (Bradley, 1958), the 2.4 m tidal range of this area allows < 12 m of vertical range of platform development along this coastline. If a slope of ~ 1° is assumed, a platform of ~ 0.8 km could be eroded. The Makran platforms are not known to extend more than ~ 60 m beyond the coastal cliff and, on the basis of fathometer profiles, the 9 m depth is reached ~ 150 m from shore. Beyond this depth the sea floor drops off markedly. A great potential for wave erosion along the coastline thus exists before the theoretical 0.8 km limit is reached.

The climate of the Makran coast is ideally suited to platform formation because tropical or monsoonal conditions will accelerate the development of platforms (King, 1957). Arabian Sea cyclones and depressions during the southwest monsoon provide a powerful potential source of wave energy.

## RECOGNITION OF TERRACE DEPOSITS

Shoreline deposits are poorly represented except in late Pleistocene or Holocene sequences because progressive folding, uplift and erosion have continuously removed the northern edges of the sedimentary prisms. Two kinds of shoreline deposits exist. One is sandy and shows typical coarsening-upward cycles of lower shoreface, surf-zone, and beach facies developed on wave-dominated prograding shorelines (Harms et al., 1975). The other is more calcareous and contains abundant remains of coral, robust clams, and other forms that thrive in shallow wave-agitated water where detrital supply does not smother their growth. These two types presumably represent reaches of shoreline that were either at or near stream mouths or between streams where longshore supply of mud or sand was limited (Haq and Milliman, 1984).

Surf-zone deposits of trough cross-stratified, medium to fine grained sand show orientations indicating strong westward longshore drift, commonly observed on the modern shoreline under the attack of large Indian Ocean waves approaching at an oblique angle. Well sorted beach beds cap these surf-zone sandstone deposits, showing delicate lamination dipping at low angles seaward in cusp-shaped sets. In some areas, aeolian dune fields overlie the beaches, identified by intricately cross-stratified large sets with high local dips and large sweeping sets. Within the shoreline deposits are limpets, cerrithid snails, and

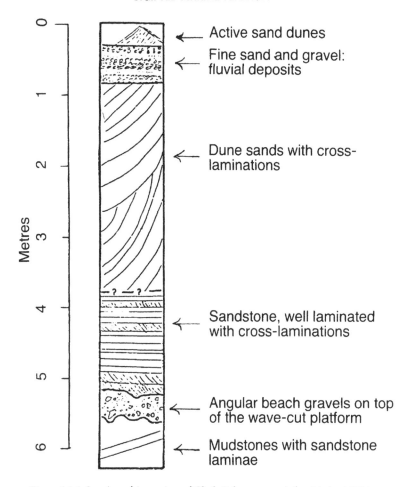

*Figure* 16.6 Stratigraphic section of Chah Bahar terrace (after Little, 1972).

large clams that live only at the shoreline today. A typical sandy shoreline sequence is late Pleistocene in age but already tilted and eroding (Figure 16.6).

Calcareous shoreline deposits contain coral, large clams, oyster colonies, and many other robust wave-resistant forms. Such deposits are relatively thin and represent areas with temporarily low rates of detrital supply. These beds, presumably young and deposited at sea level, now cap terraces rising up to 500 m above sea level. They provide excellent evidence of recent uplift rates (Haq and Milliman, 1984).

## OTHER TYPES OF TERRACES

Besides the raised marine-formed terraces discussed mainly in this chapter, three

*Plate 16.4* Paired stream terraces ~ 50 km inland from Chah Bahar, Iran, on the road to Zahedan. (Photograph: Snead, 1968)

other types of terraces occur in this coastal zone: submarine terraces, stream terraces and interior structural terraces.

### Submarine terraces

Several authors believe that irregularities on the continental shelf of the Makran area result from either faulting (Medlicott and Blanford, 1879; Hunting Survey Corp., 1960) or lower sea levels (Stiffe, 1873). Transects made by the writer along the coast of Chah Bahar, Iran, revealed several shelf breaks with a prominent terrace feature. The offshore terraces seem to bear a striking resemblance to the terraces present at the coastline, although these submarine landforms are smaller.

### Stream terraces

Fluvial terraces are prominent features along the Makran ranges which back the coastal plain, and may underlie most of the coastal plain as well. The stream terraces of the mountains occur as paired bedrock surfaces covered with only a few m of gravel. At least four levels were observed above the present stream-valley floor in the Chah Bahar area of Iran (Plate 16.4). Falcon (1947) observed that the stream terraces, like the present water courses, are graded, and they rise

350

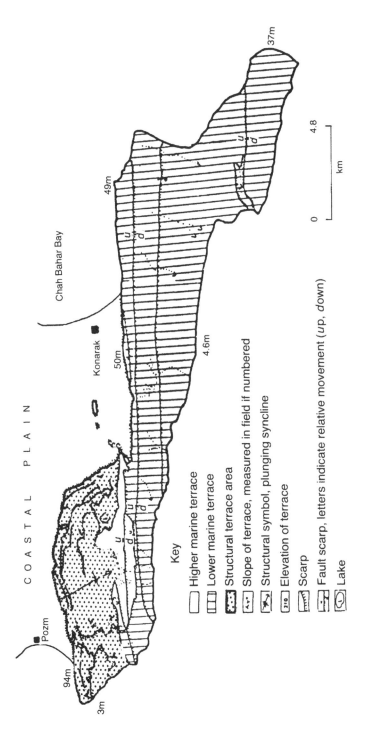

*Figure* 16.7 Landforms of the Konarak terrace region (after Little, 1972).

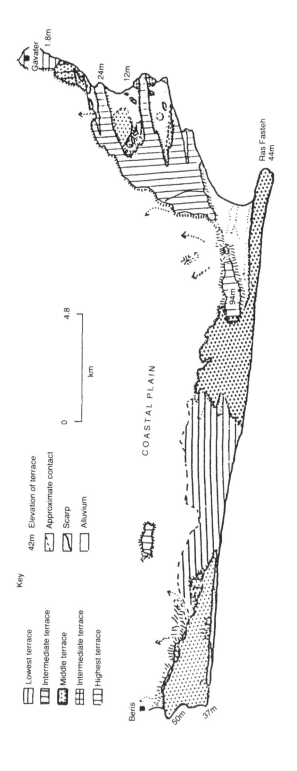

*Figure* 16.8 Beris, Ras Fasteh and Gavater terraces in eastern Iran (after Little, 1972).

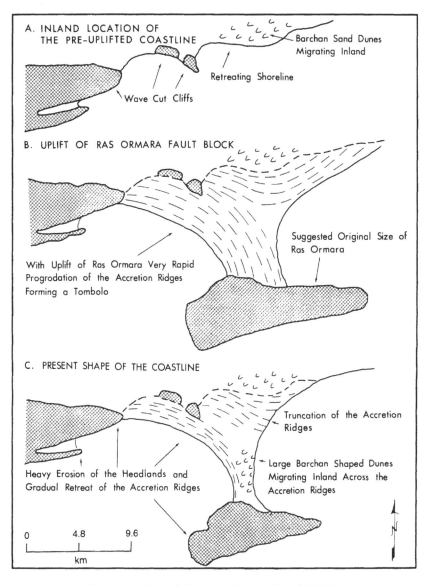

A. INLAND LOCATION OF
THE PRE-UPLIFTED COASTLINE

Barchan Sand Dunes
Migrating Inland

Retreating Shoreline

Wave Cut Cliffs

B. UPLIFT OF RAS ORMARA FAULT BLOCK

Suggested Original Size of
Ras Ormara

With Uplift of Ras Ormara Very Rapid
Progrodation of the Accretion Ridges
Forming a Tombolo

C. PRESENT SHAPE OF THE COASTLINE

Truncation of the Accretion
Ridges

Large Barchan Shaped Dunes
Migrating Inland Across the
Accretion Ridges

Heavy Erosion of the Headlands and
Gradual Retreat of the Accretion Ridges

0    4.8    9.6

km

N

*Figure 16.9* Coastal changes at Ormara (Snead, 1967).

A  Coastal faulting has uplifted terrace above sea level, but the initial shape of the coast
has not yet been affected;
B  The coast, being in the lee of the terrace, has prograded out to the island. Beach ridges
illustrate the positions of the prograding shoreline;
C  As the terraced headland is eroded, it can protect only a smaller section of coast from
wave action. The coast, therefore, begins gradually to retreat, and the more recent beach
ridges are removed.

*Plate 16.5* The Konarak platform, which is one of the more prominent terraces occurring along the Makran coast of Iran. This flat-topped platform is ~ 18 km long, 4.8 km wide, and rises 64 m above mean sea level. Step-like slumping occurs along the seaward side. (Photograph: Snead, 1968)

*Plate 16.6* The well developed Chah Bahar terrace in Iran. The cliffs rise to 50 m. More erosion has occurred on this complex terrace than on Konarak terrace, although both are the same age. (Photograph: Snead, 1968)

*Plate 16.7* Wave-cut sea arch along Hawks Bay ~ 30 km east of Cape Monze, Pakistan. The flat terrace surface is 7.6 m above mean sea level. (Photograph: Snead, 1964–8)

*Plate 16.8* Steeply dipping wave-cut platform on the west side of Hawks Bay near the eastern flanks of Cape Monze in the eastern section of the Pakistani coastal terraces. The wave-cut platform is 6–9 m above mean sea level. (Photograph: Snead, 1964–8)

gradually upstream parallel to the thalweg of the water courses. The highest terraces are ~ 75 m above the stream beds. These paired bedrock terraces are geomorphic monuments to periods of uplift in the region. The lack of alluvial deposition on the terraces indicates that minimal time passed between uplift movements. Apparently as every uplift occurred, the stream courses quickly cut back to their headwater area. Over time, some streams started to meander laterally (Falcon, 1947). Although uplift in this mountainous area is considered to be the mechanism causing periods of stream erosion, changing sea levels owing to eustatic changes during the Pleistocene also must have had some effect.

A widespread terrace with fluvial gravels occurs in the northern portion, north of Chah Bahar, where the land is extremely irregular and dissected by entrenched streams. This level appears to be cut by streams meandering across mudstones of the coastal plain. The streams deposited gravels over the cut surface, and uplift produced rejuvenation of the stream.

### Structural Terraces

Structural terraces are relatively flat uplands caused by the resistance of underlying formations to erosion. From analysis of aerial photographs they resemble wave-cut terraces, but in the field these levels conform to the attitude of the underlying strata.

## ANALYSIS OF THE INDIVIDUAL TERRACE AREAS

In Iran the Konarak (Figure 16.7; Plate 16.5), Chah Bahar (Plate 16.6) and Gavater (Figure 16.8) terraces were investigated in the field. In Pakistan the Jiwani, Gwadar, Pasni, Ormara (Figure 16.9), Manora and Cape Monze (Plates 16.7, 16.8) terrace regions were similarly studied. The remaining terraces were inaccessible because of steep cliffs, poor or non-existent roads, or general regional inaccessibility. Vertical, black and white, aerial photographs (1:50,000) and oblique colour slides (35 mm) provided most of the information about the inaccessible locations.

The individual marine terraces and structural platforms occurring near the Makran coast are outlined in Table 16.1. Many of the terraces are most prominent. They can be seen easily on maps and aerial photographs. These are, east to west, Ormara, Gwadar and Jiwani in Pakistan; and Ras Fastah, Beris, Chah Bahar, Konarak, Gurdim, Ras Tang (Figure 16.10) and Jask terraces in Iran. Several of the terrace regions are so dissected or distorted by tilting and faulting that it is difficult accurately to measure, date and map their characteristics. This is especially true for the terraces and platforms near Cape Monze, Ras Malan, and the Nishar area (Plate 16.9) where step-like marine terraces dip towards the west and disappear into the sea. The Meydani and Gohart terraces of Iran are complex because of faulting and severe erosion. The lowest, smallest and most simple structural terraces occur at Manora, the easternmost terrace, at

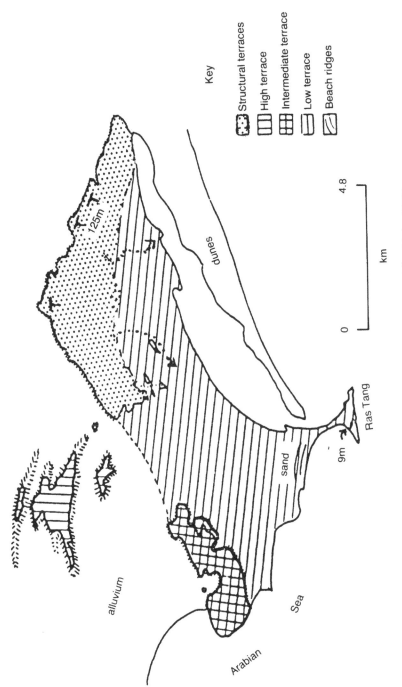

*Figure 16.10* Ras Tang terrace area in Iran (after Little, 1972).

Key

Structural terraces
High terrace
Intermediate terrace
Low terrace
Beach ridges

125m

dunes

Ras Tang

9m

sand

alluvium

Arabian

Sea

0    km    4.8

*Plate 16.9* Wave-cut terraces and platforms along the Nishar section of the Iran Makran coast. The terraces are dipping ~ 3°W. The highest terrace rises to nearly 122 m above mean sea level. (Photograph: Snead, 1968)

the entrance to Karachi harbour, and at Jask, the westernmost terrace, near the entrance to the Strait of Hormuz in Iran. The highest structural platforms and possibly very old marine terraces are in the central Pakistan coastal zone at Ras Malan. Here wave-cut cliffs and uplifted platforms reach 611 m.

## TIMING OF FORMATION OF THE MARINE TERRACES

A basic clue to the history of the terraces is provided by [14]C dating of shell material in the terrace caps, plotted against terrace elevation (Figure 16.11). In spite of the shell material having been obtained from a variety of terrace locations and elevations, all of the dates fall into two general time categories: 30,000 [14]C yr BP and 23,000 [14]C yr BP. In general, the older dates apply to the higher terraces in any given area, as would be expected.

Datable material was sampled from the various levels at each area. At Konarak a fault separates what had been thought to be two different marine-cut levels, with the result that the two dates for Konarak are similar. At Gavater the sample from the highest level was not *in situ* and provided a spurious date. But the Gwadar, Pakistan, terraces yielded two excellent localities where pelecypods and a coral were collected in growth positions. Although these terraces are somewhat higher than the Iranian localities, they seem to represent a simpler

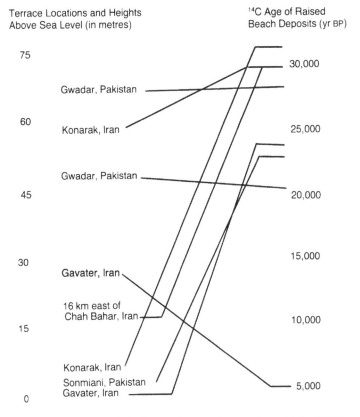

*Figure 16.11* Results of ¹⁴C dating of terrace deposits (after Little, 1972).

history which would serve as a general model for the development of the Makran terraces.

Sea level controls some terrace development, so that the elevation of sea level at 30,000 yr BP and 23,000 yr BP is critical. There is no concrete explanation of sea-level behaviour for these time periods, but the consensus seems to be that a transgression reached its peak at ~ 30,000–25,000 yr BP, followed by a regression that reached its lowest elevation at ~ 20,000–17,000 yr BP. At the height of the transgression the ocean is believed to have been approximately at today's level, and the regression at least 90 m lower than present sea level (Curray, 1965; Hoyt *et al.*, 1968; Bull, 1984). During the times represented by the dated Makran terraces, therefore, sea level reached a high about the date of the older, higher terraces and was beginning to recede about the date of the younger, lower terraces (Chappell, 1974).

While it is remarkable that only two general dates emerge from the diverse locations and levels that were analysed, it would be unwise to assume that all of

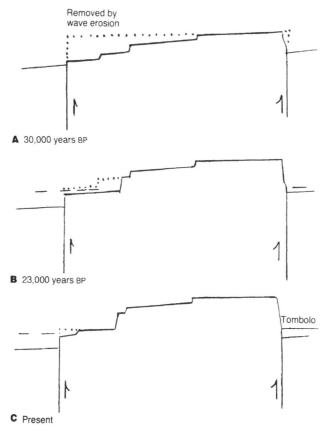

A 30,000 years BP

B 23,000 years BP

C Present

*Figure 16.12* Progressive development of marine terraces (after Little, 1972).

A  30,000 yr BP. Uplift along parallel coastal faults, combined with rising sea levels, produce a series of wave-cut levels.

B  23,000 yr BP. A possible stillstand at about this time produces another wave-cut terrace level, which is subsequently uplifted by renewed faulting.

C  Present. After the lower sea levels of the late Pleistocene, transgression of the sea again floods the terrace area and a new erosion cycle is initiated. Coastal progradation has connected most of the terraces with the coastal plain.

the terraces were formed at the indicated times. Sea-level changes have occurred throughout the Quaternary and it is likely that tectonic uplift has been active as well. Many of the higher terraces along the coast, therefore, may be much older than 30,000 yr BP. This assumption is supported by a date of 39,000 $^{14}$C yr BP which records a transgressive sequence occurring near Chah Bahar (Figure 16.12).

Tectonic activity plays a prominent role in the evolution of the Makran landscape, and uplift during historic times is well documented. Ancient seaports are now 32 km inland and ~ 60 m above sea level (Dales, 1962a,b), and fault movements have caused significant uplifts of the coast (Snead, 1967) and continental shelf (Sondhi, 1947). The terraces are monuments to tectonic uplift. Terrace levels dated at ~ 30,000 yr BP and representing a stand of the sea at about present-day levels now are uplifted to elevations > 60 m. The highest recorded marine terrace is at Ormara, whose undated flat top is 457 m above sea level (Hunting Survey Corp., 1960).

The terraces probably developed according to the following general sequence (Figure 16.12):

1   30,000 yr BP. Approximately simultaneous transgression and uplift tend to keep the shoreline at about the same position, so that a significant terrace level is developed. Probably the processes did not exactly balance, so that several levels may have developed, as at Gwadar (Table 16.1). Because the up-faulted horsts are rather narrow, it is likely that they were islands at the time.

2   23,000 yr BP. Sea level was probably receding as the land was rising, with the result that terraces should not be produced easily. Because they exist, either sea level underwent a stillstand or slight rise at this time or perhaps an interval of faulting uplifted the marine deposits left by the receding sea.

3   Late Pleistocene and Holocene events. Sea level of the late Pleistocene regression was much below the level of the previously described terraces, and probably reached to ~ 110 m below present sea level (Curray, 1965; Emery, 1967). This low stand of the sea may have cut the submerged terraces at 55–91 m. As the late Pleistocene glaciers melted, sea level gradually rose to present levels. During this final transgression the terraces were again affected by waves. These isolated areas again became islands as sea level rose and they began to be undercut by wave action, thus setting the stage for the latest phase of coastal change.

Distinct changes in coastal morphology must have begun ~ 30,000 yr BP with the rise of the offshore island horsts. In the lee of the islands, wave energy would have been greatly reduced, allowing the coastline to prograde. The receding sea levels of the late Pleistocene interrupted this process. With the Holocene rise of sea level to its present position, the process of coastal progradation in the lee of the terraces continued until the tombolos connected the terraced blocks to the mainland. Shell material from the beach ridges around Chah Bahar Bay indicates that tombolos to both Konarak and Tis terraces were present by ~ 10,000 [14]C yr BP.

Because the terraces are the controlling mechanism for these coastal changes, modification of the terrace would affect the adjacent coast. As the terraces become smaller through erosion, the adjacent coastline is being affected to a greater degree by wave action. Germann (in Snead, 1970) discovered that, in areas near the eroding terraces, the coast is now regressive; any ridges that may have been built after 7775 BC (9725 yr BP) appear to have been destroyed by erosion.

## CONCLUSIONS

Marine terraces dominate the coastal physiography and tectonic development of the Makran coast of Pakistan and Iran. Submarine, stream and structural terraces also occur, but these have not been studied in detail. The marine terraces have been uplifted, apparently as horsts along faults parallel to the coast, to as much as 457 m above present sea level. Marine terraces generally have diverse elevations which make correlation difficult but all terraces sampled were formed ~ 30,000 and 23,000 yr BP. The history and future of the Makran coast lies in the complex interrelationships of tectonism, coastal sedimentation and erosion, and eustatic sea level changes. Erosion continues to affect much of the coastline, especially the terrace areas.

# 17

# GEOGRAPHY, GEOMORPHIC PROCESS AND EFFECTS ON ARCHAEOLOGICAL SITES ON THE MAKRAN COAST

*Rodman J. Snead*

## ABSTRACT

The coastal region of Pakistan is important archaeologically because it is transitional between the Indus valley Harappan civilization and those of later times, with the well known ancient cultures of Mesopotamia, Egypt and the Middle East. The region is also one of the most tectonically active of the world. More than thirty major earthquakes have occurred there during 1939–75. In addition, there has been considerable faulting, folding and warping of coastal landforms, particularly the coastal terraces. In the last 5000 years, structural movements have been ~ 60–120 m along several sections of the coast, especially near the fishing village of Pasni. Also, this coastal region is occasionally hit by severe Arabian Sea cyclones. These are small hurricanes that can cause extensive flooding and rapid erosion of the soft marls and clays. These physical events have resulted in the loss and destruction of numerous archaeological locations. The materials used in the construction of these sites were mud, mud-brick or matting, none of which long resists accelerated processes of desert and coastal erosion. Near Karachi, archaeological sites are destroyed through transportation and urban expansion. More than 250 possible locations have been identified but few are well preserved. Considerable concern exists that the presently known sites will be lost by natural and human forces. Existing sites need further protection by good government management in order to protect the important evidence of long-term contact between Indus valley civilizations and those of the Persian Gulf and further west for over 6000 yr.

## INTRODUCTION

The coast of Pakistan is an isolated desert region that has received little attention because of its remoteness from major travel routes and its difficult terrain and climate. A number of known travellers and explorers, including Alexander the Great and Arab traders, have crossed this region but, until the British studies

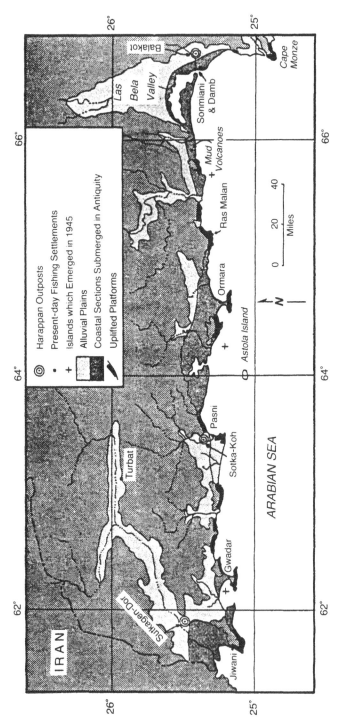

*Figure 17.1* The Makran coast of Pakistan.

during the mid-1800s, few outsiders wrote about the region. Even today, fewer than a dozen people have done field research on this area (Goldsmid, 1863; Dales, 1962a, 1962b, 1974, 1979; Khan, 1968; Snead, 1963, 1964, 1967, 1968, 1969, 1987). This region is important archaeologically, however, because it is between the well known ancient cultures of Mesopotamia and Egypt in the Middle East and the ancient Indus valley Harappan civilization and others.

This coastal zone extends for 990 km from Karachi to the Iran border and inland for as much as 30–160 km. The region has uplifted mountains and platforms separated by scalloped bays, wide sandy plains, salt marshes and lagoons (Figure 17.1). Several of the platforms represent terraces and fault blocks; for example, Ormara and Ras Malan, which have sheer cliffs > 30 m high. The region also is undergoing dramatic changes. Tectonic movements and erosion are highly active, with over thirty earthquakes in the last forty years. Uncommon Arabian Sea cyclones, or small hurricanes, occasionally cause extensive flooding and erosion (Sondhi, 1947; Snead, 1967, 1969; Snead 1987; Vita-Finzi and Ghorashi, 1978).

## ARCHAEOLOGICAL FINDINGS

The Makran coast and the Karachi region to the east offer a fascinating but frustrating area of archaeological research. It appears that, since prehistoric time, considerable change has taken place. The projection of Cape Monze into the sea (Ras Mauri) was even more pronounced in the past when much of the lower Indus delta on the east side, and part of the Las Bela plain on the west, were under the sea. Moreover, the east–west trending mountains of central and western Baluchistan change in this region, and the north–south ranges, part of the major Las Bela axial belt, become lower and merge into the extensive plains of Sind and the Indus valley.

From the strategic position of this area, it must have been a meeting place of different prehistoric cultures flourishing around it: the Kulli culture to the northwest in the Makran and parts of Las Bela, the Nal-Nundara to the north in Jhalawan, and the Indus and Amri cultures to the east and northeast. In the maritime communication between the Indus and Mesopotamian civilizations, for which we have quite good evidence (Dales, 1962a,b), the importance of this region deserves special attention. Apart from being on the highways of ancient civilizations, the Makran region, with its maritime environment, its sources of food from the sea, its genial climate and its terrain with low parallel ridges and flat intervening valleys which was watered by numerous good springs and perennial and semi-perennial streams, enjoyed highly favourable conditions for being the home of indigenous cultures. From such considerations one could expect that this region must have played an important role in the drama of human progress from the earliest prehistoric times.

The search of prehistoric sites in this area is attended by rather peculiar difficulties. The land has risen vertically, relative to sea level, several times since

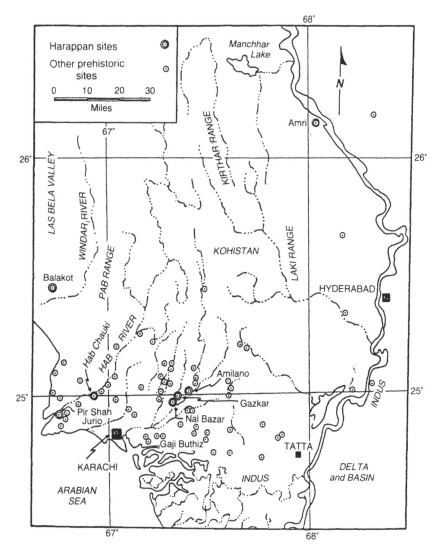

*Figure 17.2* Archaeological sites in the Karachi region of Pakistan.

the end of the Pleistocene. After every uplift of land, erosion has accelerated. As a result, the old surfaces have been destroyed to a great extent. Only limited patches remain in the form of old terraces. The old archaeological sites themselves have been eroded down and no longer occur as high mounds that could easily be detected.

The materials used for building houses were mud, mud-brick or matting. There is little evidence of fired bricks. Only at a few sites, especially the

Harappan and Amri sites, has stone been used, but merely for a few courses in the foundations. With such weak building materials, all of the sites have suffered from relentless erosion by wind, rain and sheet flow. The mud and mud-bricks have dissolved and gradually disappeared. The imperishable materials, such as pottery, stone implements, shingle (probably used in hearths), oyster shells, some gastropods (their flesh used as food or their shells for ornaments) and some bones, have collected on the surface as lag material while the finer material has been carried away by wind and water. This process has mixed up and accumulated materials from different horizons into one layer on the surface. This makes the working out of any stratigraphic sequence at any one site almost impossible. The number of sites is quite large and the period under study very long, while the land surface configuration has changed and the people have shifted their location many times. The result is that a comparative study of different sites makes their general classification only partly possible in terms of time sequence. Because the studies have been based only on surface finds and the decorated pottery is quite scarce, some of the locations remain tentative. For older sites researchers rely on the evolving technology of tool-making, which has been a universal phenomenon. There is no reason to suppose that this region has been any exception. Correlation of these sites with marine and river terraces and geological deposits provides further aid in arriving at certain acceptable conclusions regarding these settlements.

As a result of a systematic search in the Karachi region, about 250 prehistoric sites have been located, which range as far back in time as the upper Palaeolithic period. They range from small camping grounds to fair-sized rural settlements, from trading posts on the ancient highways to fortified outposts guarding the frontiers. Thus, this area offers the vestiges of the oldest cultures so far known for Pakistan after the Soan valley culture of the Potwar plateau.

At least four sites are assigned with certainty to the Harappan culture period: the Sutkagen-Dor, Sotka-Koh, Balakot and Pir Shah Jurio. The last is in some ways the most important because it is located near the mouth of the Hab river and most probably served as a port (Khan, 1968, 1973; Khuhro, 1979; Figure 17.2). The possibility of trade connections by sea between Mesopotamia and Egypt and the Harappan civilization of the Indus valley of Pakistan became likely when stamp seals of the Indus type were discovered in southern Mesopotamia and the Persian Gulf regions.

The Sutkagen-Dor site was discovered and partly excavated by a British officer in 1876, but it was not until the excavation of the site by Sir Aurel Stein in 1931 that it was identified as a 3000 yr old Harappan settlement. Sutkagen-Dor lies on the extreme eastern edge of the wide Dasht valley, ~ 48 km from the present Arabian Sea coast and ~ 56 km east of the Iran border (Figure 17.1). A second site, Sotka-Koh, was discovered and partially excavated by George Dales in 1960. This fortified settlement, ~ 13 km north of Pasni in the Shadi Kaur valley, was similar in many ways to Sutkagen-Dor in that it was built on top of a rock outcrop at a strategic location in the valley (Dales, 1962a,b).

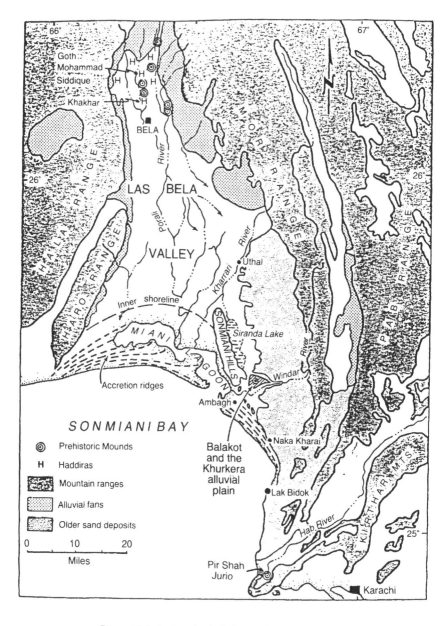

*Figure 17.3* Archaeological sites in the Las Bela valley.

The most exciting archaeological site to be uncovered to date along the Makran coast is the mound of Balakot, which lies near the middle of the Khurkera alluvial plain on the east side of the Las Bela valley, ~ 14 km inland from the sea (Figure 17.3). Balakot's present physical attributes differ sharply from its most likely condition in antiquity because of the extensive alluviation in the valley and the southward shift of the main bed of the Windar river. The small fishing village of Sonmiani, itself almost unusable now because of recent silting, is ~ 13 km west of Balakot (Dales, 1979).

The three above-mentioned Harappan sites, guarding the entrances to the three main routes from the sea to the interior of the Makran, provide important clues concerning the trading activities of the Harappans. When combined with other geological and geographical observations, their existence becomes even more significant (Dales, 1962b).

Pir Shah Jurio is probably the most important of the four Harappan sites, because it is located near the mouth of the Hab river and most likely served as a port. This site is located on a conglomerate terrace, > 15 m above sea level, on the left bank of the Hab river. The sea water at high tide reaches beyond this point even now. There are indications of stone foundations covered with wind-blown sand. The site is now occupied by a graveyard and is presented here to show what objects can actually occur on the surface when the mound itself is otherwise buried or destroyed. The objects found at the site include much plain and red pottery, some decorated with designs in black, perforated pottery, chert blades and scrapers, clay and shell bangles, copper pieces, triangular clay tablets, a polished chert weight, and part of a toy cart. All of these objects are typical of Harappan sites, but none of the characteristic terracotta figurines was found on the surface.

Whether the Harappans maintained just a few outposts along the trade route passing through the domain of an alien culture and terminating at the port-site at the Hab mouth or whether there were other Harappan sites in the area, which have completely eroded away, is not yet certain. Even the known Harappan sites, with the possible exception of Nal Bazar, have been eroded almost down to their foundations.

Apart from the prehistoric sites, many settlements and structures occur which belong to the proto-historic and historic periods. Several circular structures of stones have been located in the Hab valley west of Karachi, at Manghopir northwest of Karachi, and around the Drigh Road hills in the residential area of Karachi. Many of these occur in Baluchistan and may be Dakhma or Parsi Towers of Silence in which the dead are exposed to scavengers. It is of interest to note that Alexander's historians have recorded the fact that the people of Las Bela disposed of their dead as do the Zoroastrians (Parsi). Even Buddhist carvings and statues have been found in the Karachi and Las Bela region. Northwest of the town of Las Bela, along the cliffs of the Hala range, are Buddhist statues carved into the rock cliffs. These statues are well above the base of the cliff face. Evidence of Muslim occupation can be found in the large number of very old graveyards. The graves are rectangular platforms built with squared thin slabs of stones, with no mortar or plaster now visible. Their antiquity is proven by the fact that at many places erosion has lowered the surrounding area considerably, leaving these graves perched on higher patches of land. In many cases erosion has gone so deep that the graves have crumbled and even the bones are scattered on the surface. Near these graves, and at a number of other points at many prehistoric sites, microlithic as well as Chalcolithic evidence can be found (Khan, 1968, 1973; Khuhro, 1979).

## NATURAL FORCES THAT DESTROY ARCHAEOLOGICAL SITES

A series of natural forces has radically altered the configuration of the Makran coast since Palaeolithic times.

### Tectonics

The coastal zone and the interior Makran ranges are an active geological area and indications are that much of the region has been rising for millenia. Shells collected by the author (Snead, herein) in the uplifted marine terraces were radiocarbon dated and found to be as recent as the second millennium BC, and movements are continuing at the present time. Several major transverse faults have been identified and broad warping of the uplifted terraces can be discerned clearly. Uplift appears greatest, ~ 90–120 m, in the middle of the Makran region, near the Iran border, and decreases to ~ 1.8–2.4 m in the Indus delta.

In recent years there has been a concentration of earthquake epicentres off the Makran coast near the Pasni fishing village. This is an active anticlinal area. The November 1945 earthquake was one of the most destructive of the twenty earthquakes recorded in Baluchistan, because a tsunami (seismic sea wave), estimated to be 3–6 m high, moved over the coast and caused great loss of life and property (Figure 17.1). Local fishermen assert that a section of the coast near Pasni was uplifted ~ 4.5 m. The magnitude of this earthquake, as recorded at the seismological station in Quetta, 644 km away, was > 6.7 Richter magnitude (Sohdhi, 1947). So active is this coastal region that Hunt Petroleum Co., drilling near the Hingol river, reported that its drill became stuck permanently during an earthquake of slight intensity ~ 80 km north. When earthquakes occur, large mud volcanoes begin to erupt liquid mud down the sides of their cones (Snead, 1964). A number of archeological sites that were once close to or directly on the coast are now 24–30 km inland, and some are uplifted well above the high-tide mark.

The small archaeological site of Pir Shah Jurio probably once served as a port but it is now left isolated on a small terrace > 15 m above sea level (Rauf Khan, pers. comm., 1964–6; Figure 17.3). If this site is not investigated and excavated very soon, waves will completely undercut the terrace and the site will be lost to the sea.

The Karachi region provides valuable clues about the tectonic processes that have operated in the Holocene. A series of raised beaches and marine terraces occur along the coast, which are 6–7 m, 9–12 m and > 15 m above sea level and are matched by river terraces of roughly similar height inland. The highest gravel terrace, capped by wind-blown sand, has yielded implements of the latest phase of the upper Palaeolithic and Mesolithic periods. The middle terrace is occupied by Neolithic and Chalcolithic sites. The lowest one has so far failed to show any evidence of occupation during prehistoric times (Khan, 1968; Khuhro, 1979).

The areas at an elevation of 6–7 m above sea level, along the west side of the Indus delta and also along the Las Bela coast, are backed by old sea cliffs that are partly rock. This old shoreline corresponds very closely to the description of the coast given by Nearchus, the admiral of Alexander the Great, who took his fleet along this torturous rocky coast. Some uplift may have occurred there since 325 BC (Stein, 1943). On the lower Indus plain there is some indication that even since the Arab period there may have been uplift of 3–4 m, but this requires confirmation by other evidence. The 4 m high, wave-cut bench of Cape Monze may correspond with this uplift. Toward the end of the thirteenth century, or shortly thereafter, drastic changes seem to have occurred all over the lower Indus basin (see Flam, this volume). In general, for millenia the land has been rising intermittently in most places and sinking in a few others. Uplift has resulted in recession of the sea, erosion of river beds into their own deposits, gradual lowering of the regional water table, drying up of many springs, destruction of scrub forests previously inundated by seasonal floods but now left dry, severe soil erosion, and disruption and deterioration of irrigation works that may have been in operation before such movements. These geomorphic processes help to explain the presence of Indus silt much above the present floodplain. If any of these movements were abrupt, then earthquakes may also have destroyed settlements (Khan, 1968, 1973).

## Climatic effects

Climate in the harsh desert environment plays the most significant role in the destruction of archaeological sites. But climate varies greatly from year to year and is affected by various factors, such as proximity to the sea (more humidity) or to the drier inland, which control location of the sites (east or west sides of the Makran coastal region).

The climate of the coast is mainly determined by the southwest monsoons in summer and by western depressions in winter. According to the prevailing influence of the two wind systems, therefore, the coast can be divided into east and west sections (Figure 17.4).

In the east section, mainly Karachi and Las Bela, the effects of the southwest monsoon and occasional Arabian Sea tropical cyclone predominate, but these systems slowly decrease to the west. The winds associated with these influences produce heavy swells along the coast, especially between Karachi and Ormara.

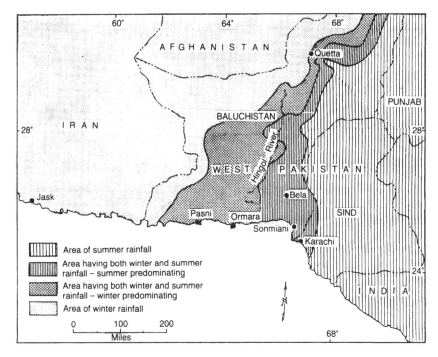

*Figure 17.4* Distribution of summer and winter rainfall in southern Pakistan (from Snead, 1968).

Rocky points, such as Ras Malan, Cape Monze and Ormara, receive a continual pounding of summer waves which causes severe erosion of the coast and of archaeological sites, as well as the deposition of plentiful new sand on the beaches. The force of the monsoons on the eastern part of the coast of Pakistan is neither as destructive to archaeological sites nor as great as along the Konkan and Malabar coasts of India. The rain-bearing southwest winds leave most of their moisture on the coast of southern India and the rest is absorbed in the hot dry air of southeastern Pakistan.

The western Pakistan coast from Ormara to Iran is less exposed to influence from the southwest winds, which bring no rain to the area in summer. Instead, the region is dominated in winter by weak westerly, poorly developed, cyclonic depressions. The intensity of the winds around these storms decreases after crossing the Arabian desert but they sometimes regain vigour over the Persian Gulf and provide occasional rain on the Makran coast. Storm intensity is thus slight and the light winter precipitation occurs only at long intervals. The line of balance between rainfall from summer monsoons and winter depressions trends northeast through Pakistan from Ormara (Figure 17.4).

## Temperature effects

Summer temperatures rise rapidly in the interior of the Makran coastal region. Mean maximum temperature for coastal Sonmiani in May and June is 29.4 °C, whereas 102 km inland at Las Bela the mean maximum temperature exceeds 37.8 °C. Freezing temperatures have been recorded rarely at Las Bela but never at Sonmiani. The more extreme interior temperatures produce more mechanical weathering there, but salt weathering near the coast appears greater.

## Humidity

Absolute humidity of 70–80 per cent is commonly high on the east coastal areas of Pakistan. Weathering increases because the chemical processes accelerate in saturated air, but research on the breakdown of different substances is needed.

## Precipitation

Summer monsoon conditions prevail in June–September and become fully developed on the east part of the Pakistan coast. Sometimes one or two strong cyclonic gales bring heavy rain, winds and floods to the coastal area from Karachi to Ormara. The monsoon barely reaches a height of a few hundred m around Karachi. Farther west it loses progressively more height until it is too low to cross the coastal mountains and so the highlands of Baluchistan and Iran remain free of its influence.

The highlands of Baluchistan generate hot dry winds that blow to the southeast. Their dryness increases continuously as they descend the steep slopes of the mountains and pass toward the coast. They resemble chinook (foehn) winds as they enter the lowlands of Las Bela and Sind. The moisture of the oceanic air, brought inland by the monsoon countercurrent, consequently is absorbed in the hot dry air of the interior not far inland from the coast.

Annual rainfall on all parts of the Pakistan coast is scanty and uncertain, but commonly intensive. It usually falls during June–September and November–February in heavy downpours. Nowhere in this area does the mean annual rainfall reach even 250 mm, and long periods of drought can occur, at times lasting as long as 2–3 yr. Summer rain is always < 220 m. Mean monthly summer rain is 216 mm in Karachi, 68 mm in Ormara and only 25 mm in Pasni, and declines toward nothing in the west. Wintertime westerly depressions provide a small amount of precipitation in the west, however. Along the Iran coast, Jask has a mean monthly winter rainfall of 170 mm, but in Pakistan Pasni gets 131 mm, Las Bela 137 mm and Karachi 55 mm. Only the middle section of the Las Bela plain receives rain in both seasons because of the surrounding mountains that encircle rain-bearing winds and force orographic precipitation (Siddiqi, 1956; Snead, 1966).

Because of the common droughts, no significant rainfall may occur but then,

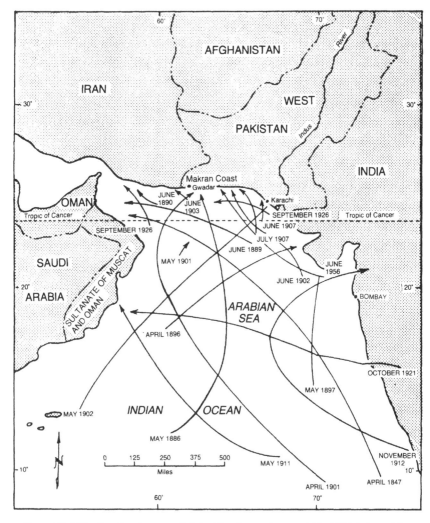

*Figure 17.5* Tracks of major recorded Arabian Sea cyclones, 1847–1956 (adapted from Normand, 1937 and Records of Pakistan Meteorological Department, Karachi).

in one year, immense erosion and alluviation may occur as a result of one or two storms. Along with this, the steady winds, often accompanied by blowing dust and sand, have slow abrasive action which wears down the soft mud walls of any site that rises above the desert landscape.

### Arabian Sea cyclones

Weather during the transition months of April–May and October–November is generally more settled than at any other time of the year. Arabian Sea cyclones may occur then, however, possibly connected to either the onset or the wane of the monsoon. They sometimes follow the path of the southwest monsoon (Figure 17.5). Some of these cyclones have the character of small hurricanes but generally lose intensity as they move north and reach the coast. Such storms create havoc and cause damage, not so much by rainfall but by causing high tides and floods blown by forceful winds toward the mainland (Snead, 1966, 1968).

## WEATHERING INFLUENCES ON ARCHAEOLOGICAL SITES

### Insolation weathering

This process is thought to involve rupture of rocks and other earth materials as a result of large diurnal temperature changes, or seasonal changes, temperature gradients from the surface into the rock, and the different coefficients of thermal expansion of various materials. The idea is old and now controversial. It was defended by reports of fusillades of heard-but-unseen warm rocks cracking in the cool desert night. It was thought to initiate splits in otherwise pristine rocks, and had the appeal of simplicity (Tricart and Cailleux, 1969; Ollier, 1963, 1965). Monthly means of diurnal air temperatures of as much as 29 °C (Mackelein, 1959) and some daily ranges as high as 54 °C were thought to be effective in the process, but whether materials react sufficiently to cause disintegration remains speculative. Nevertheless, the fabric of the material may be weakened by numerous small volume changes over long periods, and rock fatigue or hysteresis. Also, some rocks can have considerable built-in internal stress caused by original confining pressures at time of formation, or by forces of crystallization (Smalley, 1966).

### Honeycomb weathering

Honeycomb formation leading to alveoles and tafoni is based on the relocation of calcite and/or silica toward exposed surfaces while the interior stone crumbles away because of the deprivation of its interior cement. The process is often accelerated by salt action (Winkler, 1987). Honeycomb weathering was observed near the base of several buildings in Sonmiani, Pakistan. Such weathering was more pronounced on the windward side of structures owing to more salt spray from Miani Lagoon and a higher air humidity.

### Weathering of silicate minerals

Weathering of silicates is measured not in surface reduction but by thickness of alteration rims or rinds. Iron from mafic (manganese- and iron-rich) minerals such as biotite mica, amphiboles and pyroxenes is commonly mobilized first. The thickness of the weakened rind depends upon length of exposure to moisture, permeability of the material, and the degree of the iron bond in the crystal lattice (Winkler, 1987).

## Salt weathering

Salt weathering is quite common in the Makran region and can lead to rapid material disintegration of mud-brick walls and pottery. The process acts in three ways: the expansion of salts in confined spaces as crystals grow from salt solutions, stress caused by hydration of salts in confined spaces, and thermal expansion, especially under desert sun (Cooke and Smalley, 1968).

Evidence of salt corrosion or salt fretting can be seen along the upper fringe of the capillary water transport (Arnold, 1981). Sources of salt vary: from groundwater rising through salt-laden building materials of clay sand or silt, salts from ocean spray, and desert dust. The effectiveness of salt action depends on the kinds of salt present, on the size and shape of the capillary system, on the moisture content, and on the exposure to solar radiation. Exposure to the sun means that often the west and southwest sides of structures have more salt deposits because of the increased intensity of the afternoon sun's rays in the northern hemisphere.

## Wetting and drying weathering

Simple wetting and drying of some materials is sufficient to cause them to disintegrate (Goudie *et al.*, 1970). Desert surfaces in Pakistan are wetted by rainfall or dew, especially near the coast where salt spray and higher humidities condense into dew during early morning. Dew is also saline in many places so that weathering of surface material could be promoted by both wetting and drying, and salt (Yaalon and Ganor, 1968). Some salts also attract moisture through hygroscopicity without hydrating so that areas of salt concentration remain constantly moist (Arnold, 1981).

## GEOMORPHIC CHANGES

### River deposition

Large amounts of sand, silt and clay are brought down by desert rivers during floods and deposited along the lower stretches of the valleys and at river mouths. The Sonmiani fishing village, a prosperous small port in the 1800s, is now completely closed off by silt from the Windar river. A new port at Danb, 3.2 km northwest, was built because it is closer to the deeper water of Miani Lagoon (Snead, 1963; Figure 17.1). Excessive monsoonal or cyclonic rains have caused exceptional floods and severe erosion of the soft marls, clays and sandstones of the region. Poorly located or poorly constructed sites probably had to be moved or abandoned. Both erratic rainfall, droughts or floods must have existed in ancient times as they do now, with the result that sites would have had to have been abandoned from time to time.

## Coastal deposition

Constant sea wave and current action from the southwest, especially during the monsoon, deposit large amounts of new sand along the beaches. Even the Karachi harbour has to be dredged because of Indus river silts filling in the estuary (Snead, 1964, 1966, 1967, 1968, 1969, 1987). Another result is that at the Balakot site in the Las Bela area the coast has grown seaward ~ 4–7 km in the last 10,000 yr. Over 30 well defined beach ridges were created as the coast prograded (Figure 17.3). This rapid deposition has closed off a number of large lagoons, embayments and ports along the Makran coast, which could have been used by small ships in the past, and thereby has affected trading patterns. Nearly every major embayment should be investigated geomorphologically to determine not only the former size of the embayments but also how far inland the sea was in the past.

As a result of these geomorphic processes, wide expanses of barren sand desert extend along much of the Makran coast, from the foot of the Makran mountains to the present shoreline. Also, the thin spits of sand connecting the massive rock headlands with the mainland at Gwadar and Ormara, as well as the wide, sandy plain south of Pasni, have been formed by wave and current action during the Holocene, since the sea level came to its basic present position 10,000–8000 yr ago. Concentric beach ridge arcs marking successive formation of new beaches through this time are especially noticeable at Gwadar and Ormara (Figure 17.1). It is thus probable that the major headlands and other smaller rock outcrops were actually islands in antiquity, just as the island of Astola, off the coast of Pasni, represents a new uplifted terrace. With continued coastal progradation, this island may become connected to the mainland by a tombolo, as are Ormara and Gwadar (Vrendenburg, 1901; Ullah, 1954; Snead, 1967, 1969).

## DESTRUCTION OF ARCHAEOLOGICAL SITES THROUGH TRANSPORTATION AND URBAN EXPANSION

Many of the archaeological sites in the Karachi region and along the coast are in great danger of destruction through construction of buildings and roads. One mound, Gaki Buthi-2, which may have been a Neolithic settlement, has been destroyed. Haddiras (rectangular platforms and stone circles of the Zoroastrian Arhaemanian period) near the village of Goth Mohammad Siddique, which were well preserved and were visible some km from the main road, have been destroyed to obtain large stones for bridge construction. A similar fate may have befallen the largest group of Haddiras in the Las Bela valley. Stones have been dug from the walls of the Khakhar mound, 16 km north of Bela city (Figure 17.3). This is the most valuable Neolithic site in southern Pakistan, and its destruction is an enormous loss (Khan, 1973).

## METHODS TO PRESERVE PREHISTORICAL SITES

The 250 known archaeological sites must be protected from natural forces and from humans. It is difficult in an isolated coastal region, such as the Makran, to manage the remote and barren sites. It may be wiser to build large fences around the sites, but these would have to be repeatedly checked. In some places excavations should be recovered with earth, and in a few places it may be important to construct a sheltering roof.

Archaeological sites must be investigated and mapped before natural or human-caused processes destroy them further. Recent political problems in Pakistan have resulted in cancellation of archaeological work in the Makran and Baluchistan regions. Even the Balakot site, only 97 km from Karachi, has been closed to investigation. The Government of Pakistan must be encouraged to promote scientific inquiry for the betterment of knowledge and before further destruction has occurred. Local Pakistani scholars should be trained and encouraged in order to seek and investigate as yet unrecognized sites. Eventually some sites should be set aside as valuable national historical sites for education and to promote tourism.

## CONCLUSION

The 250 known archaeological sites along the Makran coast of Pakistan are being destroyed or removed by a variety of natural and human-caused processes. Archaeological findings have clearly established that there was substantial maritime penetration down the whole length of the Persian Gulf from lower Mesopotamia by the late fourth millennium BC ( ~ 6000 yr ago), and eastward to the Indus river valley civilizations 500 yr later. By the turn of the second millennium, a colony of Harappan traders from the Indus had established itself near Ur in Mesopotamia. Later periods of extensive contact across the Indian Ocean show that it was a vast concourse of meeting grounds for much of the peoples of the region. The trade connections between the Indus valley civilization and Mesopotamia and Egypt must have involved much travel through sites along the Makran coast and, therefore, these sites need much more examination before the sites are completely destroyed.

# REFERENCES

Abbasi, I.A. and Friend, P.F. (1988) 'The uplift and evolution of the Himalayan orogenic belts as recorded in the foredeep molasse sediments', *The Neogene of the Karakoram and Himalayas*, Leicester: Department of Geography, University of Leicester.

Abdel-Gawad, M. (1971) 'Wrench movements in the Baluchistan arc and relation to Himalayan–Indian Ocean tectonics', *Bulletin of the Geological Society of America* 82: 1235–50.

Abdunazarov, U. *et al.* [27 authors] (1984) 'Quaternary deposits of central Asia', *27th International Geological Congress, Uzbekistan excursions guidebook*, Tashkent: Fan Publishers, 44–71.

Abers, G., Bryan, C., Roecker, S. and McCaffrey, R. (1988) 'Thrusting of the Hindu Kush over the southeastern Tadjik basin, Afghanistan: evidence from two large earthquakes', *Tectonics* 7: 41–56.

Adams, J. (1980) 'Active tilting of the United States midcontinent: geodetic and geomorphic evidence', *Geology* 8: 442–6.

Agrawal, D.P. (1984) 'Palaeoclimatic studies in Kashmir: a summary', *East Asian Tertiary/Quaternary Newsletter* 1: 17–21.

Ahmed, S.S. (1969) 'Tertiary geology of part of south Makran, Baluchistan, West Pakistan', *Bulletin of the American Association of Petroleum Geologists* 53: 1480–99.

Ahnert, F. (1970) 'Functional relationships between denudation, relief and uplift in large mid-latitude drainage basins', *American Journal of Science* 268: 243–63.

Ahuja, and Rao, (1958)

Akhtar, G.H. (1970) *Reconnaissance Soil Survey of Khairpur*, Lahore: Soil Survey of Pakistan.

Akram, M. (1971) *Reconnaissance Soil Survey of Bahawalpur Area*, Lahore: Soil Survey of Pakistan.

Alexander, J. and Leeder, M.R. (1987) 'Active tectonic control of alluvial architecture', in F.G. Ethridge, R.M. Flores and M.D. Harvey (eds) *Recent Developments in Fluvial Sedimentology*, Tulsa, OK: Society of Economic Paleontologists and Mineralogists, Special Publication 39: 243–52.

Ali, M.A. (1967) *Reconnaissance Soil Survey of Rawalpindi Area*, Lahore: Soil Survey of Pakistan.

Ali, M.A. (1969) *Reconnaissance Soil Survey of Dera Ismail Khan*, Lahore: Soil Survey of Pakistan.

Alim, M.M. (1968) *Reconnaissance Soil Survey of Sahiwal District*, Lahore: Soil Survey of Pakistan.

Allchin, B. (1976) 'Palaeolithic sites in the plains of Sind and their geographical implication', *Geographical Journal* 142: 471–89.

Allchin, B. and Goudie, A. (1978) 'Climatic change in the Indian desert and northwest

India during the Late Pleistocene and Early Holocene', in W.C. Brice (ed.) *The Environmental History of the Near and Middle East*, London: Academic Press, 307–18.

Allchin, B., Goudie, A. and Karunarkara, Hedge (1978) *The Prehistory and Paleogeography of the Great Indian Desert*, London: Academic Press.

Allen, C. (1984) *A Mountain in Tibet*, London: Futura.

Allen, C.R. (1975) 'Geological criteria for evaluating seismicity', *Bulletin of the Geological Society of America* 86: 1041–57.

Allen, J.R.L. (1985) *Principles of Physical Sedimentology*, Boston, MA: Allen & Unwin.

Ambraseys, N., Lensen, G., Moinfar, A. and Pennington, W. (1981) 'The Pattan (Pakistan) earthquake of 28 December 1974: field observations', *Quaternary Journal of Engineering Geology, London* 14: 1–16.

Andersen, B.G. (1968) 'Glacial geology of western Troms, north Norway', *Norges Geologiske Undersøgelse* 256.

Andrews-Speed, C.P. and Brookfield, M.E. (1982) 'Middle Paleozoic to Cenozoic geology and tectonic evolution of the northwestern Himalaya', *Tectonophysics* 82: 253–75.

Ansari, A.H. (1971) *Reconnaissance Soil Survey of Bahawalnager Area*, Lahore: Soil Survey of Pakistan.

Ansari, A.H. (1972) *Reconnaissance Soil Survey: Larkana district*, Lahore: Soil Survey of Pakistan.

Ansari, A.H. (1973) *Reconnaissance Soil Survey: Dadu district*, Lahore: Soil Survey of Pakistan.

Arnold, A. (1981) 'Nature and reactions of saline materials in walls', in R. Rossi-Manaresi (ed.) *The Conservation of Stone II, Bologna 27–30 October 1981*, Centro per la Conservazione delle Sculture all'aperto, 13–23.

Arrian, F. (1879) 'First century AD, the Indika of Megasthenes and Arrian', Part 2, in trans. by J.W. McCrindle, London: Constable.

Artem'yev, M.Ye. and Belousov, T.P. (1979) 'Isostasy and neotectonics of the Pamirs and southern Tien Shan', *Doklady Akademii Nauk SSSR* 249: 51–4.

Ashmore, P. (1987) 'Bed load transfer and channel morphology in braided streams: erosion and sedimentation in the Pacific Rim', *Proceedings of the Corvallis Symposium*, International Association of Hydrological Sciences, Publication 165: 333–41.

Asrar, U. (1953) 'Physiography and structure of southwest Makran', *Pakistan Geographical Review* 9(1): 28–54.

Auden, B.A. (1974) 'Afghanistan–west Pakistan', in A.M. Spencer (ed.) *Mesozoic-Orogenic Belts*, Edinburgh: Scottish Academic Press, 235–53.

Azzaroli, A. and Napoleone, G. (1982) 'Magnetostratigraphic investigation of the Upper Siwaliks near Pinjor, India', *Rivista Italiana di Paleontologia e Stratigrafia* 87: 739–62.

Bagnold, R.A. (1941) *The Physics of Blown Sand and Desert Dunes*, New York: Morrow.

Baig, M.S. (1969) *Reconnaissance Soil Survey of Ghotki*, Lahore: Soil Survey of Pakistan.

Bakr, A.M. and Jackson, R.O. (1964) *Geological Map of Pakistan, 1:2,000,000*, Quetta: Geological Survey of Pakistan.

Bard, J.P., Maluske, H., Matte, P. and Proust, F. (1980) 'The Kohistan sequence: crust and mantle of an obducted island arc', *Geological Bulletin of the University of Peshawar, Pakistan* 13: 87–94 (Amsterdam: Elsevier).

Barndt, J., Johnson, N.M., Johnson, G.D., Opdyke, N.D., Lindsay, E.H., Pilbeam, D. and Tahirkheli, R.A.K. (1978) 'The magnetic polarity stratigraphy and age of the Siwalik group near Dhok Pathan village, Potwar Plateau, Pakistan', *Earth and Planetary Science Letters* 41: 355–64 (Amsterdam: Elsevier).

Barry, J.C., Johnson, N.M., Raza, S.M. and Jacobs, L.L. (1985) 'Neogene mammalian faunal change in southern asia: correlations with climatic, tectonic and eustatic events', *Geology* 13: 637–40.

Batura Glacier Investigation Group (1976) *Investigation Report on the Batura Glacier in the Karakoram Mountains, the Islamic Republic of Pakistan (1974–1975)*, Beijing: Engineering Headquarters of the Batura Glacier Investigation Group of Karakoram Highway.

Batura Glacier Investigation Group (1979) 'The Batura Glacier in the Karakoram mountains and its variations', *Scientia Sinica* 22(8): 959–74.

Batura Glacier Investigation Group (1980) *Karakoram Batura Glacier, Exploration and Research* (in Chinese with English abstracts), Lanzhou Institute of Glaciology and Cryopedology, Academia Sinica, Beijing: Science Press.

Becher, J. (1859) 'Letter addressed to R.H. Davis, Esq., Secretary to the Government of the Punjab and its dependencies, 1 July 1859 (on the 1858 Indus Flood)', *Journal of the Asiatic Society of Bengal* 28: 219–28.

Beck, R.A. and Burbank, D.W. (1990) 'Continental-scale diversion of rivers: a control of alluvial stratigraphy', Geological Society of America, *Abstracts with Programs* 22: A238.

Behrensmeyer, A.K. and Tauxe, L. (1982) 'Isochronous fluvial systems in Miocene deposits of northern Pakistan', *Sedimentology* 29: 331–52.

Belousov, T.P. (1976) 'Evolution of vertical movements in the Pamirs in the Pleistocene and Holocene', *Geotectonics* 1: 68–75.

Belyayevskiy, N.A. (1966) 'Principal geological features of Karakoram', *International Geological Review* 8: 127–43.

Bendefy, L., Dohnalik, J. and Mike, K. (1967) 'Nouvelles methodes de l'étude genetiques des cours d'eau', *Symposium on River Morphology*, International Association of Hydrological Sciences, Publication 75: 64–72.

Berggren, W.A. and van Couvering, J. (1974) 'The late Neogene: biostratigraphy, geochronology and palaeoclimatology of the last 15 million years in marine and continental sequences', *Palaeogeography, Palaeoclimatology, Palaeoecology* 16: 1–216 (Amsterdam: Elsevier).

Berggren, W.A., Kent, D.V. and van Couvering, J.A. (1985a) 'Neogene geochronology and chronostratigraphy', in N.J. Snelling (ed.) *The Chronology of the Geological Record*, Geological Society of London, Memoir 10: 211–60.

Berggren, W.A., Kent, D.V., Flynn, J.A. and van Couvering, J.A. (1985b) 'Cenozoic chronology', *Bulletin of the Geological Society of America* 96: 1407–18.

Bhandari, L.L., Venkatachalaya, B.S. and Patap Singh (1977) 'Stratigraphy, palynology and palaeontology of Ladakh Molasse group', *Proceedings of the 4th Colloquium on Indian Micropalaeontology and Stratigraphy 1974–75*, Dehra Dun, India: Oil and Gas Commission and Institute of Petroleum, 127–33.

Biagi, P. and Cremaschi, M. (1988) 'The early Paleolithic sites of the Rohri Hills (Sind, Pakistan) and their environmental significance', *World Archaeology* 19: 421–33.

Biswas, S.K. (1987) 'Regional tectonic framework, structure and evolution of the western marginal basins of India', *Tectonophysics* 135: 307–27.

Blanford, W.T. (1868–73) 'Coast of Baluchistan and Persia from Karachi to the head of the Persian Gulf', *Records of the Geological Survey of India* 5 (part 2).

Blanford, W.T. (1872) 'Note on the geological formation seen along the coasts of Baluchistan and Persia from Karachi to the head of Persian Gulf', *Records of the Geological Survey of India* 4: 41–5.

Blanford, W.T. (1876) 'On the physical geography of the Great Indian Desert with special reference to the former existence of the sea in the Indus valley; and on the origin and mode of formation of the sand-hills', *Journal of the Asiatic Society of Bengal* 45(2): 86–103.

Bonney, T.G. (1902) 'Moraines and mud-streams in the Alps', *Geological Magazine* (new series) 9: 9–16.

Bordet, P. (1970) 'A propos du volcanisme d'âge plioquaternaire d'Afghanistan central: ignimbrites et brèches de nuées péléenes', *Bulletin Volcanologique* (of International Volcanological Association) 33: 1220–8.

Bordet, P. (1972a) 'Le volcanisme acide récent du Dacht-e-Nawar (Afghanistan central)', *Bulletin Volcanologique* (of International Volcanological Association) 36: 289–300.

Bordet, P. (1972b) 'Le volcanisme récent du Dacht-e-Nawar Meridional (Afghanistan central)', *Revue de Géographie Physique et de Géologie Dynamique* 14(4), 427–32.

Boulton, G.S. and Eyles, N.I. (1979) 'Sedimentation by valley glaciers: a model and genetic classification', in C.H. Schluchter (ed.) *Moraines and Varves*, Rotterdam: Balkema, 11–25.

Boutière, A. and Clocchiatti, R. (1971) 'Sur les roches pyroclastiques du Nord du Dacht-e-Nawar (Afghanistan): les quartz et leurs inclusions vitreuses', *Compte Rendu Sommaire et Bulletin* 41(3): 181.

Bovis, M.J. (1982) 'Uphill-facing (antislope) scarps in the coast mountains, southwest British Columbia', *Bulletin of the Geological Society of America* 93: 804–12.

Bradley, W.C. (1958) 'Submarine abrasion and wave-cut platforms', *Bulletin of the Geological Society of America* 69: 723–35.

Brady, N.C. (1984) *The Nature and Properties of Soils*, 9th edn, New York: Macmillan.

Breed, C.S., Bryberger, S.C., Andrews, S., McCauley, C., Lennertz, F., Gebel, D. and Horstman, K. (1979) 'Regional studies of sand seas using Landsat (ERTS) imagery', in E.D. McKee (ed.) *A Study of Global Sand Seas*, United States Geological Survey, Professional Paper 1052: 305–97.

Bridge, J.S. and Leeder, M.R. (1979) 'A simulation model of alluvial stratigraphy', *Sedimentology* 26: 617–44.

Brinkman, R. (1971) 'Soil genesis in West Pakistan', *Pakistan Soil Bulletin* 4, Lahore: Soil Survey of Pakistan.

Brinkman, R.A. and Rafiq, C.M. (1971) 'Landforms and soil parent material in West Pakistan', *Pakistan Soil Bulletin* 2, Lahore: Soil Survey of Pakistan.

Broecker, W.S. and Denton, G.S. (1990) 'The role of ocean–atmosphere reorganizations in glacial cycles', *Quaternary Science Reviews* 9: 305–41.

Brookfield, M.E. (1981) 'Metamorphic distributions and events in the Ladakh range, Indus suture zone and Karakoram mountains', in P.S. Saklani (ed.) *Metamorphic Tectonites of the Himalaya*, New Delhi: Today and Tomorrows Publishing, 1–14.

Brookfield, M.E. and Andrews-Speed, C.P. (1984a) 'Sedimentology, petrography and tectonic significance of the shelf, flysch and molasse clastic deposits across the Indus suture zone, Ladakh, northwest India', *Sedimentary Geology* 40: 249–86.

Brookfield, M.E. and Andrews-Speed, C.P. (1984b) 'Sedimentation in a high-altitude intermontane basin – the Wakka Chu Molasse (mid-Tertiary, northwestern India)', *Bulletin of the Indian Geological Association* 17: 176–93.

Bruckner, E. (1886) 'Die vergletscherung des salsachgebiets', *Geografische Abhandlungen* 1: 1–183.

Brunsden, D. and Jones, D.K.C. (1984) 'The geomorphology of high magnitude–low frequency events in the Karakoram mountains', in K.J. Miller (ed.) *The International Karakoram Project* 1, Cambridge: Cambridge University Press, 383–8.

Brunsden, D., Jones, D.K.C. and Goudie, A.S. (1984) 'Particle size distribution on the debris slopes of the Hunza valley', in K.J. Miller (ed.) *The International Karakoram Project* 2, Cambridge: Cambridge University Press, 536–80.

Bryan, R. and Yair, A. (eds) (1982) *Badland Geomorphology and Piping*, Norwich: Geobooks.

Bryson, R.A. and Swain, A.M. (1981) 'Holocene variations of monsoon rainfall in Rajasthan', *Quaternary Research* 16: 135–45.

Buhl, H. (1986) *Nanga Parbat Pilgrimage*, Harmondsworth, UK: Penguin Books.

Bull, W.B. (1972) 'Recognition of alluvial-fan deposits in the stratigraphic record', in J.K. Rigby and W.K. Hamblin (eds) *Recognition of Ancient Sedimentary Environments*, Tulsa, OK: Society of Economic Paleontologists and Mineralogists, Special Publication 16: 63–83.

Bull, W.B. (1984) 'Correlation of flights of global marine terraces', in M. Morisawa and J. Hack (eds) *Tectonic Geomorphology*, Proceedings of the 15th Annual Geomorphology Symposium at Binghamton, Boston, MA: Allen & Unwin, 129–52.

Burbank, D.W. (1982) 'The chronologic and stratigraphic evolution of the Kashmir and Peshawar intermontane basins, northwest Himalaya', Unpublished Ph.D. thesis, Dartmouth College, Hanover, NH.

Burbank, D.W. (1983a) 'The chronology of intermontane-basin development in the northwestern Himalaya and the evolution of the northwest syntaxis', *Earth and Planetary Science Letters* 64: 77–92 (Amsterdam: Elsevier).

Burbank, D.W. (1983b) 'Multiple episode of catastrophic flooding in the Peshawar Basin during the past 700,000 years', *Geological Bulletin of the University of Peshawar, Pakistan* 16: 43–9 (Amsterdam: Elsevier).

Burbank, D.W. and Beck, R.A. (1991) 'Rapid, long-term rates of denudation', *Geology* 19: 1169–72.

Burbank, D.W. and Fort, M.B. (1985) 'Bedrock control on glacial limits: examples from the Ladakh and Zanskar Ranges, northwestern Himalaya, India', *Journal of Glaciology* 31(108): 143–9.

Burbank, D.W. and Johnson, G.D. (1982) 'Intermontane basin development in the past 4 myr in the northwest Himalaya', *Nature* 298: 432–6.

Burbank, D.W. and Johnson, G.D. (1983) 'The late Cenozoic chronologic and stratigraphic development of the Kashmir intermontane basin, northwestern Himalaya', *Palaeogeography, Palaeoclimatology, Palaeoecology* 43: 205–35 (Amsterdam: Elsevier).

Burbank, D.W. and Raynolds, R.G.H. (1984) 'Sequential late Cenozoic disruption of the northern Himalayan foredeep', *Nature* 311: 114–18.

Burbank, D.W. and Tahirkheli, R.A.K. (1985) 'The magnetostratigraphy, fission-track dating, and stratigraphic evolution of the Peshawar intermontane basin, northern Pakistan', *Bulletin of the Geological Society of America* 96: 539–52.

Burbank, D.W., Raynolds, R.H.G. and Johnson, G.D. (1986) 'Late Cenozoic tectonics and sedimentation in the northwestern Himalayan foredeep, II: Eastern limb of the northwest syntaxis and regional syntexis', in P.A. Allen and P. Homewood (eds) *Foreland Basins*, International Association of Sedimentologists, Special Publication 8: 293–306.

Burbank, D.W., Beck, R.A., Raynolds, R.G.H., Hobbs, R. and Tahirkheli, R.A.K. (1988) 'Thrusting and gravel progradation in foreland basins: a test of post-thrusting gravel dispersal', *Geology* 16: 1143–6.

Burgisser, H.M., Gansser, A. and Pika, J. (1982) 'Late glacial lake sediments of the Indus valley area, northwestern Himalayas', *Ecologae Geologische Helvetica* 75: 51–63.

Burnes, A. (1835) 'Memoir on the eastern branch of the River Indus', *Transactions of the Royal Asiatic Society* 3: 550–8.

Burnett, A.W. and Schumm, S.A. (1983) 'Active tectonics and river response in Louisiana and Mississippi', *Science* 222: 49–50.

Burrard, S.G. and Hayden, H.H. (1907) 'Geography and geology of Himalayan Mountains and Tibet', *The Rivers of the Himalaya and Tibet*, Part 3, Calcutta: Government of India Press, 119–230.

Butler, B.E. (1950) 'A theory of prior streams as a causal factor of soil occurrence in the riverine plain of southeastern Australia', *Australian Journal of Agricultural Research* 1: 231–52.

Butler, R.W.H. and Prior, D.J. (1988a) 'Anatomy of a continental subduction zone: the main mantle thrust in northern Pakistan', *Geologische Rundschau* 77(1): 239–55.

Butler, R.W.H. and Prior, D.J. (1988b) 'Tectonic controls on the uplift of the Nanga Parbat Massif, Pakistan Himalayas', *Nature* 333: 247–50.

Butler, R.W.H., Owen, L.A. and Prior, D.J. (1988) 'Flooding, earthquakes and uplift in the Pakistan Himalayas', *Geology Today*, Nov.–Dec.: 197–201.

Butler, R.W.H., Prior, D.J. and Knipe, R.J. (1989) 'Neotectonics of the Nanga Parbat syntaxis, Pakistan, and crustal stacking in the northwest Himalayas', *Earth and Planetary Science Letters* 94: 329–43 (Amsterdam: Elsevier).

Butz, D.A.O. (1987) 'Irrigation agriculture in high mountain communities: the example of Hopar villages, Pakistan', Unpublished Master's thesis, Department of Geography, Wilfrid Laurier University, Waterloo, Ontario.

Calkins, J.A., Offield, T.W., Abdullah, S.K.M. and Ali, S.T. (1975) *Geology of the Southern Himalaya in Hazara, Pakistan, and Adjacent Areas*, United States Geological Survey, Professional Paper 716-C.

Carson, M.A. (1984) 'The meandering–braided river threshold: a reappraisal', *Journal of Hydrology* 73: 315–34.

Casnedi, R. (1976) 'Geological notes on the junction between the Haramosh–Nanga Parbat structure and the Karakoram Range', *Accademia Nazionale dei Lincei, Rendiconti* 61: 631–3.

Casnedi, R. and Ebblin, C. (1977) 'Geological notes on the area between Astor and Skardu, Kashmir', *Academia Nazionale dei Lincei, Rendiconti* 62: 662–8.

Cerling, T.E. (1984) 'The stable isotopic composition of modern soil carbonate and its relationship to climate', *Earth and Planetary Science Letters* 71: 229–40 (Amsterdam: Elsevier).

Cerling, T.E., Quade, J., Wang, Y. and Bowman, J.R. (1989) 'Soil and paleosols as ecologic and paleoecologic indicators', *Nature* 341: 138–9.

Cerveny, P.F., Naeser, N.D., Zeitler, P.K., Naeser, C.W. and Johnson, N.M. (1988) 'History of uplift and relief of the Himalaya during the past 18 million years: evidence from fission track ages of detrital zircons from sandstones of the Siwalik group', in K.L. Kleinspehn and C. Paola (eds) *New Perspectives in Basin Analysis*, New York: Springer Verlag, 43–61.

Chang Chengfa *et al.* [26 authors] (1986) 'Preliminary conclusions of the Royal Society and Academia Sinica 1985 geotraverse of Tibet', *Nature* 323: 501–7.

Chappell, J. (1974) 'Geology of coral terraces, Huon Peninsula, New Guinea: a study of Quaternary tectonic movements and sea level changes', *Bulletin of the Geological Society of America* 85: 553–70.

Chaudhri, R.S. (1975) 'Sedimentology and genesis of the Cenozoic sediments of northwestern Himalayas (India)', *Geologische Rundschau* 64: 958–77.

Chi-yen Wang and Yaolin Shui (1982) 'On the tectonics of the Himalaya and the Tibet Plateau', *Journal of Geophysical Research* 87: 2949–57.

Church, M. and Ryder, J.M. (1972) 'Paraglacial sedimentation: a consideration of fluvial processes conditioned by glaciation', *Bulletin of the Geological Society of America* 83: 3059–72.

Clemens, S.C. and Prell, W.L. (1989) 'Terrigenous palaeoclimate indicators of continental aridity and summer monsoon strength: northwest Arabian Sea', *The Late Cenozoic Ice Age Symposium*, Edinburgh.

Clemens, S.C. and Prell, W.L. (1990) 'Late Pleistocene variability of Arabian Sea summer monsoon winds and continental aridity: Eolian records from the lithogenic component of deep sea cores', *Paleoceanography* 5: 109–45.

Code, J.A. and Sirhindi, S. (1986) 'Engineering implications of impoundment of the Indus River by an earthquake-induced landslide', in R.L. Schuster (ed.) *Landslide Dams:*

*Processes, Risk, and Mitigation*, American Society of Civil Engineers, Special Publication, 97–110.

COHMAP members [Co-operative Holocene Mapping Project – 33 members] (1988) 'Climatic changes of the last 18,000 years: observations and model simulations', *Science* 241: 1043–52.

Coleman, J.M. (1969) 'Brahmaputra River: channel processes and sedimentation', *Sedimentary Geology* 3: 129–239.

Collins, D.N. (1988) 'Meltwater characteristics as indicators of glacial and hydrological processes beneath large valley glaciers in the Karakoram', Paper handed out and presented but not published at the symposium on the Neogene of the Karakoram and Himalaya, Department of Geography, University of Leicester.

Cooke, R.U. and Smalley, I.J. (1968) 'Salt weathering in deserts', *Nature* 220: 1226–7.

Corbel, J. (1959) 'Vitesse de l'erosion', *Zeitschrift für Geomorphologie* 1: 1–28.

Cotter, G. de P. (1929) 'The Erratics of the Punjab', *Records of the Geological Survey of India* 61: 327–36.

Cotter, G. de P. (1933) 'The geology of the part of the Attock District west of longitude 72 degrees 45 minutes E', *Memoirs of the Geological Survey of India* 55: 63–161.

Coulson, A.L. (1938) 'Pleistocene glaciation in north-western India with special reference to erratics of the Punjab', *Records of the Geological Survey of India* 72: 422–39.

Coumes, F. and Kolla, V. (1984) 'Indus fan: seismic structure, channel migration and sediment thickness in the upper fan', in B.U. Haq and J.D. Milliman (eds) *Marine Geology and Oceanography of Arabian Sea and Coastal Pakistan*, New York: Van Nostrand Reinhold, 101–10.

Coward, M.P. and Butler, R.W.H. (1985) 'Thrust tectonics and the deep structure of the Pakistan Himalaya', *Geology* 13: 417–20.

Coward, M.P., Windley, B.F., Broughton, R.D., Luff, I.W., Petterson, M.G., Pudsey, C.J., Rex, D.C. and Khan, M.A. (1986) 'Collision tectonics in the northwest Himalaya', in M.P. Coward and A.R. Ries (eds) *Collision Tectonics*, Geological Society of London, Special Publication 19: 203–19.

Crawford, A.R. (1979) 'Gondwanaland and the Pakistan Region', in A. Farah and K.A. DeJong (eds) *Geodynamics of Pakistan*, Quetta: Geological Survey of Pakistan, 103–10.

Cronin, V.S. (1982) 'The physical and magnetic polarity stratigraphy of the Skardu Basin, Baltistan, northern Pakistan', Unpublished Master's thesis, Dartmouth College, Hanover, NH.

Cronin, V.S. (1989) 'Structural setting of Skardu intermontane basin, Karakoram Himalaya, Pakistan', in L.L. Malinconico, jun. and R.J. Lillie (eds) *Tectonics and Geophysics of the Western Himalaya*, Geological Society of America, Special Paper 232: 183–201.

Cronin, V.S., Johnson, W.P., Johnson, N.M. and Johnson, G.P. (1989) 'Chronostratigraphy of the upper Cenozoic Bunthang sequence and possible mechanisms controlling base level in Skardu intermontane basin, Karakoram Himalaya, Pakistan', in L.L. Malinconico, jun. and R.J. Lillie (eds) *Tectonics and Geophysics of the Western Himalaya*, Geological Society of America, Special Paper 232: 295–309.

Cronin, V.S., Sverdrup, K.A. and Schurter, G. (1990) 'Landsat drainage lineaments, seismicity and uplift of the Nanga Parbat–Haramosh Massif, northwest Himalaya', Geological Society of America, *Abstracts with programs* 22: A232.

Cullen, J.L. (1981) 'Microfossil evidence for changing salinity patterns in the Bay of Bengal over the last 20,000 years', *Palaeogeography, Palaeoclimatology, Palaeoecology* 35: 315–56 (Amsterdam: Elsevier).

Curray, J.R. (1965) 'Late Quaternary history, continental shelves of the United States', in H.E. Wright, jun. and D.G. Frey (eds) *The Quaternary of the United States*, Princeton,

NJ: Princeton University Press, 723–35.

Curray, J.R. and Moore, D.G. (1971) 'Growth of the Bengal deep-sea fan and denudation in the Himalayas', *Bulletin of the Geological Society of America* 82: 563–72.

Dainelli, G. (1922) *Studi sul Glaciale: Spedizione Italiana De Filippi nell'Himalaia, Caracorum e. Turchestan Chinese (1913–1914)*, Series II, Vol. 3, Bologna: Zanichelli, 658.

Dainelli, G. (1924–35) *Relazioni Scientifiche della Spedizione Italiana De Filippi nell'Himalaia, Caracorum e. Turchestan Chinese (1913–1914)*, Series II, 10 volumes, Bologna: Zanichelli.

Dainelli, G. (1934) *La Esplorazione della Regione fra l'Himalaia Occidentale e il Caracorum: Spedizione Italiana De Filippi nell'Himalaia, Caracorum e. Turchestan Chinese (1913–1914)*, Series II, Vol 1, Bologna: Zanichelli, 430.

Dainelli, G. (1935) *La Serie dei Terreni: Spedizione Italiana De Filippi nell'Himalaia, Caracorum e. Turchestan Chinese (1913–1914)*, Series II, Vol. 2, Bologna: Zanichelli, 230.

Dales, G. (1962a) 'A search for ancient seaports', *Expedition* 4: 2–11.

Dales, G. (1962b) 'Harappan outposts on the Makran coast', *Antiquity* 36(142): 86–92.

Dales, G.F. (1964) 'The mythical massacre at Moenjo-daro', *Expedition* 6(3): 36–43.

Dales, G.F. (1965) 'New investigations at Moenjo-daro', *Archaeology* 18(2): 145–50.

Dales, G.F. (1972) 'The decline of the Harappans', *Old World Archaeology: Foundations of Civilization*; reading from *Scientific American*, New York: W.H. Freeman.

Dales, G.F. (1974) 'Excavations at Balakot, Pakistan, 1973', *Journal of Field Archaeology* 1: 3–22.

Dales, G.F. (1979) 'The Balakot Project: summary of four years of excavations in Pakistan', in M. Taddei (ed.) *South Asian Archaeology 1977*, Naples: Instituto Universitario Orientale, 241–74.

Dales, G.F. and Raikes, R.L. (1977) 'The Moenjo-daro floods: a rejoinder', *American Anthropologist* 70: 957–61.

Danilchik, W. and Shah, S.M.I. (1976) *Stratigraphy and coal resources of the Makarwal area, Trans-Indus Mountains, Mianwali District, Pakistan*, United States Geological Society Project Report, Pakistan Investigations (IR), PK 60.

Davies, D.D. (1967) 'Origin of friable sandstone–calcareous sandstone rhythms in the upper Lias of England', *Journal of Sedimentary Petrology* 37(4): 1179–88.

Davis, J.C. (1986) *Statistics and Data Analysis in Geology*, New York: Wiley.

de Terra, H. (1932) *Geologische Forschungen im Westlichen K'un-lun und Karakorum Himalaya*, Berlin: Reimer & Vohsen.

de Terra, H. and Paterson, T.T. (1939) *Studies on the Ice Age in India*, Washington, DC: Carnegie Institute.

de Terra, H. and Teilhard de Chardin, P. (1936) 'Observations on the Upper Siwalik formations and later Pleistocene deposits in India', *American Philosophical Society Proceedings* 76: 791–822.

Derbyshire, E. (1981) 'Glacial regime and glacial sediment facies: a hypothetical framework for the Qinghai–Xizang Plateau', *Proceedings of Symposium on Qinghai–Xizang (Tibet) Plateau* 2, Beijing: Science Press; New York: Gordon and Breach Science Publishers, 1649–56.

Derbyshire, E. and Owen, L.A. (1990) 'Quaternary alluvial fans in the Karakoram Mountains', in A.H. Rachocki and M. Church (eds) *Alluvial Fans: a Field Approach*, Chichester: Wiley, 27–53.

Derbyshire, E., Li, J., Perott, F.A., Xu, S. and Waters, R.S. (1984) 'Quaternary glacial history of the Hunza valley, Karakoram Mountains, Pakistan', in K.J. Miller (ed.) *The International Karakoram Project* 2, Cambridge: Cambridge University Press, 456–95.

Derbyshire, E., Owen, L.A. and Fort, M. (1987) 'Fabric and the problem of distinguishing glacial and non-glacial diamictons in the Karakoram and Himalaya', *Seminaire Himalaya–Karakorum*, Abstracts Volume, Centre National de Recherche Scientifique, Nancy, France.

Desio, A. (1964) *Geological Tenative Map of Western Karakorum – 1:5000,000*. Geological Institute, University of Milan.

Desio, A. (1980) *Geology of the Shaksgam Valley*, Leiden: Brill.

Desio, A. and Orombelli, G. (1971) 'Notizie preliminari sulla presenza diun grande ghlacciaio vallivo nella media valle dell' Indo (Pakistan) durante il Pleistocene', *Atti della Accademia Nazionale dei Lincei*, 53: 387–92.

Desio, A. and Orombelli, G. (1983) 'The Punjab erratics and the maximum extent of the glaciers in the middle Indus valley (Pakistan) during the Pleistocene', *Atti della Accademia Nazionale dei Lincei*, Series 8(17): 179.

Desio, A., Tongiorgi, E. and Ferra, G. (1964) 'On the geological age of some granites of the Karakoram, Hindu Kush, and Badakshan, Central Asia', *Proceedings of the 22nd International Geological Congress, New Delhi*, 11: 479–96.

Deutsch, M. and Ruggles, F.H., jun. (1978) 'Hydrological applications of Landsat imagery used in the study of the 1973 Indus River flood, Pakistan', *Water Resource Bulletin* 14: 261–74.

Dongol, G.M.S. (1985) 'Geology of the Kathmandu fluviatile lacustrine sediments in the light of new vertebrate fossil occurrences', *Journal of the Nepal Geological Society* 3: 43–57.

Dongol, G.M.S. (1987) 'The stratigraphic significance of vertebrate fossils from the Quaternary deposits of the Kathmandu Basin, Nepal', *Newsletters in Stratigraphy (Nepal)* 18: 21–9.

Drew, F. (1873) 'Alluvial and lacustrine deposits and glacial records of the Upper-Indus Basin', *Quarterly Journal of the Geological Society of London* 29: 441–71.

Drew, F. (1875) 'The Jumoo and Kashmir territories. A geographical account', *Quarterly Journal of the Geological Society of London* (republished 1980 by Indus Publications, Karachi).

Duplessy, J.C. (1982) 'Glacial to interglacial contrasts in the northern Indian Ocean', *Nature* 295: 494–8.

Ebblin, C. (1976) 'Tectonic lineaments in Karakorum, Pamir, and Hindu Kush from ERTS imageries', *Accademia Nazionale dei Lincei, Rendiconti* 60: 245–53.

Edwards, A.B. (1941) 'Storm-wave platforms', *Journal of Geomorphology* 2: 233–6.

Ehleringer, J.R. (1978) 'Implications of quantum yield differences on the distributions of C4 and C3 grasses', *Oecologia* 31: 255–67.

Eisbacher, G.H. and Clague, J.J. (1984) *Destructive Mass Movements in High Mountains: hazard and management*, Geological Survey of Canada, 84–16.

Emery, A.B. (1967) 'Some stages in the development of knowledge of the Atlantic coast of the United States', *The Quaternary History of Ocean Basins, Progress in Oceanography* 4, New York: Pergamon, 307–32.

England, P. and Molnar, P. (1990) 'Surface uplift, uplift of rocks, and exhumation of rocks', *Geology* 18: 1173–7.

Eyles, N., Eyles, C.H. and Miall, A.D. (1983) 'Lithofacies types and vertical profile models: an alternative approach to the description and environmental interpretation of glacial diamict and diamictite sequences', *Sedimentology* 30: 393–410.

Fairservis, W.A., jun. (1967) 'The origin, character, and decline of an early civilization', *American Museum Novitates* 2302, American Museum of Natural History, New York.

Falcon, N.L. (1947) 'Raised beaches and terraces of the Iranian Makran coast', *Geographical Journal* 109–10: 149–51.

Farah, A. and DeJong, K.A. (eds) (1979) *Geodynamics of Pakistan*, Quetta: Geological Survey of Pakistan.

Farah, A., Abbas, G., DeJong, K.A. and Lawrence, R.D. (1984a) 'The evolution of lithosphere in Pakistan', *Tectonophysics* 105: 207–27.

Farah, A., Lawrence, R.D. and DeJong, K.A. (1984b) 'An overview of the tectonics of Pakistan', in B.U. Haq and J.D. Milliman (eds) *Marine Geology and Oceanography of Arabian Sea and Coastal Pakistan*, New York: Van Nostrand Reinhold, 161–76.

Farhoudi, G. and Karig, D.E. (1977) 'Makran of Iran and Pakistan as an active arc system', *Geology* 5: 664–8.

Farrell, K.M. (1987) 'Sedimentology and facies architecture of overbank deposits of the Mississippi River, False River region, Louisiana', in F.G. Ethridge, R.M. Flores and M.D. Harvey (eds) *Recent Developments in Fluvial Sedimentology*, Tulsa, OK: Society of Economic Paleontologists and Mineralogists, Special Publication 39: 111–20.

Fatmi, A.N. (1974) 'Lithostratigraphic units of the Kohat–Potwar Province, Indus Basin, Pakistan', *Memoirs of the Geological Survey of Pakistan* 10: 1–80.

Fedden, F. (1884) 'The geology of the Kathiawar peninsula in Guzerat', *Memoirs of the Geological Survey of India* 21: 73–136.

Ferguson, R.I. (1984) 'Sediment load of the Hunza River', in K.J. Miller (ed.) *The International Karakoram Project* 2, Cambridge: Cambridge University Press, 581–98.

Ferguson, R.I., Collins, D.N. and Whalley, W.B. (1984) 'Techniques for investigating meltwater runoff and erosion', in K.J. Miller (ed.) *The International Karakoram Project* 1, Cambridge: Cambridge University Press, 374–82.

Ferguson, R.T. (1984) 'The threshold between meandering and braiding', in L.V.H. Smith (ed.) *Channels and Channel Control Structures*, Berlin: Springer Verlag, 6–29.

Fillon, R.H. and Williams, D.F. (1983) 'Glacial evolution of the Plio–Pleistocene: role of continental and Arctic Ocean ice sheets', *Palaeogeography, Palaeoclimatology, Palaeoecology* 42: 7–33 (Amsterdam: Elsevier).

Finko, Ye.A. and Enman, V.B. (1971) 'Present surface movements in the Surkhob fault zone', *Geotectonics* 5: 330–4.

Finsterwalder, R. (1935) 'The scientific work of the German Himalayan expedition to Nanga Parbat, 1934', *Himalayan Journal* 7: 44–52.

Finsterwalder, R. (1937) 'Die Gletscher des Nanga Parbat: Glaziologische Arbeiten der Deutschen Himalaya-Expedition 1934 und ihre Ergebnisse', *Zeitschrift für Gletscherkunde* 25: 57–108.

Finsterwalder, R. (1950) 'Some comments on glacier flow', *Journal of Glaciology* 1: 383–8.

Fisher, R.S. (1953) 'Dispersion on a sphere', *Quarterly Journal of the Geological Society of London*, A217: 295–303.

Fisk, H.N. (1947) *Fine Grained Alluvial Deposits and their Effects on Mississippi River Activity*, Vicksburg, MS: United States Army Engineer Waterways Experiment Station.

Flam, Louis (1976) 'Settlement, subsistence and population: a dynamic approach to the development of the Indus valley civilization', in K.A.R. Kennedy and G.L. Possehl (eds), *Ecological Backgrounds of South Asian Prehistory*, Ithaca, NY: Cornell University Press, 76–93.

Flam, Louis (1981a) 'Toward an ecological analysis of prehistoric settlement patterns in Sind, Pakistan', *Man and Environment* 5: 52–8.

Flam, Louis (1981b) 'The paleogeography and prehistoric settlement patterns in Sind, Pakistan (*c.*4000–2000 BC)', Unpublished Ph.D. dissertation, University of Pennsylvania, Philadelphia, PA.

Flam, Louis (1986a) 'Recent explorations in Sind: paleogeography, regional ecology, and prehistoric settlement patterns', in J. Jacobson (ed.) *Studies in the Archaeology of India and Pakistan*, New Delhi: Oxford & IBH Publishing, 65–89.

Flam, Louis (1986b) 'The Indus River and the Arab period in Sind', *Sindological Studies*, 5–14.

Flint, R.F. (1971) *Glacial and Quaternary Geology*, New York: Wiley.

Food and Agricultural Organization (1971) *Soil Resources of West Pakistan and Their Development Possibilities*, Soil Survey Project, Pakistan: United Nations Development Programme.

Fort, M. (1982) 'Apport de la télédetection à la connaissance des formations superficielles et des structures dans le bassin de Leh (Vallée de l'Indus, Himalaya du Ladakh)', *Bulletin Société Géologique de France* 7(4): 97–104.

Fort, M.B. (1983) 'Geomorphological observations in the Ladakh area (Himalayas): Quaternary evolution and present dynamics', in V.J. Gupta (ed.) *Stratigraphy and Structure of Kashmir and Ladakh Himalaya*, Delhi: Hindustan Publishing, 39–58.

Fort, M. (1987) 'Sporadic morphogenesis in a continental subduction setting: An example from the Annapurna Range, Nepal, Himalaya', *Zeitschrift für Geomorphologie, Neue Folge, Suppelmentband* 63: 9–36.

Fort, M., Freytet, P. and Colchen, M. (1982a) 'Structural and sedimentological evolution of the Thakkhola–Mustang graben (Nepal Himalayas)', *Zeitschrift für Geomorphologie, Neue Folge, Supplementband* 42: 75–98.

Fort, M., Freytet, P. and Colchen, M. (1982b) 'The structural and sedimentological evolution of the Thakkhola–Mustang graben (Nepal Himalaya) in relation to the uplift of the Himalayan range', *Proceedings of Symposium on Qinghai–Xizang (Tibet) Plateau* 1, Beijing: Science Press, 1981; New York: Gordon & Breach, 307–13.

Fraser, I.S. (1958) *Report on a Reconnaissance Survey of the Landforms, Soils and Present Land Use of the Indus Plains, West Pakistan*, Toronto: Colombo Plan Cooperative Project.

Galloway, W.E. and Hobday, D.K. (1983) *Terrigenous Clastic Depositional Systems: Applications to Petroleum, Coal and Uranium Exploration*, New York: Springer Verlag.

Gansser, A. (1964) *Geology of the Himalayas*, London: Wiley Interscience Publications.

Gansser, A. (1980) 'The significance of the Himalayan suture zone', *Tectonophysics* 62: 37–52.

Gansser, A. (1983) 'The morphogenetic phase of mountain building', in K.J. Hsu (ed.) *Mountain Building Processes*, London: Academic Press, 221–8.

Gardner, J.S. (1986) 'Recent fluctuations of Rakhiot Glacier, Nanga Parbat, Punjab Himalaya, Pakistan', *Journal of Glaciology* 32(112): 527–9.

Gardner, J.S. and Hewitt, K. (1990) 'A surge of Bualtar Glacier, Karakoram Range, Pakistan: a possible landslide trigger', *Journal of Glaciology* 36(123): 159–62.

Gardner, T.W., Jorgensen, D.W., Shurman, C. and Leimieux, C.R. (1987) 'Geomorphic and tectonic process rates: effects of measured time interval', *Geology* 15: 259–61.

Germanoski, D. (1989) 'The effect of sediment load and gradient on braided river morphology', Unpublished Ph.D. thesis, Colorado State University, Fort Collins, CO.

Ghose, B., Kar, A. and Husain, Z. (1979) 'The lost courses of the Sarawati River in the Great Indian desert: new evidence from Landsat imagery', *Geographical Journal* 145: 446–51.

Giardino, J.R., Shroder, J.F. jr. and Vitek, J.D. (eds) (1987) *Rock glaciers*, London: Allen and Unwin.

Giardino, J.R., Vitek, J.D., Johnson, P.G., Marston, R.A. and Shroder, J.F. jr. (1988) 'Development of an alpine landform continuum model', Geological Society of America, *Abstracts with Programs* 20: A284.

Gibbons, A.B., Megeath, J.D. and Pierce, K.L. (1984) 'Probability of moraine survival in a succession of glacial advances', *Geology* 12: 327–30.

Gibbs, R.J. (1981) 'Sites of river-derived sedimentation in the ocean', *Geology* 9: 77–80.

Gilbert, R.O. (1987) *Statistical Methods for Environmental Pollution Monitoring*, New York: Van Nostrand Reinhold.

Gill, W.D. (1952) 'The stratigraphy of the Siwalik Series in the northern Potwar, Punjab,

Pakistan', *Quarterly Journal of the Geological Society of London* 107(4): 375–97.

Giu Dongzhou (1987) 'Sedimentary models of gypsum-bearing clastic rocks and prospects for associated hydrocarbons west of the Tarim basin (China) in Miocene', in T.M. Peryt (ed.) *Evaporite Basins*, New York: Springer Verlag, 123–32.

Glennie, E.A. (1956) 'Gravity data and crustal warping in northwest Pakistan and adjacent parts of India', Royal Astronomical Society Monthly Notices, Geophysics Supplement 7(4): 162–75.

Glikman, L.S. and Ishckenko, V.V. (1967) 'Marine Miocene sediments in central Asia', *Doklady Akademii Nauk SSSR* 117: 78–81.

Godwin-Austen, H. (1864) 'On the glaciers of the Mustagh Range', *Journal of the Royal Geographical Society of London* 34: 19–56.

Goldsmid, F.G. (1863) 'Diary of proceedings of the mission into Makran for political and survey purposes from the 12th to the 19th of December, 1861', *Journal of the Royal Geographical Society of London* 33: 181–213.

Gole, C.V. and Chitale, S.V. (1966) 'Inland delta building activity of Kosi River', *Proceedings of the American Society of Civil Engineers, Journal of the Hydraulics Division* HY-2: 111–26.

Gornitz, V. and Seeber, L. (1981) 'Morphotectonic analysis of the Hazara arc region of the Himalayas, north Pakistan and northwest India', *Tectonophysics* 74: 263–82.

Goudie, A., Cooke, R. and Evans, I. (1970) 'Experimental investigation of rock weathering by salts', *Area*, 42–8.

Goudie, A.S. (1984) 'Salt efflorescences and salt weathering in the Hunza Valley, Karakoram Mountains, Pakistan', in K.J. Miller (ed.) *The International Karakoram Project* 2, Cambridge: Cambridge University Press, 607–15.

Goudie, A.S., Brunsden, D., Collins, D.N., Derbyshire, E., Ferguson, R.I., Jones, D.K.C., Perrott, F.A., Said, M., Waters, R.S. and Whalley, W.B. (1984a) 'The geomorphology of the Hunza Valley, Karakoram Mountains, Pakistan', in K.J. Miller (ed.) *The International Karakoram Project* 2, Cambridge: Cambridge University Press, 359–410.

Goudie, A.S., Jones, D.K.C. and Brunsden, D. (1984b) 'Recent fluctuations in some glaciers of the western Karakoram Mountains, Pakistan', in K.J. Miller (ed.) *The International Karakoram Project* 2, Cambridge: Cambridge University Press, 411–55.

Grinlinton, J.L. (1928) 'The former glaciation of the east Lidar Valley, Kashmir', *Memoirs of the Geological Survey of India* 49(2): 289–381.

Guo Shuang-Xing (1982) 'On the elevation and climatic changes of the Qinghai–Xizang plateau based on fossil angiosperms', *Proceedings of Symposium on Qinghai–Xizang (Tibet) Plateau* 1, Beijing: Science Press, 1981; New York: Gordon & Breach, 201–6.

Haghipour, A., Ghorashi, M. and Kadjar, M.H. (1984) *Seismotectonic Map of Iran, Afghanistan, and Pakistan*, Tehran: Geological Survey of Iran.

Haneef, M., Jan, M.Q. and Rabbi, F. (1986) 'Fracture fills of "sedimentary dykes" in the lake sediments of Jalala N.W.F.P.: a preliminary report', *Geological Bulletin of the University of Peshawar, Pakistan* 19: 151–6 (Amsterdam: Elsevier).

Haq, B.U. and Milliman, J.D. (eds) (1984) *Marine Geology and Oceanography of Arabian Sea and Coastal Pakistan*, New York: Van Nostrand Reinhold.

Harms, J.C., Southard, J.B., Spearing, D.R. and Walker, R.G. (1975) *Depositional Environments as interpreted from Primary Sedimentary Structures and Stratification Sequences*, Tulsa, OK: Society of Economic Paleontologists and Mineralogists, SEPM Short Course 2.

Harms, J.C., Cappel, H.N. and Francis, D.C. (1984) 'The Makran Coast of Pakistan: its stratigraphy and hydrocarbon potential', in B.U. Haq and J.D. Milliman (eds) *Marine Geology and Oceanography of Arabian Sea and Coastal Pakistan*, New York: Van Nostrand Reinhold, 3–26.

Harrison, J.V. (1941) 'Coastal Makran', *Geographical Journal* XCVII: 1–17.

Harrison, J.V. (1944) 'Mud volcanoes on the Makran coast', *Geographical Journal* 103: 180–1.

Harrison, T.M., Copeland, P., Kidd, W.S. and An Yin (1992) 'Raising Tibet', *Science* 255: 1663–70.

Harvey, A. (1982) 'The role of piping in the development of badlands and gully systems in southeast Spain', in R. Bryan and A. Yair (eds) *Badland Geomorphology and Piping*, Norwich: Geobooks, 317–35.

Harvey, M.D. (1989) 'Meanderbelt dynamics of Sacramento River, California', *Proceedings of the California Riparian Systems Conference, General Technical Report, PSW 110*, Washington, DC: US Department of Agriculture, Forest Service, 54–9.

Harvey, M.D., Pranger, H.H. II, Biedenharn, D.S. and Combs, P. (1988) 'Morphologic and hydraulic adjustments of Red River from Shreveport, LA to Fulton, AK, between 1886 and 1980', in S.R. Abt and J. Gessler (eds), *1988 National Conference at Williamsburg, VA*, New York: American Society of Civil Engineers, Hydraulics Division, 120–5.

Haserodt, K. (1984) 'Abflussverhalten der Flusse mit Bezugen zur Sonnenscheindauer und zum Niederschlag zwischen Hindukusch (Chitral) und Hunza—Karakorum (Gilgit, Nordpakistan)', *Mitteilungen Geographischen Gesellschaft in Munchen*, 129–61.

Hashimi, N.H. and Nair, R.R. (1986) 'Climatic aridity over India 11,000 years ago: evidence from feldspar distribution in shelf sediments', *Palaeogreography, Palaeoclimatology, Palaeoecology* 53: 309–19 (Amsterdam: Elsevier).

Heller, F. and Lui Tungsheng (1984) 'Magnetism of Chinese loess deposits', *Geophysical Journal of the Royal Astronomical Society* 77: 125–41.

Hewitt, K. (1961) 'Karakoram glaciers and the Indus', *Indus* (Lahore) 2: 4–14.

Hewitt, K. (1964) 'A Karakoram ice dam', *Indus* (Lahore) 5: 18–30.

Hewitt, K. (1967) 'Ice-front sedimentation and the seasonal effect: a Himalayan example', *Transactions of the Institute of British Geographers* 42: 93–106.

Hewitt, K. (1968a) 'Studies of the geomorphology of the Mountains Regions of the Upper Indus Basin', Unpublished Ph.D. thesis, 2 vols, University of London.

Hewitt, K. (1968b) 'The freeze–thaw environment of the Karakoram Himalaya', *Canadian Geographer* 12: 85–98.

Hewitt, K. (1969) 'Glacier surges in the Karakoram Himalaya (Central Asia)', *Canadian Journal of Earth Sciences* 6: 1009–18.

Hewitt, K. (1982) 'Natural dams and outburst floods of the Karakoram Himalaya', in J. Glen (ed.) *Hydrological Aspects of Alpine and High Mountain Areas, International Hydrological Association (I.A.H.S.)*, 138: 259–69.

Hewitt, K. (1985) 'Snow and ice hydrology in remote, high mountain regions: the Himalayan sources of the River Indus', *Snow and Ice Hydrology Project*, Working Paper 1, Waterloo, Ontario: Wilfrid Laurier University.

Hewitt, K. (ed.) (1986) *Annual Report 1985: Snow and Ice Hydrology Project (Upper Indus Basin)*, Waterloo, Ontario: Wilfrid Laurier University.

Hewitt, K. (1987) *Snow and Ice Hydrology Project (Upper Indus Basin): Annual Report*, Waterloo, Ontario: Wilfrid Laurier University.

Hewitt, K. (1988) 'Catastrophic landslide deposits in the Karakoram Himalaya', *Science* 242: 64–77.

Hewitt, K. (1989a) 'European science in high Asia: geomorphology in the Karakoram Himalaya to 1939', in E.J. Tinkler (ed.) *History of Geomorphology: Hutton to Hack*, Binghamton Geomorphology Series, Boston: Unwin Hyman, 19: 165–203.

Hewitt, K. (1989b) 'The altitudinal organization of Karakoram geomorphic processes and depositional environments', *Zeitschrift für Geomorphologie, Supplementband* 76: 9–32.

Higgins, G.M., Ahmad, M. and Brinkman, R. (1973) 'The Thal interfluve, Pakistan geomorphology and history', *Geologie en Mijnbouw* 52(3): 147–55.

Higgins, G.M., Baig, S. and Brinkman, R. (1974) 'The sands of Thal: wind regimes and sand ridge formations, *Zeitschrift für Geomorphologie, Neue Folge* 18(3): 272–90.

Higgins, S.M. (1986) 'A SEM investigation of quartz grain surface textures: Evidence for the origin of Jalipur and Bain diamictites', Unpublished B.S. thesis, Department of Geography, University of Nebraska at Omaha.

Hobbs, B.E., Means, W.D. and Williams, P.F. (1976) *An Outline of Structural Geology*, New York: Wiley.

Holmes, D.A. (1968) 'The recent history of the Indus', *Geographical Journal* 134: 367–81.

Holmes, J.A. (1988) 'Pliocene and Quaternary environmental change in Kashmir, northwest Himalayas', Unpublished Ph.D. thesis, University of Oxford.

Honegger, K., Cietrich, V., Frank, W., Gansser, A., Thoni, M. and Trommsdorff, V. (1982) 'Magmatism and metamorphism in the Ladakh Himalayas (the Indus–Tsangpo suture zone)', *Earth and Planetary Science Letters* 60: 253–92 (Amsterdam: Elsevier).

Hoyt, T.H., Henry, V.J. and Weimer, R.T. (1968) 'Age of Late Pleistocene shoreline deposits, coastal Georgia', *Means of Correlation of Quaternary Successions*, International Association for Quaternary Research, 7th congress, Salt Lake City, Utah, 381–93.

Hsu, K.J. (1975) 'On sturzstroms–catastrophic debris streams generated by rockfalls', *Bulletin of the Geological Society of America* 86: 129–40.

Hsu, K.J. (1988) 'Relict back-arc basins: principles of recognition and possible new examples from China', in K.L. Kleinspehn and C. Paola (eds), *New Perspectives in Basin Analysis*, New York: Springer Verlag, 245–63.

Hunting Survey Corporation (1960) *A Reconnaissance Geology of Part of West Pakistan*, Ottawa: Maracle Press.

Hussain, S.T., Munthe, J., West, R.M. and Lucace, J.R. (1977) 'The Daud Khel local fauna: a Neogene small-mammal assemblage from the Trans-Indus Siwaliks, Pakistan', Milwaukee Public Museum Special Publications in Biology and Geology, 16: 1–16.

Inglis, C.C. (1949) 'The behaviour and control of rivers and canals (with the aid of models)', *Poona Research Publication* 13, Central Water Power, Irrigation and Navigation Research Station, Govt of India.

Irrigation and Power Department (1978) *Bund Manual*, Government of Sind, Pakistan.

Isachenko, A.G. (1965) *Principles of Landscape Science and Physico–Geographical Zonation* (trans. R.J. Zatorski, 1975), Melbourne: Melbourne University Press.

Jackson, M.L. (1979) *Soil Chemical Analysis Advanced Course*, 2nd edn, published by the Author, Madison, Wisconsin 53705.

Jacob, K.H. and Quittmeyer, R.C. (1979) 'The Makran region of Pakistan and Iran: trench-arc system with active plate subduction', in A. Farah and K.A. DeJong (eds), *Geodynamics of Pakistan*, Quetta: Geological Survey of Pakistan, 305–18.

Jalal-ud-din, Ch., Brinkman, R. and Rafiq, Ch.M. (1970a) 'Landforms of the Indus delta', *Pakistan Geographical Review* 25: 12–22.

Jalal-ud-din, Ch., Brinkman, R. and Rafiq, Ch.M. (1970b) 'Soils of the Indus delta: their nature, genesis and classification', *Pakistan Geographical Review* 25: 71–85.

Jan, M.Q., Khattack, M.U.K., Parvez, M.K. and Windley, B.F. (1984) 'The Chilas stratiform complex: field and mineralogical aspects', *Geological Bulletin of the University of Peshawar, Pakistan* 17: 153–69 (Amsterdam: Elsevier).

Jenny, H. (1941) *Factors in Soil Formation*, New York: McGraw-Hill.

Johnson, G.D., Johnson, N.M., Opdyke, D.W. and Tahirkheli, R.A.K. (1979) 'Magnetic reversal stratigraphy and sedimentary tectonic history of the Upper Siwalik group, eastern Salt Range and southwestern Kashmir', in A. Farah and K.A. DeJong (eds)

*Geodynamics of Pakistan*, Quetta: Geological Survey of Pakistan, 149–65.

Johnson, G.D., Rey, P.H., Ardrey, R.H., Visser, C.F., Opdyke, N.D. and Tahirkheli, R.A.K. (1981) 'Paleoenvironments of the Siwalik Group, Pakistan and India', in G. Rapp, jun. and C.F. Vondra (eds) *Hominid Sites: Their Geologic Settings*, American Association for the Advancement of Science, Selected Symposium, Washington, DC, 63: 197–254.

Johnson, G.D., Zeitler, P., Naeser, C.W., Johnson, N.M., Summers, D.M., Frost, C.D., Opdyke, N.D. and Tahirkheli, R.A.K. (1982) 'The occurrence and fission-track ages of Late Neogene and Quaternary volcanic sediments, Siwalik group, northern Pakistan', *Palaeogeography, Palaeoclimatology, Palaeoecology* 37: 63–93 (Amsterdam: Elsevier).

Johnson, G.D., Raynolds, R.G.A. and Burbank, D.W. (1986) 'Late Cenozoic tectonics and sedimentation in the northwestern Himalayan foredeep, 1: Thrust ramping and associated deformation in the Potwar Region', in P.A. Allen and P. Homewood (eds) *Foreland Basins*, International Association of Sedimentologists, Special Publication 8: 273–91.

Johnson, N.M. and McGee, V.E. (1983) 'Magnetic polarity stratigraphy: stochastic properties of data, sampling problems and the evaluations of interpretations', *Journal of Geophysical Research* 88: 1213–21.

Johnson, N.M., Opdyke, N.D. and Lindsay, E.H. (1975) 'Magnetic polarity stratigraphy of Pliocene/Pleistocene terrestrial deposits and vertebrate faunas, San Pedro Valley, Arizona', *Bulletin of the Geological Society of America* 86: 5–12.

Johnson, N.M., Opdyke, N.D., Johnson, G.D., Lindsay, E.M. and Tahirkheli, R.A.K. (1982) 'Magnetic polarity stratigraphy and ages of Siwalik Group rocks of the Potwar Plateau, Pakistan', *Palaeogeography, Palaeoclimatology, Palaeoecology* 37: 17–42 (Amsterdam: Elsevier).

Johnson, W.P. (1986) 'The physical and magnetic polarity stratigraphy of the Bunthang sequence, Skardu Basin, northern Pakistan', Unpublished M.S. thesis, Dartmouth College, Hanover, NH.

Jorgensen, D.W. (1989) 'Channel migration in response to tectonically influenced sediment transport and deposition' (Abs.), *Transactions of the American Geophysical Union (EOS)* 70: 333.

Kamb, B., Raymond, C.F., Harrison, W.D., Englehardt, H., Echelmeyer, K.A., Humphrey, N., Brugman, M.A. and Pfeffer, T. (1985) 'Glacier surge mechanism: 1982–1983 surge of variegated glacier, Alaska', *Science* 227: 469–79.

Kar, Amal (1990) 'Megabarchanoids of the Thar: their environment, morphology and relationship with longitudinal dunes', *Geographical Journal* 156(1): 51–61.

Kazmi, A.H. (1979) 'Active faults in Pakistan', in A. Farah and K.A. DeJong (eds) *Geodynamics of Pakistan*, Quetta: Geological Survey of Pakistan, 258–94.

Kazmi, A.H. (1984) 'Geology of the Indus Delta', in B.U. Haq and J.D. Milliman (eds) *Marine Geology and Oceanography of Arabian Sea and Coastal Pakistan*, New York: Van Nostrand Reinhold, 71–84.

Kazmi, A.H. and Rana, R.A. (1982) *Tectonic Map of Pakistan (Scale 1:2 million)*, Quetta: Geological Survey of Pakistan.

Kazmi, A.K., Lawrence, R.D., Anwar, J., Snee, L.W. and Hussain, S.S. (1984) 'Geology of the Indus suture zone in Mingora–Shangla area of Swat', *Geological Bulletin of the University of Peshawar, Pakistan*, 17: 127–44 (Amsterdam: Elsevier).

Kaz'min, V.G. and Faradzhev, V.A. (1963) 'Tectonic development of the Yarkand sector of the Kun Lun Shan', *International Geological Review* 5: 180–8.

Keller, H.M., Tahirkheli, R.A.K., Mirza, M.M., Johnson, G.D., Johnson, N.M. and Opdyke, N.D. (1977) 'Magnetic polarity stratigraphy of the upper Siwalik deposits, Pabbi Hills, Pakistan', *Earth and Planetary Science Letters* 36: 187–201 (Amsterdam: Elsevier).

Kent, D.V. and Gradstein, F.M. (1986) 'A Jurassic to Recent chronology', in P.R. Vogt and B.E. Tucholke (eds) *The Geology of North America*, vol. M: *The western North Atlantic Region*, Geological Society of America, 45–50.

Kesel, R.H., Danne, K.C., McDonald, R.C. and Allison, K.R. (1974) 'Lateral erosion and overbank deposition on the Mississippi River in Louisiana caused by 1973 flooding', *Geology* 28: 461–4.

Khan, A.R. (1968) 'Ancient settlements in the Karachi region', *Dawn*, Sunday Magazine Section, Karachi, 21/28 July.

Khan, A.R. (1973) *New Archaeological Sites in Las Bela*, Department of Geography, Karachi University, Karachi, 1–17.

Khan, F.A. (1965) 'Excavations at Kot Diji', *Pakistan Archaeology* 2: 11–85.

Khan, M.A., Ahmed, R., Raza, H.A. and Kemal, A. (1986) 'Geology of petroleum in Kohat–Potwar depression, Pakistan', *Bulletin of the American Association of Petroleum Geologists* 70: 396–414.

Khan, M.J. (1983) 'Magnetostratigraphy of Neogene and Quaternary Siwalik Group Sediments of the Trans-Indus Salt Range, Northwestern Pakistan', Unpublished Ph.D. dissertation, Columbia University, New York.

Khan, M.J., Opdyke, N.D. and Shroder, J.F. jun. (1985) 'Bain diamictite: lithology, age and the origin', *Geological Bulletin of the University of Peshawar, Pakistan* 18: 53–64 (Amsterdam: Elsevier).

Khan, M.J., Opdyke, N.D. and Tahirkheli, R.A.K. (1988) 'Magnetic polarity stratigraphy of the Siwalik Group Bhittani, Marwat, and Khasor Ranges, northwestern Pakistan', *Journal of Geophysical Research* 93: 11,733–90.

Khosla, A.N. (1953) *Silting of Reservoirs*, Central Board of Irrigation and Power (India) Publication 51.

Khuhro, H. (1979) 'Studies in geomorphology and prehistory of Sind', *Grassroots* 3(2): 112.

Kick, W. (1964) 'Der Chogo Lungma gletscher im Karakoram', *Zeitschrift für Gletscherkunde und Glazialgeologie* 5: 19–24.

Kick, W. (1975) 'Application of geology, photogrammetry, history and geography to the study of long term mass balances of central Asiatic glaciers', *Union de Géodésie et Géophysics*, International Association of Hydrological Sciences, Publication 104: 150–60.

Kick, W. (1980) 'Material for a glacier inventory of the Indus drainage basin – the Nanga Parbat massif', *World Glacier Inventory, Proceedings of the Riederalp Workshop*, International Association of Hydrological Sciences, Publication 126: 105–9.

Kick, W. (1986) 'Glacier mapping for an inventory of the Indus drainage Basin: current state and future possibilities', *Annals of Glaciology* 8: 102–5.

King, C.A.M. (1957) *Beaches and Coasts*, London: Edward Arnold.

King, G.C.P. and Vita-Finzi, C. (1980) 'Active folding in the Algerian earthquake of 10 October, 1980', *Nature* 292: 22–6.

Klimek, K. and Starkel, L. (eds) (1984) *Vertical Zonality in the Southern Khangai Mountains (Mongolia): result of the Polish–Mongolian Physico-Geographical Expedition*. Prace Geograficzne 136, Wroclaw: PAN Instytut Geografii.

Kowalkowski, A. and Starkel, L. (1984) 'Altitudinal belts of geomorphic processes in the southern Khangai Mountains (Mongolia)', *Studia Geomorphologica Carpatha–Balcanica (Krakow)* 17: 95–116.

Krestinov, V.N. (1963) 'History of the geological development of the Pamirs and adjacent regions of Asia in the Mesozoic–Cenozoic (Upper Cretaceous–Quaternary)', *International Geological Review* 5: 38–62.

Kureshy, K.U. (1977) *A Geography of Pakistan*, Oxford: Oxford University Press.

Lakhanpal, R.N., Sah, S.C.D., Sharma, K.K. and Guleria, J.S. (1983) 'Occurrence of

Livistona in the Hemis conglomerate horizon of Ladakh', in V.C. Thakur and K.K. Sharma (eds) *Geology of the Indus Suture Zone of Ladakh*, Dehra Dun, India: Wadia Institute of Himalayan Geology, 179–85.

Lambrick, H.T. (1967) 'The Indus floodplain and the "Indus" Civilization', *Geographical Journal* 133: 483–95.

Lambrick, H.T. (1971) 'Stratigraphy at Moenjo-Daro', *Journal of the Oriental Institute* 20(4): 363–9.

Lambrick, H.T. (1975) *Sind: a Generation Introduction*, Hyderabad: Sindhi Adabi Board.

Lane, E.W. (1957) *A Study of the Shape of Channels Formed by Natural Streams Flowing in Erodible Materials*, Omaha, NE: United States Army Engineer, Missouri River Division, MRD Sediment Series 9.

Lawrence, R.D. and Ghauri, A.A.K. (1983) 'Evidence of active faulting in Chilas district', *Geological Bulletin of the University of Peshawar, Pakistan* 16: 185–6 (Amsterdam: Elsevier).

Lawrence, R.D. and Shroder, J.F., jun. (1984) 'Active fault northwest of Nanga Parbat', *First Pakistan Geological Congress Volume of Abstracts*, Lahore: Punjab University, 50–1.

Lawrence, R.D. and Shroder, J.F., jun. (1985) 'Tectonic geomorphology between Thakot and Mansehra, northern Pakistan', *Geological Bulletin of the University of Peshawar, Pakistan* 18: 153–61 (Amsterdam: Elsevier).

Leeder, M.R. (1978) 'A quantitative stratigraphic model for alluvium, with special reference to channel deposit density and interconnectedness', in A.D. Miall (ed.) *Fluvial Sedimentology*, Calgary, Alberta: Canadian Society of Petroleum Geologists, Memoir 5: 587–98.

Lees, G.M. (1928) 'The geology and tectonics of Oman', *Quarterly Journal of the Geological Society of London* 84: 585–670.

Le Fort, P. (1975) 'Himalayas: the collided range: present knowledge of the continental arc', *American Journal of Science* 275-A: 1–44.

Le Fort, P. (1986) 'Metamorphism and magmatism during the Himalayan collision', in M.P. Coward and A.C. Ries (eds) *Collision Tectonics*, Geological Society of London, Special Publication 19: 159–72.

Leggett, J.K. and Platt, J. (1984) 'Structural features of the Makran fore-arc on Landsat imagery', in B.U. Haq and J.D. Milliman (eds) *Marine Geology and Oceanography of Arabian Sea and Coastal Pakistan*, New York: Van Nostrand Reinhold, 33–43.

Leith, W. (1985) 'A mid-Mesozoic extension across central Asia', *Nature* 313: 567–70.

Leopold, L.B. and Wolman, M.G. (1957) 'River channel patterns – braided, meandering and straight', *Physiographic and Hydraulic Studies of Rivers*, United States Geological Survey, Professional Paper 282-B: 39–85.

Li Jijun, Wen Shixuan, Zhong Qingsong, Wong Fubao, Zheng Benxing, and Li Bingyuan (1979) 'A discussion on the period, amplitude and type of the uplift of the Qinghai–Xizang Plateau', *Scientia Sinica* 22, 1314–27.

Li Jijun, Derbyshire, E. and Shuyung, Xu (1984) 'Glacial and paraglacial sediments of the Hunza Valley, Karakoram Pakistan: a preliminary analysis', in K.J. Miller (ed.) *The Internatinoal Karakoram Project* 2, Cambridge: Cambridge University Press, 496–535.

Little, R.D. (1972) 'Terraces of the Makran Coast of Iran and parts of west Pakistan', Unpublished M.A. thesis, University of Southern California, Los Angeles, CA.

Lower Indus Project (1965a) *Lower Indus Report: Physical Resources*, supplemental reports: *1 Climate, 2 Geomorphology, Soils and Watertable*, Karachi: Feroz Sons.

Lower Indus Project (1965b) *Lower Indus Report: Physical Resources*, supplemental reports: *3 River Indus, 4 Torrents, 5 Surface Water Storage*, Karachi: Feroz Sons.

Lower Indus Project (1966) *Lower Indus Report: Main Report*, 2 vols, London: Lion House.

Loziyev, V.P. (1976) 'Present structure and types of local deformations in the south Tadzhik depression', *Geotectonics* 10: 291–6.

Lyon-Caen, H. and Molnar, P. (1983) 'Constraints on the structure of the Himalaya from an analysis of gravity anomalies and a flexural model of the lithosphere', *Journal of Geophysical Research* 88: 8171–91.

McDougall, J.W. (1987) 'Tectonic map and interpretation of Kalabagh tear fault, Himalayan foreland fold–thrust belt, western Salt Range area, Pakistan', Geological Society of America, *Abstracts with Programs* 19: 765.

McDougall, J.W. (1989) 'Tectonically-induced diversion of the Indus River west of the Salt Range, Pakistan fault, Himalayan foreland fold–thrust belt, western salt Range area, Pakistan', *Palaeogeography, Palaeoclimatology, Palaeoecology* 71: 301–7 (Amsterdam: Elsevier).

Mackay, E.J.H. (1938) *Further Excavations at Mohenjo-Daro*, Delhi: Manager of Publications, Government of India.

Mackay, E.J.H. (1943) *Moenjo-Daro Excavations: 1935–1936*, New Haven, Conn.: American Oriental Society.

Mackelein, W. (1959) *Forschungen in der Zentralen Sahara*, Braunschweig: Westermann.

Macklin, M.G. and Lewin, J. (1989) 'Sediment transfer and transformation of an alluvial valley floor: the River South Tyne, Northumbria, UK', *Earth Surface Processes and Landforms* 14: 233–46.

Madin, I. (1986) 'Structure and neotectonics of the northwestern Nanga Parbat–Haramosh Massif', Unpublished M.S. thesis, Oregon State University, Corvallis, Oregon.

Madin, I.P., Lawrence, R.D. and Rehman, S.U. (1989) 'The northwestern Nanga Parbat Haramosh Massif: evidence for crustal uplift at the northwestern corner of the Indian Craton', in L.L. Malinconico, jun. and R.J. Lillie (eds) *Tectonics and Geophysics of the Western Himalaya*, Geological Society of America, Special Paper 232: 169–82.

Maluski, H. and Matte, P. (1983) 'Ages of alpine tectonometamorphic events in the northwestern Himalaya (northern Pakistan) by $^{39}Ar/^{40}Ar$ method', *Tectonics* 3: 1–18.

Maluski, H., Matte, P. and Brunel, M. (1988) 'Argon 39–Argon 40 dating of metamorphic and plutonic events in the north and high Himalaya belts (Southern Tibet-China)', *Tectonics* 7: 299–326.

Manabe, S. and Hahn, D.G. (1977) 'Simulation of the tropical climate of an ice age', *Journal of Geophysical Research* 82: 3889–911.

Mani, M.S. (1974) 'Biogeography of the Himalaya', in M.S. Mani (ed.) *Ecology and Biogeography in India*, The Hague: W. Junk, 664–81.

Mankinen, E.A. and Dalrymple, G.B. (1979) 'Revised geomagnetic polarity time scale for the interval 0–5 m.y.b.p.', *Journal of Geophysical Research* 84: 615–27.

Marshall, Sir John (1931) *Mohenjo-Daro and the Indus Civilization*, London: Arthur Probsthain.

Mason, K. (1929) 'Indus floods and the Shyok glaciers', *Himalayan Journal* 1: 10–29.

Mason, K. (1930) 'The glaciers of the Karakoram and neighbourhood', *Records of the Geological Survey of India* 63(2): 214–79.

Mason, K. (1935) 'The study of threatening glaciers', *Geography Journal* 85: 24–41.

Mathur, Y.K. (1984) 'Cenozoic palynofossils, vegetation, ecology and climate of the north and northwestern sub-Himalayan region, India', in R.O. White (ed.) *The Evolution of the East Asian Environment*, Hong Kong: Centre of Asian Studies, 2: 504–49.

Mattauer, M. (1986) 'Intracontinental subduction, crust-mantle decollement and crustal-stacking wedge in the Himalaya and other collision belts', in M.P. Coward and A.R. Ries (eds) *Collision Tectonics*, Geological Society of London, Special Publication 19: 37–50.

Mattson, L.E. and Gardner, J.S. (1989) 'Energy exchanges and ablation rates on the debris-covered Rakhiot Glacier, Pakistan', *Zeitschrift für Gletscherkunde und Glazialgeolgie* 25(1): 17–32.

Mayewski, P.A. and Jeschke, P.A. (1979) 'Himalayan and trans-Himalayan glacier fluctuations since AD 1812', *Arctic and Alpine Research* 11(3): 267–87.

Mayewski, P.A., Pergent, G.P., Jeschke, P.A. and Ahemad, N. (1980) 'Himalayan and Trans-Himalayan glacier fluctuations and the south Asian monsoon record', *Arctic and Alpine Research* 12(2): 171–82.

Medlicott, H.B. and Blanford, W.T. (1879) *Manual of the Geology of India* 1, Calcutta: Geological Survey of India.

Mehta, P.K. (1980) 'Tectonic significance of the young mineral dates and the rates of cooling and uplift in the Himalaya', *Tectonophysics* 62: 205–17.

Meier, M.F. and Post, A.S. (1962) 'Recent variations in net mass budget of glaciers in western North America', International Association of Hydrological Sciences, Publication 58: 63–77.

Meierding, T.C. (1982) 'Late Pleistocene equilibrium-line altitudes in the Colorado Front Range: a comparison of methods', *Quaternary Research* 18: 289–310.

Meissner, C.R., Master, J.M., Rashid, M.A. and Hussain, M. (1974) *Stratigraphy of the Kohat Quadrangle, Pakistan*, United States Geological Survey, Professional Paper 716-D.

Melamed, Ya.R. (1966) 'Quantitative characteristics of tectonic movements taking the Afghan–Tadzhik depression as an example', *Doklady Akademii Nauk SSRR* 171: 81–3.

Memon, M.M. (1969) 'Alluvial morphology of the lower Indus plain and its relation to land use', *Pakistan Geographical Review* 24: 1–34.

Menard, H.W. (1961) 'Some rates of regional erosion', *Journal of Geology* 69: 154–61.

Mercer, J.H. (1975) 'Glaciers of the Karakoram', in W.O. Field (ed.) *Mountain Glaciers of the Northern Hemisphere* 1, United States Army Cold Regions Research and Engineering Laboratory, Hanover, NH, 371–409.

Mercer, J.H., Fleck, R.J. and Mankiner, E.A. (1975) 'Southern Patagonia: glacial events between 4 M.Y. and 1 M.Y. ago', in D.P. Suggate and M.M. Cresswell (eds) *Quaternary Studies* 13, Bulletin of the Royal Society of New Zealand, 223–30.

Middlemiss, C.S. (1896) 'The geology o Hazara and the Black Mountains', *Memoirs of the Geological Survey of India* 26: 1–302.

Middleton, G.V. (1965) *Primary Sedimentary Structures and their Hydrodynamic Interpretation*, Tulsa, OK: Society of Economic Paleontologists and Mineralogists, Special Publication 12.

Miller, K.J. (ed.) (1984) *The International Karakoram Project* (2 vols), Cambridge: Cambridge University Press.

Milliman, J.D., Quraishee, G.S. and Beg, M.A.A. (1984) 'Sediment discharge from the Indus River to the ocean: past, present and future', in B.U. Haq and J.D. Milliman (eds) *Marine Geology and Oceanography of Arabian Sea and Coastal Pakistan*, New York: Van Nostrand Reinhold, 65–70.

Misch, P. (1935) 'Ein gefalteter junger Sandstein im Nordwest-Himalaya und sein Gefuge', *Festschrift zum 60 geburstag von Hans Stille*, Stuttgart: F. Enke Verlag.

Misch, P. (1949) 'Metasomatic granitization of batholithic dimensions', *American Journal of Science*, 209–45.

Mithal, R.S. (1968) 'The physiographical and structural evolution of the Himalaya', in B.C. Law (ed.) *Mountains and Rivers of India: National Committee for Geography, Calcutta, 21st International Geography Congress, India*, 41–81.

Molnar, P. and Tapponnier, P. (1975) 'Cenozoic tectonics of Asia: effects of a continental collision', *Science* 189(4201): 419–25.

Molnar, R. and England, P. (1990) 'Late Cenozoic uplift of mountain ranges and global climatic change: chicken or egg?', *Nature* 346: 29–34.

Moorcroft, W. and Trebeck, G. (1841) *Travels in the Himalaya Provinces of Hindoostan and the Punjab* (2 vols), London: John Murray.

Moosvi, A.T., Haque, S.M. and Muslim, M. (1974) 'Geology and china clay deposits, Shah Dheri (Swat) N.W.F.P.', *Records of the Geological Survey of Pakistan* 26: 28.

Moralev, V.M., Skotarenko, V.V. and Fokina, N.A. (1967) 'Eocene paleogeography of the northern Pamirs', *Doklady Akademii Nauk SSSR* 175: 100–1.

Morris, T.O. (1938) 'Bain boulder-bed: a glacial episode in the Siwalik series of the Marwat Kundi Range and Shekh Budin, Northwest Frontier Province, India', *Quarterly Journal of the Geological Society of London*, 94: 385–421.

Morrison, R.B. (1964) 'Soil stratigraphy: principles, applications to differentiation and correlation of Quaternary deposits and landforms and applications to soil science', Unpublished Ph.D. dissertation, University of Nevada, Reno.

Mughal, M.R. (1971) 'The early Harappan period in the greater Indus Valley and northern Baluchistan (c. 3000–2400 BC)', Unpublished Ph.D. dissertation, University of Pennsylvania, Philadelphia, PA.

Mughal, M.R. (1973) *Present State of Research on the Indus Valley Civilizations*, Karachi: Government of Pakistan.

Mughal, M.R. (1981) 'New archaeological evidence from Bahawalpur', in A.H. Dani (ed.) *Indus Civilization: New Perspectives*, Islamabad: Quaid-i-Azam University, 33–42.

Mughal, M.R. (1982) 'Recent archaeological research in the Cholistan Desert', in G.L. Possehl (ed.) *Harappan Civilization: a contemporary perspective*, New Delhi: Oxford & IBH Publishing, 85–96.

Mughal, M.R. (1988) 'Genesis of the Indus Valley civilization', *Lahore Museum Bulletin* 1: 45–54.

Nanson, G.C. (1980) 'A regional trend to meander migration', *Journal of Geology* 88: 100–7.

Nazir, A. (1974) *Ground Water Resources of Pakistan*, Lahore: S. Nazir.

Ni, J. and Barazangi, M. (1984) 'Seismotectonics of the Himalayan collision zone: geometry of the underthrusting Indian plate beneath the Himalaya', *Journal of Geophysical Research* 89: 1147–63.

Nikonov, A.A. (1970) 'Evolution of river valleys in the southern part of central Asia in the Anthropogene', *Doklady Akademii Nauk SSSR*, 195, 29–31.

Nikonov, A.A. (1981) Dating of seismotectonic movements and old earthquakes in the mountains of Soviet Central Asia by means of radiocarbon analysis and archaeologic data, *Doklady Akademii Nauk SSSR* 257: 62–5.

Norin, E. (1925) 'Preliminary notes on the late Quaternary glaciation of the northwestern Himalaya', *Geografiska Annaler* 7: 165–94.

Norin, E. (1932) 'Quaternary climatic changes within the Tarim basin', *Geographical Review* 22: 591–8.

Norin, E. (1946) 'Geological explorations in western Tibet: Reports of the scientific expedition to the northwestern provinces of China under the leadership of Dr Sven Hedin', *Geology* 7(29): 214.

Normand, C.W. (1937) *The Weather of India*, Calcutta: Indian Science Congress Association, 1–16.

Oestreich, K. (1906) 'Die Taler des nordwestlichen Himalaya', *Erganzungsheft zu Petermanns Mitteilungen* 155, Gotha: Justus Perthes.

Oldham, R.D. (1893) 'The river valleys of the Himalayas', *Journal of Manchester Geographical Society* 9: 112–25.

Oldham, R.D. (1926) 'The Cutch earthquake of 16 June 1819 with a revision of the great earthquake of 12 June 1897', *Memoirs of the Geological Survey of India* 46: 1–77.

Ollier, C.D. (1963) 'Insolation weathering: examples from central Australia', *American Journal of Science* 261: 376–8.

Ollier, C.D. (1965) 'Dirt-cracking – a type of insolation weathering', *Australian Journal of Science* 27: 236–7.

Olson, T.M. (1982) 'Sedimentary tectonics of the Jalipur Sequence, northwest Himalaya, Pakistan', Unpublished M.A. thesis, Dartmouth College, Hanover, NH.

Opdyke, N.D., Lindsay, E., Johnson, G.D., Tahirkheli, R.A.K. and Mirza, M.A. (1979) 'Magnetic polairty stratigraphy and vertebrate paleontology of the Upper Siwalik subgroup of northern Pakistan', *Palaeogeography, Palaeoclimatology, Palaeoecology* 27: 1–34 (Amsterdam: Elsevier).

Opdyke, N.D., Johnson, N.M., Johnson, G.D., Lindsay, E.H. and Tahirkheli, R.A.K. (1982) 'Paleomagnetism of the Middle Siwalik formations of northern Pakistan and rotation of the Salt Range decollement', *Palaeogeography, Palaeoclimatology, Palaeoecology* 37: 1–15 (Amsterdam: Elsevier).

Ori, G.G. and Friend, P.F. (1984) 'Sedimentary basins formed and carried piggyback on active thrust sheets', *Geology* 12: 475–8.

Osmaston, H.A. (1975) 'Models for the estimation of firnlines of present and Pleistocene glaciers', in R. Peel, M. Chisholm and P. Haggett (eds) *Process in Physical and Human Geography*, London: Heinemann Educational, 218–45.

Osmaston, H.A. (in press) 'The geology, geomorphology and Quaternary history of Zangskar', in J.H. Crook and H.A. Osmaston (eds) *Himalayan Buddhist villages: a study of communities in Zangskar, Ladakh*, Central Asian Studies Series, Warminster, UK: Aris & Phillips.

Østrem, G. (1961) 'The height of the glaciation limit in southern British Columbia and Alberta', *Geografiska Annaler* 48A: 126–38.

Ouchi, S. (1985) 'Response of alluvial rivers to slow active tectonics', *Bulletin of the Geological Society of America* 96: 504–15.

Owen, L.A. (1988a) 'Terraces, uplift and climate, Karakoram mountains, northern Pakistan', Unpublished Ph.D. thesis, University of Leicester.

Owen, L.A. (1988b) 'Wet-sediment deformation of Quaternary and recent sediments in the Skardu Basin, Karakoram mountains, Pakistan', in D.G. Croot (ed.) *Glaciotectonics: Forms and Processes*, Rotterdam: Balkema, 123–47.

Owen, L.A. (1989a) 'Neotectonics and glacial deformation in the Karakoram mountains, and Nanga Parbat Himalaya', *Tectonophysics* 163: 227–65.

Owen, L.A. (1989b) 'Terraces, uplift and climate in the Karakoram Mountains, Northern Pakistan: Karakoram intermontane basin evolution', *Zeitschrift für Geomorphologie, Neue Folge, Supplementband* 76: 117–46.

Owen, L.A. (1991) 'Mass movement deposits in the Karakoram Mountains: their sedimentary characteristics, recognition and role in Karakoram landform evolution', *Zeitschrift für Geomorphologie, Neue Folge* 35: 401–24.

Owen, L.A. and Derbyshire, E. (1988) 'Glacially deformed diamictons in the Karakoram Mountains, northern Pakistan', in D.G. Croot (ed.) *Glaciotectonics: Forms and Processes*, Rotterdam: Balkema, 149–76.

Owen, L.A., White, B., Rendell, H. and Derbyshire, E. (in press) 'Loessic silt deposits in the western Himalayas: their sedimentology, generics and age', *Catena* 19.

Paffen, K.H., Pillewizer, W. and Schneider, H.J. (1956) Forschungen im Hunza-Karakorum, *Erdkunde* 10(1): 1–33.

Pakhomov, M.M. and Nikonov, A.A. (1983) 'Quantitative evaluation of the recent uplifting of a mountain system based on palynological data: case of the Pamirs', *Doklady Akademii Nauk SSSR, Earth Science* 269: 52–6.

Panhwar, M.H. (1969) *Ground Water in Hyderabad and Khairpur Divisions*, Khairpur:

Directorate of Agriculture.

Pascoe, E.H. (1920a) 'The early history of the Indus, Brahmaputra, and Ganges', *Quarterly Journal of the Geological Society of London* 75: 138–57.

Pascoe, E.H. (1920b) 'The geotectonics of the oil belt, petroleum in the Punjab and northwest Frontier', *Memoirs of the Geological Survey of India* 40: 450–73.

Pels, S. (1964) 'The present and ancestral Murray River system', *Australian Geography Studies* 2: 111–19.

Pen'kov, A.V., Nikonov, A.A. and Pakhomov, M.M. (1976) 'First data on the paleomagnetic properties of Pliocene and Quaternary rocks of the Pamirs', *Doklady Akademii Nauk SSSR, Earth Science* 229: 86–8.

Pen'kov, A.V., Gamov, L.N. and Dodonov, A.Ye. (1977) 'A composite paleomagnetic section for the upper Pliocene–Pleistocene deposits of the Kyzylsu River basin (southern Tadzhikistan)', *International Geological Review* 19: 1207–15.

Perrot, F.A. and Goudie, A.S. (1984) 'Techniques for the study of glacial fluctuations', in K.J. Miller (ed.) *The International Karakoram Project* 1, Cambridge: Cambridge University Press, 94–100.

Peterson, A. and Robinson, G. (1969) 'Trend-surface mapping of cirque-floor levels', *Nature* 222: 75–6.

Petterson, M.G. and Windley, D.F. (1985) 'Rb–Sr dating of the Kohistan arc-batholith in the trans-Himalaya of north Pakistan, and tectonic implications', *Earth and Planetary Science Letters* 74: 45–57 (Amsterdam: Elsevier).

Pettijohn, F.J. (1962) 'Paleocurrents and paleogeography', *Bulletin of the American Association of Petroleum Geologists* 46: 1468–93.

Piggott, S. (1953) 'A forgotten empire of antiquity', *Scientific American* 189(5): 42–8.

Pilbeam, D. (1972) 'Evolutionary changes in hominoid dentition through geological time', in W.W. Bishop and J.A. Miller (eds) *Calibration of hominoid evolution*, Edinburgh: Scottish Academic Press, 369–80.

Pilgrim, Guy E. (1919) 'Suggestions concerning the history of the drainage of northern India arising out of a study of the Siwalik boulder conglomerates', *Journal of the Asiatic Society of Bengal* 15: 81–99.

Pillewizer, W. (1956) 'Der Rakhiot-Gletscher am Nanga Parbat im Jahre 1954', *Zeitschrift für Gletscherkunde und Glazialgeologie* 3(2): 181–94.

Pithawala, M.B. (1936) 'A geographical analysis of the lower Indus basin (Sind)', *Proceedings of the Indian Academy of Sciences* 4: 283–355.

Pithawala, M.B. (1959) *A Physical and Economic Geography of Sind*, Hyderabad: Sindhi Adabi Board.

Pivnik, D.A. (1988) 'Magnetostratigraphy and sedimentology of the Hazara intermontane basin, Pakistan', *Geological Bulletin of the University of Peshawar, Pakistan* 21: 85–104 (Amsterdam: Elsevier).

Popp, N. (1971) 'Hydrogeographische und geomorphologische Geseichspunkte zum Problem der rezenten verikalen krustenbewegungen in Rumanian', *Zeitschrift für Geomorphologie* 15: 445–59.

Porter, S.C. (1970) 'Quaternary glacial record in Swat, Kohistan, and west Pakistan', *Bulletin of the Geological Society of America* 81: 1421–46.

Porter, S.C. (1977) 'Present and past glaciation threshold in the Cascade Range, Washington, USA: topographic and climatic controls and paleoclimatic implications', *Journal of Glaciology* 18: 101–16.

Possehl, G.L. (1967) 'The Moenjo-daro floods: a reply', *American Anthropologist* 69: 32–40.

Potter, P.E. (1978) 'Significance and origin of big rivers', *Journal of Geology* 86: 13–33.

Powell, C.McA. (1979) 'A speculative tectonic history of Pakistan and surroundings: some constraints from the Indian Ocean', in A. Farah and K.A. DeJong (eds)

*Geodynamics of Pakistan*, Quetta: Geological Survey of Pakistan, 5–25.

Powell, C.McA. (1986) 'Continental underplating model for the rise of the Tibetan Plateau', *Earth and Planetary Science Letters* 81: 79–94 (Amsterdam: Elsevier).

Powell, C.M. and Conaghan, P.J. (1973) 'Plate tectonics and the Himalayas', *Earth and Planetary Science Letters* 20: 1–12 (Amsterdam: Elsevier).

Prell, W.L. (1984) 'Monsoonal climate of the Arabian Sea during the Late Quaternary: a response to changing solar radiation', in A. Berger, J. Imbrie, J. Hays, G. Kukla and B. Saltzman (eds), *Milankovitch and Climate* 1, Palisades, NY: NATO Advanced Research Workshop, 349–66.

Prell, W.L., Hutson, W.H., Williams, D.F., Be, A.W.H., Geitzenauer, K. and Molfino, B. (1980) 'Surface circulation of the Indian Ocean during the last glacial maximum, approximately 18,000 yr BP', *Quaternary Research* 14: 309–36.

Price, L.W. (1981) *Mountains and Man: A Study of Process and Environment*, Berkeley: University of California Press.

Qiang Fang and Ma Xing-Hua (1979) 'A preliminary discussion on the period, amplitude and type of the uplift of the Qinghai–Xizang Plateau', *Scientia Sinica* 22: 1314–28.

Quade, J., Cerling, T.E. and Bowman, J.R. (1989) 'Systematic variations in the carbon and oxygen isotopic composition of pedogenic carbonate along elevation transects in the southern Great Basin, USA', *Bulletin of the Geological Society of America* 101: 464–75.

Quittmeyer, R.C., Farah, A. and Jacob, K.H. (1979) 'The seismicity of Pakistan and its relation to surface faults', in A. Farah and K.A. DeJong (eds) *Geodynamics of Pakistan*, Quetta: Geological Survey of Pakistan, 271–84.

Rabot, C. (1905) 'Glacial reservoirs and their outbursts', *Geographical Journal* 25: 545–8.

Rafiq, Ch.M. (1971) 'Soil classification', *Soil Survey Project, Pakistan: Soil Resources in West Pakistan and their Development Possibilities*, Rome: United Nations Development Programme, 93–110.

Raikes, R.L. (1964) 'The end of the ancient cities of the Indus', *American Anthropologist* 66: 284–99.

Raikes, R.L. (1965) 'The Moenjo-daro floods', *Antiquity* 155: 196–203.

Raikes, R.L. and Dales, G.F. (1977) 'The Moenjo-daro floods reconsidered', *Journal of the Palaeontological Society of India* 20: 251–60.

Ramsay, W.J.H. (1985) 'Erosion in the Middle Himalaya, Nepal, with a case study of the Phewa Valley', Unpublished M.Sc. thesis, Department of Forest Resources Management, University of British Columbia, Vancouver, BC.

Ranga Rao, A., Agarwal, R.P., Sharma, U.N. and Bhalla, M.S. (1988) 'Magnetic polarity stratigraphy and vertebrate palaeontology of the Upper Siwalik subgroup of Jammu hills, India', *Journal of the Geological Society of India* 31: 361–85.

Raufi, F. and Sickenberg, O. (1973) 'Zur Geologie und Palaontologie der Becken von Lagman und Jalalabad', *Geologisches Jahrbuch* B3: 63–99.

Raverty, H.G. (1895) 'The Mihran of Sind and its tributaries: a geographical and historical study', *Journal of the Asiatic Society of Bengal* 61 (Extra): 155–508.

Raymo, M.E., Ruddiman, W.F. and Froelich, P.N. (1988) 'Influence of Late Cenozoic mountain building on ocean geochemical cycles', *Geology* 16: 649–53.

Raynolds, R.G.H. (1981) 'Did the ancestral Indus flow into the Ganges drainage?' *Geological Bulletin of the University of Peshawar, Pakistan* 14: 141–50 (Amsterdam: Elsevier).

Raynolds, R.G.H. and Johnson, G.D. (1985) 'Rates of Neogene depositional and deformational processes, northwest Himalayan foredeep margin, Pakistan', in N.J. Snelling (ed.) *The Chronology of the Geological Record*, Geological Society of London, Memoir 10: 297–311.

Raza, H.A. and Alam, S. (1983) 'Pakistan's Makran region merits extensive oil hunt', *Oil and Gas Journal* 81(29): 170–4.

Raza, H.A., Alam, S., Ali, S.M., Elahi, N. and Anwar, M. (1981) 'Hydrocarbon potential of Makran region of Baluchistan basin', *Seminar on Mineral Policy of Baluchistan, Quetta*, 29–30 April, Hydrocarbon Development Institute, Pakistan.

Reading, H.G. (1978) *Sedimentary Environments and Facies*, New York: Elsevier.

Reconnaissance Soil Survey, Rawalpindi Area (1967) Lahore: Directorate of Soil Survey, West Pakistan preliminary edn.

Reconnaissance Soil Survey, Sahiwal District (1968) Lahore: Directorate of Soil Survey, West Pakistan preliminary edn.

Reconnaissance Soil Survey, Gujrat District (1968) Lahore: Directorate of Soil Survey, West Pakistan preliminary edn.

Reconnaissance Soil Survey, Gujranwala Area (1965) Lahore: Directorate of Soil Survey, West Pakistan preliminary edn.

Reconnaissance Soil Survey, Lahore Area (1968) Lahore: Directorate of Soil Survey, West Pakistan preliminary edn.

Reconnaissance Soil Survey, Multan North (1969) Lahore: Directorate of Soil Survey, West Pakistan preliminary edn.

Reconnaissance Soil Survey, Hyderabad (1970) Lahore: Directorate of Soil Survey, West Pakistan preliminary edn.

Reconnaissance Soil Survey, Thatta West (1979) Lahore: Directorate of Soil Survey, West Pakistan preliminary edn.

Rendell, H.M. (1988) 'Palaeoenvironmental change during the Pleistocene in northern Pakistan', *Proceedings of the Indian National Science Academy* 54A: 392–400.

Rendell, H.M. (1989) 'Loess deposition during the Late Pleistocene in northern Pakistan', *Zeitschrift für Geomorphologie, Neue Folge* 76: 247–55.

Rendell, H.M. and Dennell, R.W. (1987) 'Thermoluminescence dating of an upper Pleistocene site, northern Pakistan', *Geoarchaeology* 2: 63–7.

Rendell, H.M. and Townsend, P.D. (1988) 'Thermoluminescence dating of a 10 m loess profile in Pakistan', *Quaternary Science Reviews* 7: 251–5.

Rendell, H.M., Dennell, R.W. and Halim, M.A. (1989) 'Pleistocene and palaeolithic investigations in the Soan Valley, northern Pakistan', *British Archaeological Reports, International Series* (supplementary) 544, 346.

Repenning, C.A. (1984) 'Quaternary rodent biochronology and its correlation with climate and magnetic stratigraphies', in W.C. Mahaney (ed.) *Correlation of Quaternary Chronologies*, Norwich: Geo Books, 105–18.

Revelle, R. (1964) *Report on Land and Water Development in the Indus Plain*, Washington, DC: White House Department of the Interior Panel on Waterlogging and Salinity in West Pakistan.

Rickmers, W.R. (1913) *The Duab of Turkestan*, Cambridge: Cambridge University Press.

Rigby, J.K. and Hamblin, W.K. (eds) (1972) *Recognition of Ancient Sedimentary Environments*, Tulsa, OK: Society of Economic Paleontologists and Mineralogists, Special Publication 16.

Ritter, D.F. (1978) *Process Geomorphology*, Dubuque, IA: W.M. Brown.

Rowlands, D. (1978) 'The structure and seismicity of portion of the southern Sulaiman Range, Pakistan', *Tectonophysics* 51: 41–56.

Royden, L.H. and Burchfiel, B.C. (1987) 'Thin-skinned N–S extension with the convergent Himalyan region: gravitational collapse of Miocene topographic front', in M.P. Coward, J.F. Dewey and P.L. Hancock (eds) *Continental Extension Tectonics*, Geological Society of London (Oxford: Blackwell), 611–19.

Ruddiman, W.F. and Kutzbach, J.E. (1989) 'Forcing of Late Cenozoic Northern Hemisphere climate by plateau uplift in southern Asia and the American west', *Journal of Geophysical Research* 94: 18,409–27.

Ruddiman, W.F., Prell, W.L. and Raymo, M.E. (1989) 'Late Cenozoic uplift in southern Asia and the American West: rationale for general circulation modeling experiments', *Journal of Geophysical Research* 94: 18,379–91.

Ruhe, R.V. (1983) 'Depositional environments of Late Wisconsin loess in midcontinental United States', in S.C. Porter (ed.) *The Late Pleistocene Environments of the United States*, Minneapolis, MN: University of Minnesota Press, 130–7.

Rundel, P.W. (1980) 'The ecological distribution of Carbon 4 and Carbon 3 plants in the Hawaiian Islands', *Oecologia* 45: 354–9.

Saeed, K.M. (1974) 'Watershed Management and the Mangla Watershed Project', *Proceedings of the National Seminar on Ecology, Environment and Afforestation*, Islamabad: Environment and Urban Affairs Division, 94.

Sahni, A. and Chandra, M. (1980) 'Lower Miocene (Aquitanian–Burdigalian) palaeobiogeography of the Indian subcontinent', *Geologische Rundschau* 69: 824–48.

Sahni, M.R. (1956) 'Bio-geological evidence bearing on the decline of the Indus Valley Civilization', *Journal of the Palaeolontological Society of India* 1(1): 101–7.

Sankhla, N., Ziegler, H., Vyas, O.P., Stichler, W. and Trimborn, P. (1975) 'Eco-physiological studies on Indian arid zone plants', *Oecologia* 21: 123–9.

Sarwar, G. and DeJong, K.A. (1979) 'Arcs, oroclines, syntaxis: the curvature of mountain belts in Pakistan', in A. Farah and K.A. DeJong (eds) *Geodynamics of Pakistan*, Quetta: Geological Survey of Pakistan, 341–50.

Scheidegger, A.E. (1979) 'The principle of Antagonism in the Earth's evolution', *Tectonophysics* 55: 7–10.

Schneider, H.J. (1959) 'Zur diluvialen Geschichte des NW Karakorum', *Mitteilungen Geographische Gesellschaft Munchen* 444: 201–17.

Schumann, R.R. (1989) 'Morphology of Red Creek, Wyoming, an arid-region anastomosing channel system', *Earth Surface Processes and Landforms* 14: 277–88.

Schumm, S.A. (1963) *The Disparity between Rates of Denudation and Orogeny*, United States Geological Survey, Professional Paper 454-H.

Schumm, S.A. (1965) 'Quaternary paleohydrology', in H.E. Wright and D.G. Frey (eds) *Quaternary of the United States*, Princeton, NJ: Princeton University Press, 783–94.

Schumm, S.A. (1968) *River Adjustment to Altered Hydrologic Regimen: Murrumbidgee River and Paleochannels, Australia*, United States Geological Survey, Professional Paper 598.

Schumm, S.A. (1969) 'River metamorphosis', *Proceedings of the American Society of Civil Engineers, Journal of the Hydraulics Division* 95(HY-1): 255–73.

Schumm, S.A. (1972) 'Fluvial paleochannels: recognition of ancient sedimentary environments', in J.K. Rigby and W.K. Hamblin (eds) *Recognition of Ancient Sedimentary Environments*, Tulsa, OK: Society of Economic Paleontologists and Mineralogists, Special Publication 16: 98–107.

Schumm, S.A. (1977) *The Fluvial System*, New York: Wiley.

Schumm, S.A. (1985) 'Patterns of alluvial rivers', *Annual Review of Earth and Planetary Sciences* 13: 5–27.

Schumm, S.A. (1986) 'Alluvial river response to active tectonics', *Active Tectonics*, Washington, DC: National Academy Press, 80–94.

Schumm, S.A. and Erskine, W.D. (1989) 'Anastomosing streams or anastomosing patterns', Geological Society of America, *Abstracts with Programs* 21: A153.

Schumm, S.A. and Kahn, H.R. (1972) 'Experimental study of channel patterns', *Bulletin of the Geological Society of America* 83: 1755–70.

Schumm, S.A., Khan, H.R., Winkley, B.R. and Robbins, L.G. (1972) 'Variability of river patterns', *Nature* 237: 75–6.

Schweinfurth, U. (1956) 'Uber klimatische trockentaler im Himalaya', *Erdkunde* 10: 297–302.

Searle, M.P. (1983) 'Stratigraphy, structure and evolution of Tibetan–Tethys zone in Zanskar and the Indus suture zone in the Ladakh Himalaya', *Transactions of the Royal Society of Edinburgh, Earth Science* 73: 205–19.

Searle, M.P. (1991) *Geology and Tectonics of the Karakoram Mountains*, Chichester: Wiley.

Searle, M.P., Windley, B.F., Coward, M.P., Cooper, D.J.W., Rex, A.J., Rex, D., Li, T., Xiao, X., Jan, M.Q., Thakur, V.C. and Kumar, S. (1987) 'The closing of Tethys and the tectonics of the Himalaya', *Bulletin of the Geological Survey of America* 98: 678–701.

Searle, M.P., Rex, A.J., Tirrul, R., Rex, D.C., Barnicoat, A. and Windley, B.F. (1989) 'Metamorphic, magmatic and tectonic evolution of the central Karakoram in the Biafo–Baltoro–Hushe regions of northern Pakistan', in L.L. Malinconico, jun. and R.J. Lillie (eds) *Tectonics and Geophysics of the Western Himalaya*, Geological Survey of America, Special Paper 232: 47–53.

Seeber, L. and Armbruster, J.G. (1979) 'Seismicity of the Hazara arc in northern Pakistan: decollement vs. basement faulting', in A. Farah and K.A. DeJong (eds) *Geodynamics of Pakistan*, Quetta: Geological Survey of Pakistan, 131–284.

Seeber, L. and Armbruster, J.G. (1981) 'Great detachment earthquakes along the Himalayan arc and long-term forecasting', in D.W. Simpson and P.G. Richards (eds) *Earthquake Prediction: an international review*, American Geophysical Union, Maurice Ewing Series 4: 259–79.

Seeber, L. and Gornitz, V. (1983) 'River profiles along the Himalayan arc as indicators of active tectonics', *Tectonophysics* 92: 335–367.

Seeber, L., Armbruster, J. and Quittmeyer, R.C. (1981) 'Seismicity and continental subduction in the Himalayan arc', *Zagros–Hindu Kush–Himalaya Geodynamic Evolution*, American Geophysical Union Series 3: 215–42.

Selley, R.C. (1978) *Ancient Sedimentary Environments*, 2nd edn, Ithaca, NY: Cornell University Press.

Seth, S.K. (1978) 'The desiccation of the Thar Desert and its environs during the proto-historical and historical periods', in W.C. Brice (ed.) *The Environmental History of the Near and Middle East*, London: Academic Press, 279–305.

Shackleton, N.J. and Opdyke, N.D. (1973) 'Oxygen isotope and palaeomagnetic stratigraphy of equatorial Pacific core V28–238: oxygen isotope temperatures and ice volumes on a 10,000 and 100,000 year scale', *Quaternary Research* 3: 39–55.

Shaffer, J.G. (1982) 'Harappan commerce: an alternate perspective', in S. Pastner and L. Flam (eds) *Anthropology in Pakistan: Recent Sociocultural and Archaeological Perspectives*, Ithaca, NY: Cornell University Press, 166–210.

Shah, S.M.I. (1977) 'Stratigraphy of Pakistan', *Memoirs of the Geological Survey of Pakistan* 12: 138.

Shareq, A., Chmyriov, W.M., Stazhilo-Alekseev, K.F., Dronov, V.I., Gannon, P.J., Lubemov, B.K., Kafarskiy, A.Kh., Malyarov, E.P. and Rossovskiy, L.N. (1977) *Mineral Resources of Afghanistan*, United Nations Development Program Project (Ed. 2), 1–419.

Sharma, K.K. (1983) 'Ladakh–Deosai Batholith and its surrounding rocks', in V.J. Gupta (ed.) *Stratigraphy and Structure of Kashmir and Ladakh Himalaya*, Delhi: Hindustan Publishing, 180–7.

Sharma, K.K. (1984) 'The sequence of phased uplift of the Himalaya', in R.O. White (ed.) *The Evolution of the East Asian Environment*, Hong Kong: Centre of Asian Studies, 2: 56–70.

Sharma, K.K. and Gupta, K.R. (1983) 'Northern Ladakh, a scene of explosive volcanic activity in early Cenozoic', in V.J. Gupta (ed.) *Stratigraphy and Structure of Kashmir and Ladakh Himalaya*, Delhi: Hindustan Publishing, 87–95.

Shearman, D.J. (1976) 'The geological evolution of southern Iran', *Geographical Journal* 142(3): 393–410.

Shi Yafeng and Wang Jingtai (1981) 'The fluctuations of climate, glaciers and sea level since late Pleistocene in China', *Sea Level, Ice and Climatic Change*, International Association Hydrological Sciences, Publication 131: 281–93.

Shi Yafeng and Zhang Xiangsong (1984) 'Some studies of the Batura Glacier in the Karakoram Mountains', in K.J. Miller (ed.) *The International Karakoram Project 1*, Cambridge: Cambridge University Press, 51–63.

Shroder, J.F., jr. (1984) 'Comparison of tectonic and metallogenic provinces of Afghanistan to Pakistan', *Geological Bulletin of the University of Peshawar, Pakistan* 17: 87–109 (Amsterdam: Elsevier).

Shroder, J.F., jr. (1985) 'Quaternary chronology and geomorphology of Indus–Gilgit–Hunza valleys, Pakistan', Association of American Geographers, *Program Abstracts* 295.

Shroder, J.F., jr. (1987) 'Himalayan jökulhlaups and Punjab erratics', Association of American Geographers, *Program Abstracts*, 93.

Shroder, J.F., jr. (1989a) 'Geomorphic development of the western Himalaya', *Geological Bulletin of the University of Peshawar, Pakistan* 22: 127–51.

Shroder, J.F., jr. (1989b) 'Slope failure: extent and economic significance in Afghanistan and Pakistan', in E.E. Brabb and B.L. Harrod (eds) *Landslides: Extent and Economic Significance*, Rotterdam: Balkema, 325–41.

Shroder, J.F., jr. (1989c) 'Hazards of the Himalaya', *American Scientist* 77: 564–73.

Shroder, J.F., jr. (in press a) 'Satellite-image analysis of glaciers of northern Pakistan', in R.S. Williams, jr. and J.R. Ferrigno (eds) *Satellite Image Atlas of Glaciers of the World*, United States Geological Survey, Professional Paper 1386.

Shroder, J.F., jr. (in press b) 'Satellite glacier inventory of Afghanistan', in R.S. Williams, jun. and J.R. Ferrigno (eds) *Satellite Image Atlas of Glaciers of the World*, United States Geological Survey, Professional Paper 1386.

Shroder, J.F., jr. and Khan, M.S. (1988) 'High magnitude geomorphic processes and Quaternary chronology, Indus valley and Nanga Parbat Himalaya, Pakistan (Abstract)', *The Neogene of the Karakoram and Himalayas*, Leicester: University of Leicester, 21–3.

Shroder, J.F., jr., Johnson, R., Khan, M.S. and Spencer, M. (1984) 'Batura Glacier Terminus', *Geological Bulletin of the University of Peshawar, Pakistan* 17: 119–26 (Amsterdam: Elsevier).

Shroder, J.F., jr., Khan, M.S., Lawrence, R.D., Madin, I. and Higgins, S.M. (1986) 'Chronology and deformation of Quaternary sediments, middle and upper Indus, Pakistan', Geological Society of America, *Abstracts with Programs*, 18: 749.

Shroder, J.F., jr., Khan, M.S., Lawrence, R.D., Madin, I. and Higgins, S.M. (1989) 'Quaternary glacial chronology and neotectonics in the Himalaya of northern Pakistan', in L.L. Malinconico, jun. and R.J. Lillie (eds) *Tectonics and Geophysics of the Western Himalaya*, Geological Society of America, Special Paper 232: 275–93.

Siddiqi, M.I. (1956) 'The fishermen's settlements on the coast of West Pakistan', *Selbstverlag des Geographischen Institute der Universität Kiel* 7–92.

Siddiqi, M.I. (1959) 'Geology and physiography of the coast of West Pakistan', *Scientist* (University of Karachi) 3(1): 1–6.

Siveright, R. (1907) 'Cutch and the Ran', *Geographical Journal* 29: 518–39.

Smalley, I.J. (1966) 'Contraction crack networks in basalt flows', *Geological Magazine* 103: 110–14.

Smalley, I.J. and Smalley, V. (1983) 'Loess material and loess deposits: formation, distribution and consequences', in M.E. Brookfield and T.S. Ahlbrandt (eds) *Eolian Sediments and Processes*, Amsterdam: Elsevier, 51–68.

Smith, D.G. (1983) 'Anastomosed fluvial deposits: modern examples from western Canada', in J.D. Collinson and J. Lewin (eds) *Modern and Ancient Fluvial Systems*, International Association of Sedimentologists, Special Publication 6: 155–68.

Smith, D.G. (1986) 'Anastomosing river deposits, sedimentation rates and basin subsidence, Magdalena River, northwestern Colombia, South America', *Sedimentary Geolog* 46: 177–96.

Snead, R.E. (1963) 'Disappearing town of Sonmiani, West Pakistan', *Geografia* 1(2): 26–30.

Snead, R.E. (1964) 'Active mud volcanoes of Baluchistan, West Pakistan', *Geographical Review* 54: 546–60.

Snead, R.E. (1966) 'Physical geography reconnaissance: Las Bela Coastal Plain, West Pakistan', Coastal Studies Series 13, Baton Rouge, LA: Louisiana State University Press, 118.

Snead, R.E. (1967) 'Recent morphological changes along the coast of West Pakistan', *Annals of the Association of American Geographers* 57(3): 549–65.

Snead, R.E. (1968) 'Weather patterns in southern West Pakistan', *Archiv für Meteorologie, Geophysik und Bio Klimatologie B* 16(4): 316–46.

Snead, R.E. (1969) *Physical Geography Reconnaissance: West Pakistan Coastal Zone*, Albuquerque, NM: University of New Mexico Press.

Snead, R.E. (1970) *Physical Geography of the Makran Coastal Plain of Iran*, Springfield, VA: National Technical Information Service, US Department of Commerce, Publication AD707745.

Snead, R.E. (1987) 'Man's response to change in the coastal zone of Pakistan', *Journal of Resource Management and Optimization* 4: 371–401.

Snead, R.E. and Erickson, R. (1977) 'Morphological changes in the Balakot region of Pakistan', *Field Research Projects*, 1–47.

Snead, R.E. and Frishman, S.A. (1968) 'Origin of sands on the east side of the Las Bela Valley, West Pakistan', *Bulletin of the Geological Society of America* 79: 1671–6.

Snelgrove, A.K. (1967) *Geohydrology of the Indus River, West Pakistan*, Hyderabad: Sind University Press.

Sondhi, V.P. (1947) 'The Makran earthquake, 28th November, 1945, the birth of new islands', *Indian Minerals, India Geological Survey* 1: 147–54.

Song Zhi-Chen and Liu Geng-Wu (1982) 'Tertiary palynological assemblages from Xizang with reference to their paleogeographical significance', *Proceedings of Symposium on Qinghai–Xizang (Tibet) Plateau* 1, Beijing: Science Press, 1981; New York: Gordon & Breach, 207–14.

Spate, O.H.K. and Learmonth, A.T.A. (1967) *India and Pakistan: a General and Regional Geography*, London: Methuen.

Stein, A. (1943) 'On Alexander's route into Gedrosia: an archaeological tour in Las Bela', *Geographical Journal* C2(5–6): 193–227.

Stiffe, A.W. (1873) 'On the mud craters and geological structure of the Makran coast', *Quarterly Journal of the Geological Society of London* 30: 50–3.

Stuart, M. (1922) 'The geology of the Takki Zam valley and the Kaniguram Makin area, Waziristan', *Records of the Geological Survey of India* 54(1): 87–102.

Subramanian, V., Sitasawad, R., Abbas, N. and Jha, P.K. (1987) 'Environmental geology of the Ganga River basin', *Journal of the Geological Society of India* 30: 335–55.

Subrahmanyam, V.P. (1956) 'The water balance of India according to Thornthwaite's concept of potential evapotranspiration', *Annals of the Association of American Geographers* 46: 300–11.

Sun Dianqing and Wu Xihae (1986) 'Preliminary study of Quaternary tectonoclimatic cycles in China', in V. Sibrava, D.Q. Bowen and G.M. Richmond (eds) *Quaternary Glaciation in Northern Hemisphere*, Report of International Geologic Correlation

Program Project 24, Oxford: Pergamon Press, 497–501.

Tahirkheli, R.A.K. (1982) 'Geology of the Himalaya, Karakoram and Hindukush in Pakistan', *Geological Bulletin of the University of Peshawar, Pakistan* 15: 51 (Amsterdam: Elsevier).

Tahirkheli, R.A.K. and Jan, M.Q. (1979) 'Geology of Kohistan, Karakorum Himalayas, northern Pakistan', *Geological Bulletin of the University of Peshawar, Pakistan* 11: 187 (Amsterdam: Elsevier).

Tahirkheli, R.A.K., Mattauer, M., Proust, F. and Tapponnier, P. (1979) 'The India–Eurasia suture zone in northern Pakistan: synthesis and interpretation of recent data at plate scale', in A. Farah and K.A. DeJong (eds) *Geodynamics of Pakistan*, Quetta: Geological Survey of Pakistan, 125–30.

Tamburi, A.J. (1974) 'Geology and the water resources of the Indus plains', Unpublished Ph.D. dissertation, Colorado State University, Fort Collins, CO.

Tandon, S.K. and Kumar, R. (1984) 'Active intra-basinal highs and palaeodrainage reversals in the late orogenic hominoid-bearing Siwalik basin', *Nature* 308: 635–7.

Tang Tianfu, Yang Hengren, Lan Xiu, Hu Lanyang, Yu Congliu, Zhong Shilan, Zhang Yiyong and Wei Jingming (1984) 'Upper Cretaceous–Lower Tertiary transgression and sedimentation of western Tarim basin, China', in Su Zongwei (ed.) *Development of Geoscience*, Beijing: Science Press, 203–14.

Tanish, H.R., Stringer, K.V. and Azad, J. (1959) 'Major gas fields of west Pakistan', *Bulletin of the American Association of Petroleum Geologists* 43: 2675–700.

Tapponnier, P., Peltzer, G. and Armijo, R. (1986) 'On the mechanics of the collision between India and Asia', in M.P. Coward and A.C. Ries (eds) *Collision Tectonics*, Geological Society of London, Special Publication 19: 115–57.

Tarling, D.H. (1971) *Principle and Applications of Paleomagnetism*, Norfolk: Cox & Wyman.

Tauxe, L. and Opdyke, N.D. (1982) 'A time framework based on magnetostratigraphy for the Siwalik sediments of the Khaur area, northern Pakistan', *Palaeogeography, Palaeoclimatology, Palaeoecology* 37: 43–61 (Amsterdam: Elsevier).

Teeri, J.A. and Stowe, L.G. (1976) 'Climatic patterns and distributions of C4 grasses in North America', *Oecologia* 23: 1–12.

Tewari, B.S. and Dixit, P.C. (1972) 'A new terrestrial gastropod from freshwater beds of Kargil, Ladakh, J and K [Jammu and Kashmir] State', *Bulletin of the Indian Geological Association* 4: 61–7.

Theobald, W. (1877) 'Occurrence of erratics in the Potwar', *Records of the Geological Survey of India* 10: 140–3.

Thompson, A.B. and England, P.C. (1984) 'Pressure temperature time paths of regional metamorphism, 2: Their inference and interpretation using mineral assemblages in metamorphic rocks', *Journal of Petrology* 25: 929–55.

Treloar, P.J., Rex, D.C., Guise, P.G., Coward, M.P., Searle, M.P., Windley, B.F., Petterson, M.G., Jan, M.Q. and Luff, I.W. (1989) K–Ar and Ar–Ar geochronology of the Himalayan collision in NW Pakistan: constraints on the timing of suturing, deformation, metamorphism and uplift', *Tectonics* 8(4): 881–909.

Tricart, J. and Cailleux, A. (1969) *Le Modèle des Regions Sèches*, Paris: SEDES.

Trifonov, V.G. (1978) 'Late Quaternary tectonic movements of western and central Asia', *Bulletin of the Geological Society of America* 89: 1059–72.

Troll, K. (1938) 'Der Nanga Parbat als Ziel deutscher Forschung', *Zeitschrift die Gessell-schaft für Erdkunde* 1(2): 1–26.

Ullah, A. (1954) 'Physiography and structure of southwest Makran', Fourth Pakistan Science Conference, Peshawar, *Pakistan Geographical Review* 9: 28–54.

Van Campo, E. (1986) 'Monsoon fluctuations in two 20,000-yr BP oxygen-isotope/pollen records off southwest India', *Quaternary Research* 26: 376–88.

Velichko, A.A. and Lebedeva, I.M. (1973) 'Reconstruction of the upper Pleistocene glaciation of east Pamir', *Geoforum* 16: 67–74.

Verstappen, H.Th. (1970) 'Aeolian geomorphology of the Thar desert and palaeoclimates', *Zeitschrift für Geomorphologie, Supplementband* 10: 104–20.

Vigne, G.T. (1842) *Travels in Kashmir, Ladakh, Iskardo: the Countries Adjoining the Mountain-Course of the Indus, and the Himalaya North of the Punjab*, London: Colburn.

Visser, Ph.C. (1928) 'Von den gletschern am obersten Indus', *Zeitschrift für Gletscherkunde* 16: 169–229.

Visser, Ph.C. and Visser-Hooft, J. (1935–8) *Karakorum: Wissenschaftliche ergebnisse der Niederlandischen Expeditionen den Karakorum und die Angrenzenden Gebiete in den Jahren 1922, 1925, 1929–30, und 1935*, Leiden: E.J. Brill.

Vita-Finzi, C. and Ghorashi, M. (1978) 'A recent faulting episode in the Iranian Makran', *Tectonophysics* 44: 21–5.

Volkov, N.G., Sokolovsky, I.L. and Subbotiu, A.I. (1967) 'Effect of recent crustal movements on the shape and longitudinal profiles and water levels in rivers', *Symposium on River Morphology*, International Association of Hydrological Sciences, Publication 75: 105–16.

von Humboldt, Freiherr H.A. (1820) Sur la limite inférieure des neiges perpétuelles dans les montagnes de l'Himalaya et des régions équatoriales', *Annales de Chimie et de Physique* 14: 5.

von Klebelsberg, R. (1925–6) 'Der Turkestanische Gletscherypus', *Zeitschrift für Gletscherkunde* 14: 193–209.

von Schlagintweit, H.A. and von Schlagintweit, R. (1860–6) *Results of a Scientific Mission to India and High Asia*, 4 vols, London: Atlas, Trubner.

von Wissmann, H. (1959) 'Die Heutige Vergletscherung und Schneegrenze in Hoch Asien', *Abhandlung der Mathematisch–Naturwissenschaftlichen klasse* 14, Akademie der Wissenschaften und der literatur in Mainz, Wiesbaden: Steiner Verlag, 1103–431.

Voskresenkiy, I.A., Kravchenko, K.N. and Sokolov, B.A. (1968) 'The tectonics of west Pakistan', *Geotectonics* 2: 93–9.

Vrendenburg, E.W. (1901) 'Geological sketch of the Baluchistan desert and part of eastern Persia', *Memoirs of the Geological Survey of India* 21(2): 179–302.

Vuichard, D. and Zimmermann, M. (1987) 'The 1985 catastrophic drainage of a moraine-dammed lake, Khumbu Himal, Nepal: cause and consequences', *Mountain Research and Development* 7: 91–110.

Wadia, D.N. (1928) 'The geology of Poonch State (Kashmir) and adjacent portions of the northern Punjab', *Memoirs of the Geological Survey of India* 51: 185–370.

Wadia, D.N. (1931) 'The syntaxis of the northwest Himalaya: its rocks, tectonics, and orogeny', *Records of the Geological Survey of India* 65: 189–220.

Wadia, D.N. (1939) *Geology of India for Students*, London: Macmillan.

Wahraftig, C. and Birman, J. (1965) 'The Quaternary of the Pacific mountain system in California', in H.E. Wright and D.G. Frey (eds) *The Quaternary of the United States*, Princeton, NJ: Princeton University Press, 299–340.

Wake, C.P. (1987a) 'Snow accumulation studies in the central Karakoram, Pakistan', *Proceedings of the Eastern Snow Conference*, 44th Annual Meeting, Fredricton, NB, 19–33.

Wake, C.P. (1987b) 'Spatial and temporal variation of snow accumulation in the central Karakoram, northern Pakistan', Unpublished Master's thesis, Department of Geography, Wilfrid Laurier University, Waterloo, Ontario.

Wang, C-Y., Shi, Y-L. and Khou, W-H. (1982) 'Dynamic uplift of the Himalaya', *Nature* 298: 553–6.

Wang Wenying, Huang Maohuan and Chen Jianming (1984) 'A surging advance of Balt

Bare Glacier, Karakoram mountains', in K.J. Miller (ed.) *The International Karakoram Project* 1, Cambridge: Cambridge University Press, 76–83.

Wells, J.T. and Coleman, J.M. (1984) 'Deltaic morphology and sedimentology, with special reference to the Indus River delta', in B.U. Haq and J.D. Milliman (eds) *Marine Geology and Oceanography of Arabian Sea and Coastal Pakistan*, New York: Van Nostrand Reinhold, 85–100.

Whalley, W.B., McGreevy, J.P. and Ferguson, R.I. (1984) 'Rock temperature observations and chemical weathering in the Hunza region, Karakoram: preliminary data', in K.J. Miller (ed.) *The International Karakoram Project* 2, Cambridge: Cambridge University Press, 616–33.

Wheeler, Sir R.E.M. (1968) *The Indus Civilization*, Cambridge: Cambridge University Press.

Whiteman, P.T.S. (1985) *Mountain Oases: a Technical Report of Agricultural Studies (1982–1984) in Gilgit District, Northern Areas, Pakistan*, Gilgit: UN Food and Agriculture Organization, UN Development Programme, Pak/80/009.

Wiche, K. (1959a) 'Klimamorphologische untersuchungen im Westlichen Karakorum', *Verhandlungen des Deutschen geographentages* 32: 190–203.

Wiche, K. (1959b) 'Die Osterreichische Karakorum Expedition 1958', *Mitteilungen Geographische Gesellschaft in Wien* 100: 280–94.

Wien, K. (1936) 'Weather conditions on Nanga Parbat, July 1934', *Himalayan Journal* 8: 78–85.

Wilhelmy, H. (1969) 'Das Urstomtal am Ostrand der Indusbene und das Sarasvati Problem', *Zeitschrift für Geomorphologie, Supplementband* 8: 76–93.

Williams, V.S. (1977) 'Neotectonic implications of the alluvial record in the Sapta Kosi drainage basin, Nepalese Himalayas', Unpublished Ph.D. dissertation, University of Washington, Seattle, WA.

Winkler, W.M. (1987) 'Weathering and weathering rates of natural stone', *Environmental Geology Water Science* 9(2): 85–92.

Woldstedt, P. (1965) *Das Eiszeitalter*, Stuttgart: Ferdinand Enke.

Wu-ling Zhao and Morgan, W.J. (1985) 'Uplift of the Tibetan Plateau', *Tectonics* 4: 359–69.

Wynne, A.B. (1879) 'Further notes on geology of the upper Punjab', *Records of the Geological Survey of India* 12: 114–33.

Wynne, A.B. (1881) 'Travelled blocks of the Punjab', *Records of the Geological Survey of India* 14: 153–4.

Xu Ren (1982) 'Vegetational changes in the past and the uplift of Qinghai–Xizang plateau', *Proceedings of Symposium on Qinghai–Xizang (Tibet) Plateau* 1, Beijing: Science Press, 1981; New York: Gordon & Breach, 139–44.

Yaalon, D.H. and Ganor, E. (1968) 'Chemical composition of dew and dry fallout in Jerusalem, Israel', *Nature* 217: 1139–40.

Yeats, R.S. and Lawrence, R.D. (1984) 'Tectonics of the Himalayan thrust belt in the northern Pakistan', in B.U. Haq and J.D. Milliman (eds) *Marine Geology and Oceanography of Arabian Sea and Coastal Pakistan*, New York: Van Nostrand Reinhold, 178–98.

Yeats, R.S., Khan, S.H. and Akhtar, M. (1984) 'Late Quaternary deformation of the Salt Range of Pakistan', *Bulletin of the Geological Society of America* 95: 958–66.

Yoshida, M. and Igarashi, Y. (1984) 'Neogene to Quaternary lacustrine sediments in the Kathmandu valley, Nepal', *Journal of the Nepal Geological Society* 4: 73–100.

Yoshida, M., Igarashi, Y., Arita, K., Hayashi, D. and Sharma, T. (1984) 'Magnetostratigraphic and pollen analytic studies of the takmar series, Nepal Himalaya', *Journal of the Nepal Geological Society* 4: 101–20.

Zannetin, B. (1964) 'Geology and petrology of Haramosh–Mango Gusor area: Italian expedition to the Karakorum (K2) and Hindu Kush', *Scientific Reports*, Series 3(1), Leiden: E.J. Brill.

Zeitler, P.K. (1983) 'Uplift and cooling history of the northwest Himalaya, north Pakistan: evidence from fission-track and $Ar^{40}/Ar^{39}$ cooling ages', Unpublished Ph.D. thesis, Dartmouth College, Hanover, NH.

Zeitler, P.K. (1985) 'Cooling history of the northwest Himalaya, Pakistan', *Tectonics* 4(1): 127–51.

Zeitler, P.K. and Chamberlin, C.P. (1991) 'Petrogenetic and tectonic significance of young leucogranites from the northwestern Himalaya, Pakistan', *Tectonics* 10: 729–41.

Zeitler, P.K., Johnson, N.M., Naeser, C.W. and Tahirkheli, R.A.K. (1982a) 'Fission-track evidence for Quaternary uplift of the Nanga Parbat region, Pakistan', *Nature* 298: 255–7.

Zeitler, P.K., Tahirkheli, R.A.K., Naeser, C.W. and Johnson, N.M. (1982b) 'Unroofing history of a suture zone in the Himalaya of Pakistan by means of fission-track annealing ages', *Earth and Planetary Science Letters* 57: 227–40 (Amsterdam: Elsevier).

Zeitler, P.K., Sutter, J.F., Williams, I.S., Zartman, R.E. and Tahirkheli, R.A.K. (1986) 'The Nanga Parbat–Haramosh massif, Pakistan: geochronology and cooling history', Geological Society of America, *Abstracts with Programs* 18: 800.

Zeitler, P.K., Sutter, J.F., Williams, I.S., Zartman, R.E. and Tahirkheli, R.A.K. (1989) 'Geochronology and temperature history of the Nanga Parbat–Haramosh Massif, Pakistan', in L.L. Malinconico, jun. and R.J. Lillie (eds) *Tectonics and Geophysics of the Western Himalaya*, Geological Society of America, Special Paper 232: 1–22.

Zhang Xiang-song (1984) 'Recent variations of some glaciers in the Karakoram mountains', in K.J. Miller (ed.) *The International Karakoram Project* 1, Cambridge: Cambridge University Press, 38–50.

Zhang Xiang-song and Shi Yafeng (1980) 'Changes of the Batura glacier in the Quaternary and recent times', *Karakoram Batura Glacier, Exploration and Research* (in Chinese with English abstracts), Lanzhou Institute of Glaciology and Cryopedology, *Academia Sinica*, Beijing: Science Press, 173–90.

Zhang Xiang-song, Zheng Ben-xing and Xie Zi-chi (1981) 'Recent variations of the existing glaciers on the Quinghai–Xizang plateau', *Geological and Ecological Studies of the Qinghai–Xizang Plateau* 2, Beijing: Science Press; New York: Gordon & Breach, 1625–9.

Zijderveld, J.D.A. (1967) 'A.C. demagnetization of rocks: analysis of results', in D.W. Collinson, K.M. Creer and S.K. Runcorn (eds) *Methods in Paleomagnetism*, Amsterdam: Elsevier, 254–86.

# INDEX